AF358806

Environmental Protection by Aerobic Granular Sludge Process

Environmental Protection by Aerobic Granular Sludge Process

Editor

Yongqiang Liu

Basel • Beijing • Wuhan • Barcelona • Belgrade • Novi Sad • Cluj • Manchester

Editor
Yongqiang Liu
The University of Southampton
Southampton
UK

Editorial Office
MDPI
St. Alban-Anlage 66
4052 Basel, Switzerland

This is a reprint of articles from the Special Issue published online in the open access journal *Processes* (ISSN 2227-9717) (available at: https://www.mdpi.com/journal/processes/special_issues/Aerobic_Granular_Sludge).

For citation purposes, cite each article independently as indicated on the article page online and as indicated below:

Lastname, A.A.; Lastname, B.B. Article Title. *Journal Name* **Year**, *Volume Number*, Page Range.

ISBN 978-3-7258-0301-9 (Hbk)
ISBN 978-3-7258-0302-6 (PDF)
doi.org/10.3390/books978-3-7258-0302-6

Cover image courtesy of Yongqiang Liu

Contents

About the Editor

Yongqiang Liu

Yongqiang Liu serves as an associate professor at the University of Southampton, UK. Her research spans a broad spectrum of topics, including biological wastewater treatment, microbial fermentation, and resource recovery. Additionally, she is involved in membrane technology and the life cycle assessment of processes for wastewater treatment and resource recovery. A key focus of her research is aerobic granule technology, which she has explored in multiple contexts, from laboratory-scale to large-scale reactors for demonstration purposes. Liu has significantly contributed to our mechanistic understanding of aerobic granulation and has developed strategies for rapid granulation to facilitate quick start-up processes. In recent years, her primary work has been centered on the removal and recovery of nutrients from wastewater; microbial fermentation for the production of volatile fatty acids, succinic acid, and polyhydroxyalkanoates; and the valorization of organic waste for bioenergy. She has co-authored over 90 peer-reviewed scientific research articles.

Editorial

Environmental Protection through Aerobic Granular Sludge Process

Yong-Qiang Liu

Faculty of Engineering and Physical Sciences, University of Southampton, Southampton SO17 1BJ, UK; y.liu@soton.ac.uk

Citation: Liu, Y.-Q. Environmental Protection through Aerobic Granular Sludge Process. *Processes* **2024**, *12*, 243. https://doi.org/10.3390/pr12020243

Received: 4 January 2024
Revised: 22 January 2024
Accepted: 22 January 2024
Published: 24 January 2024

Aerobic granular sludge (AGS) represents a significant advancement in wastewater treatment technology. This innovative approach differs from traditional methods by utilising self-immobilised biofilms without the need for carrier media. The distinct features of AGS, such as its compact three-dimensional structure and larger size, endow it with remarkable sludge settling capabilities and enhanced biomass retention within reactors. One of the key benefits of AGS is its resilience against toxic compounds and its ability to adapt to fluctuating organic loading rates. This robustness is largely due to the unique physical and biological properties of the granular sludge. Additionally, the use of aerobic granular sludge in treatment plants offers considerable operational advantages by a significant reduction in spatial footprint—up to 75%—and lower energy consumption by 30–60% compared to conventional activated sludge (CAS) systems. The combination of these advantages posits aerobic granular sludge as a highly promising technology as either an alternative to CAS for wastewater treatment or to intensify CAS for better performance.

Aerobic granular sludge technology was first reported in 1991 for municipal sewage treatment in an aerobic upflow sludge blanket reactor [1], similar to an upflow anaerobic sludge blanket reactor for anaerobic granules, but this paper did not gain much attention due to the instability of the formed granules. The granulation of sludge in sequential batch reactors (SBRs) reported in 1997 [2] ignited intensive research and rapid commercialisation of AGS branded as Nereda® by Royal HaskningDHV (Amersfoort, The Netherlands). Despite this, adapting existing continuous-flow CAS systems for SBR-based AGS remains challenging. To address this, sludge intensification processes were developed in continuous-flow reactors by adding additional hydrocyclone separation to enhance sludge settleability through partial sludge granulation. Currently, approximately 60 Nereda® wastewater treatment plants and a few intensified plants exist; however, the full-scale adoption of AGS technology has not reached the anticipated levels after 25 years of intensive research and development. This underscores the need for further advancements and wider application in diverse scenarios, which is the focus of the Special Issue titled "Environmental Protection through Aerobic Granular Sludge Process".

The contents of this Special Issue cover a variety of studies on a range of topics of aerobic granular sludge, which can be classified into categories based on the types of wastewater for treatment, nutrient removal, long-term stability of granule-based reactors, and full-scale granular sludge application. In addition, the Special Issue includes a review paper on the waste granular sludge treatment that full-scale WWTPs have to deal with.

Treatment of Various Types of Wastewater with Aerobic Granules

Aerobic granular sludge offers advantages over suspended sludge in retaining slow-growing bacteria such as nitrifying bacteria and polyphosphate accumulating bacteria (PAO) for nutrient removal due to its high biomass retention and high actual sludge retention time (SRT). This is crucial for efficient nutrient removal. Zhang et al. reported on the treat of low-strength wastewater with 400 mg/L COD and 100 mg/L NH_4^+-N, respectively, i.e., a COD/N (C/N) ratio of 4:1, for granulation and nutrient removal. It was found that NH_4^+-N removal efficiency reached almost 100%, indicating excellent nitrifying performance. Liu et al. [3] introduced a novel and effective method to rapidly form nitrifying granules, which was to form heterotrophic granules with a fast granulation strategy [4] followed by converting heterotrophic granules into nitrifying granules by reducing COD concentration to 0 while simultaneously increasing NH_4^+-N concentration from 50 to 850 mg/L. Consequently, granules capable of treating ammonia-rich wastewater with 850 mg/L NH_4^+-N were formed within 37 to 49 days at temperatures of 21 °C and 32 °C, respectively. Morgan et al. explored the treatment of nutrient rich side-stream using synthetic centrate or diluted centrate with low COD and a low COD/N ratio for nitrification and denitrification. By adding acetate as the external carbon source, aerobic granules achieved over 90% COD and over 80% P removal without NOx-N accumulation when COD/N ratio was higher than 11 mg sCOD/mg NH_4^+-N. An et al. conducted nitrification in GMBR fed with wastewater containing 50 mg/L NH_4^+-N and without COD, and it was found that excellent nitrification was achieved. Iorhemen et al. focused on phosphorus removal via enhanced biological phosphorus removal (EBPR) using granular sludge for wastewater with an influent COD of 2000 (day 1–27) and 1000 mg/L (day 28–84) and a COD/N/P ratio of 100:5:1. Three different feeding strategies, anaerobic slow feeding (R1), pulse feeding followed by anaerobic mixing (R2), and pulse feeding (R3), were studied. The results showed that R1 and R2 achieved significantly higher phosphorus removal (97.6% for R1 and 98.3% for R2) than R3 (55%), suggesting the effects of feeding strategy on EBPR. Wang et al. also found that a 70% removal of total inorganic nitrogen (TIN) was achieved by granular sludge in a continuously plug–flow bioreactor fed with real domestic wastewater, meeting the discharge limit of TIN < 10 mg/L in. These studies collectively demonstrate that aerobic granules are highly effective in removing nitrogen and phosphorus from domestic wastewater, side streams, and nitrogen-rich wastewater. However, achieving optimal results depends on the optimisation of both the process and the operation for the specific wastewater with different reactor and process configurations such as the feeding strategy (Mady et al.) and the nitrate recycling location. In addition, a rapid conversion of heterotrophic granules into nitrifying granules [4] has practical implications for quick-start-up nitrifying granules to treat nitrogen-rich wastewater.

In addition to nutrients, it is crucial to address other pollutants since municipal wastewater treatment plants often receive a portion of industrial wastewater when necessary. Ortega et al. investigated the effects of increased particulate COD (pCOD) on the aerobic granular sludge process in a full-scale Nereda® plant receiving (30%) raw industrial wastewater mainly from slaughterhouses with an increased pCOD load from 0.5 to 1.3 g COD/L. Remarkably, no deterioration in granule properties or reactor performance, such as nutrient removal, was observed. This underscores the robustness of the granular sludge process in handling more challenging wastewater types.

Aerobic granules are also effective in treating industrial wastewater. This Special Issue includes a study on berberine wastewater from the pharmaceutical industry. By developing a hybrid anaerobic baffled reactor–aerobic granules process (ABR–AGS) system fed with wastewater containing a berberine concentration of 122 mg/L, Wang et al. achieved 92.2% and 94.8% overall removals of berberine and COD, respectively. Due to their compact structure and large size, aerobic granules are capable of withstanding toxic and inhibitory substances such as berberine, making them a more promising solution for treating harsher wastewater compared to suspended sludge. Experimental evidence from Tay et al. [5] demonstrated that unlike activated sludge reactors, aerobic granules quickly adapted and efficiently reduced influent phenol concentrations from 600 mg/L to 0.3 mg/L within three days after a substrate change from acetate to phenol. When the algal–bacterial aerobic granular sludge was used as biosorbent for treating wastewater containing heavy metals such as 2 mg/L Cr(VI) at a 1% salinity, the removals of total inorganic nitrogen (TIN) and total phosphorus (TP) were noticeably impacted, and they were further decreased with the co-existence of 2 mg/L Cr(VI), but the Cr(VI) removal efficiency was only influenced slightly by salinity exposure. Nitrifying bacteria are sensitive to unfavourable environmental conditions such as high salinity and high heavy-metal concentrations. Although granules' structure can protect bacteria to a certain extent, achieving the removal of both nutrients and heavy metals at a high salinity is challenging, but based on this study, using waste granular sludge as a biosorbent for Cr(VI) removal at high salinity seems viable.

For various types of wastewater mentioned above, such as that with a low COD/N ratio, side streams with high nutrient concentrations, high-pCOD wastewater, and pharmaceutical wastewater containing berberine, the aerobic granular sludge process generally demonstrates high robustness and good performance, implying its versatility for treating various types of wastewater. However, optimisation has to be conducted for each type of wastewater to ensure a good performance, and the treatment mechanism needs to be understood for better control and operation of aerobic granular process.

Long-Term Stability and Relevant Performance of Aerobic Granules in Various Treatment Processes

The long-term stability of aerobic granular sludge and reactor performance is another important factor affecting the adoption of this technology and its practical implementation. In this Special Issue, three types of reactor operation were studied: conventional SBRs, continuously plug–flow reactors for process intensification, and continuous-flow granular membrane bioreactors (GMBRs). Two pilot-scale plug–flow reactor (PFR) systems, based on a modified Ludzack–Ettinger (MLE) configuration with nitrate recycling on continuous-flow aerobic granulation fed with real domestic wastewater, were used to study the effects of nitrate recycle location on granulation. The results showed that the change in nitrate recycle location can be advantageous to drive sludge densification without a radical washout of the sludge, and it had no effects on the chemical oxygen demand (COD) and nitrogen removal efficiencies. Thus, it could be further developed as an alternative to hydrocyclones for full-scale, greenfield, continuous sludge densification applications. In conventional lab-scale SBRs, duplicate reactor operation with a low COD/N ratio (4/1) showed a slight difference for granulation. Additionally, the SVI_{30}, granule size, and size distributions of sludge in two reactors were different. However, the granulation and reactor performance in terms of nutrient removal were generally reproducible. This study validated the reproducibility of aerobic graduation and relevant treatment performance in duplicates, but meanwhile, it suggests that slight differences between different reactors and granule properties are unavoidable. This result is meaningful because the vast majority of studies on aerobic granules had no duplicate reactor operation and the comparison between different conditions, especially when the differences are not significant, could be better interpreted. Since there is typically no deliberate sludge wasting or control of biomass concentration in lab-scale reactors as in full-scale wastewater treatment plants, the biomass increase due to the granulation and biomass accumulation results in a reduced F/M ratio, which is believed to be critical for the long-term stability of aerobic granules [6]. Controlling F/M ratios in a certain range would thus benefit long-term stability. Liu et al. observed a serious calcification of partial nitrifying granules during the long-term operation with ash contents varying from 50% to 85% at stable state at 21 °C and 32 °C, fed with wastewater with high hardness (Ca^{2+} concentration > 120 mg/L). The precipitation and accumulation of hydroxyapatite and calcium carbonate in granules resulted in a significantly reduced specific nitrogen oxidation rate, and the higher temperature exacerbated this phenomenon. It was concluded that high mineral content in granules led to the instability of nitrifying granules at a higher temperature in the short or long term. The effects of water hardness on granule stability are not well investigated; thus, this study provided insights for further investigation of this aspect in the future, especially when considering adopting aerobic granule technology in areas with high water hardness. It needs to be pointed out that water hardness is not the only factor that causes the calcification of aerobic granules. The high mineral content in aerobic granules has been reported under other conditions as well. In this Special Issue, An et al. found that in GMBR without sludge wasting, inoculated nitrifying granules disintegrated and re-established over a 300-day operation, suggesting that membrane module and indefinite SRT did not affect granulation although there was granule disintegration first. However, the ash content in granules was as high as 64% after a 305-day operation with only the top 60–80 μm layer of these nitrifying granules being active, which is similar to a biofilm with an active layer on the surface of the inert carrier media. The rate of membrane fouling did not change in this GMBR, thus defeating the original purpose of GMBR development. It is noteworthy that SRT in this GMBR was indefinite because there was no sludge wasting and the reactor was operated in continuous mode. Other conditions of GMBR such as controlling SRT need further investigation to fully explore the potential of GMBR for the alleviation of membrane fouling. Generally, retaining granules in reactors for a long period, i.e., long SRT, could cause granule calcification, leading to the increase in ash content for reduced microbial activity [7], possible granule disintegration, and reactor operation instability. Thus, controlling the SRT of granules (sometimes F/M ratios) could be an effective strategy to minimise the increase in ash content and maintain the long-term stability of granules in terms of structure as well as microbial activity.

Digestibility of Waste Aerobic Granules

Just like conventional activated sludge-based plants, anaerobic digestion is a common approach to treating waste sludge in aerobic-granule-based WWTPs. The physical and chemical characteristics of granular sludge are different from those of activated sludge, which could lead to different digestibility and biogas production potential. A review paper by Zaghloul et al. [8] discussed the biological and chemical characteristics of waste AGSs and their digestibility. In addition, the comparison between aerobic granules, activated sludge, and primary sludge in terms of digestibility was conducted. It concluded that the AGS morphology and structural polymers slowed down the biomethane production rates, but the overall yield was on par with that of activated sludge. So far, the available literature on the digestibility of waste AGS is still scarce compared to WAS. Thus, further research, especially on waste AGS from full-scale plants, was suggested. In addition, aerobic granules could be used for various types of wastewater treatment in different types of reactors, from which characteristics of waste AGS and digestibility should be investigated to draw a solid conclusion.

In summary, this Special Issue addresses the key aspects of aerobic granular sludge in environmental protection. It encompasses a variety of topics including different types of wastewater characterised by diverse influent qualities, particularly varying nutrient levels. The Special Issue also explores different operation modes such as sequential-batch and continuous-flow operation modes, with different process configurations such as SBRs, GMBRs, and continuous plug–flow reactors. Significantly, the studies extend beyond synthetic wastewater to include real domestic wastewater, municipal wastewater mixed with industrial water (such as slaughterhouse water), and actual industrial wastewater like berberine wastewater. These studies, conducted in lab-scale, pilot-scale, and full-scale reactors, deepen our understanding of the practical application of aerobic granules in real-world scenarios. They demonstrate the robustness of aerobic granules in treating a wide array of wastewater types, highlighting their promising potential for widespread application. Furthermore, some key factors in different process configurations with different wastewater were investigated to understand the long-term stability of both granule structure and reactor performance. SRT is one of the most crucial operating parameters which might affect granule calcification and disintegration. Lastly, a review paper discussed an interesting finding of comparable digestibility and biogas yields between aerobic granules and activated sludge, although the methane production rate of granules could be lower due to different physical and chemical properties. This area, still under-researched, calls for further studies. Moreover, many papers in this Special Issue utilised 16S RNA gene sequencing to analyse the microbial community structure of aerobic granules under different treatment conditions. However, a detailed correlation between the microbial community structure and operating conditions or granule properties was not established. Future research in this area could provide valuable insights for improved process control and operation. Overall, this Special Issue serves as a microcosm of the current research on aerobic granules, offering a quick glimpse into the state of the art in aerobic granules for environmental protection.

Conflicts of Interest: The author declares no conflicts of interest.

List of Contributions:

1. Zhang, H.; Liu, Y.Q.; Mao, S.; Steinberg, C.E.W.; Duan, W.; Chen, F. Reproducibility of aerobic granules in treating Low-strength and low-C/N-ratio wastewater and associated microbial community structure. *Processes* **2022**, *10*, 444.
2. Liu, Y.Q.; Cinquepalmi S. Hydroxyapatite precipitation and accumulation in granules and its effects on activity and stability of partial nitrifying granules at moderate and high temperatures. *Processes* **2021**, *9*, 1710.
3. Morgan, G.; Hamza, R.A. Cultivation of Nitrifying and Nitrifying-Denitrifying Aerobic Granular Sludge for Sidestream Treatment of Anaerobically Digested Sludge Centrate. *Processes* **2022**, *10*, 1687.
4. An, Z.; Zhang, X.; Bott, C.B.; Wang, Z.W. Long-term stability of nitrifying granules in a membrane bioreactor without hydraulic selection pressure. *Processes* **2021**, *9*, 1024. https://doi.org/10.3390/pr9061024.
5. Iorhemen, O.T.; Ukaigwe, S.; Dang, H.; Liu, Y. Phosphorus removal from aerobic granular sludge: Proliferation of polyphosphate-accumulating organisms (PAOs) under different feeding strategies. *Processes* **2022**, *10*, 1339.
6. Wang, J.F.; An, Z.H.; Zhang, X.Y.; Angelotti, B.; Brooks, M.; Wang, Z.W. Effects of Nitrate Recycle on the Sludge Densification in Plug-Flow Bioreactors Fed with Real Domestic Wastewater. *Processes* **2023**, *11*, 1876.
7. Ortega, S.T.; Pronk, M.; de Kreuk, M.K. Effect of an increased particulate cod load on the aerobic granular sludge process: A full scale study. *Processes* **2021**, *9*, 1472. https://doi.org/10.3390/pr9081472.
8. Wang, Y.; Liu, Y.; Li, J.; Ma, R.; Zeng, P.; Ng, C.A.; Liu, F. The dynamic shift of bacterial communities in hybrid anaerobic baffled reactor (ABR)—Aerobic granules process for Berberine pharmaceutical wastewater treatment. *Processes* **2022**, *10*, 2506. https://doi.org/10.3390/pr10122506.
9. Van Nguyen, B.; Yang, X.; Hirayama, S.; Wang, J.; Zhao, Z.; Lei, Z. Effect of salinity on cr(Vi) bioremediation by algal-bacterial aerobic granular sludge treating synthetic wastewater. *Processes* **2021**, *9*, 1400.
10. Zaghloul, M.S.; Halbas, A.M.; Hamza, R.A.; Elbeshbishy, E. Review on Digestibility of Aerobic Granular Sludge. *Processes* **2023**, *11*, 326.
11. Mady, E.; Oleszkiewicz, J.; Yuan, Q. Feasibility Study of Applying Anaerobic Step-Feeding Mode for the Treatment of High-Strength Wastewater in Granular Sequencing Batch Reactors (GSBRs). *Processes* **2023**, *11*, 75.

References

1. Mishima, K.; Nakamura, M. Self-immobilization of aerobic activated sludge—A pilot study of the Aerobic Upflow Sludge Blanket Process in municipal sewage treatment. *Water Sci. Technol.* **1991**, *23*, 981–990. [CrossRef]
2. Morgenroth, E.; Sherden, T.; Van Loosdrecht, M.C.M.; Heijnen, J.J.; Wilderer, P.A. Aerobic granular sludge in a sequencing batch reactor. *Wat. Res.* **1997**, *31*, 3191. [CrossRef]
3. Liu, Y.Q.; Cinquepalmi, S. Hydroxyapatite precipitation and accumulation in granules and its effects on activity and stability of partial nitrifying granules at moderate and high temperatures. *Processes* **2021**, *9*, 1710. [CrossRef]
4. Liu, Y.Q.; Tay, J.H. Fast formation of aerobic granules by combining strong hydraulic selection pressure with overstressed organic loading rate. *Water Res.* **2015**, *80*, 256–266. [CrossRef] [PubMed]
5. Tay, S.T.L.; Moy, B.Y.P.; Maszenan, A.M.; Tay, J.H. Comparing activated sludge and aerobic granules as microbial inocula for phenol biodegradation. *Appl. Microbiol. Biotechnol.* **2005**, *67*, 708–713. [CrossRef] [PubMed]
6. Franca, R.D.G.; Pinheiro, H.M.; van Loosdrecht, M.C.M.; Lourenço, N.D. Stability of aerobic granules during long-term bioreactor operation. *Biotech. Adv.* **2018**, *36*, 228.
7. Yuan, S.; Gao, M.; Zhu, F.; Afzal, M.Z.; Wang, Y.-K.; Xu, H.; Wang, M.; Wang, S.-G.; Wang, X.-H. Disintegration of aerobic granules during prolonged operation. *Water Res. Technol.* **2017**, *3*, 757. [CrossRef]
8. Zaghloul, M.S.; Halbas, A.M.; Hamza, R.A.; Elbeshbishy, E. Review on Digestibility of Aerobic Granular Sludge. *Processes* **2023**, *11*, 326. [CrossRef]

 *processes*

Article

Effects of Nitrate Recycle on the Sludge Densification in Plug-Flow Bioreactors Fed with Real Domestic Wastewater

Jie-Fu Wang [1], Zhao-Hui An [1], Xue-Yao Zhang [1], Bob Angelotti [2], Matt Brooks [2] and Zhi-Wu Wang [1],*

[1] Department of Biological Systems Engineering, Virginia Tech, Blacksburg, VA 24060, USA; jiefuwang@vt.edu (J.-F.W.); za1@vt.edu (Z.-H.A.); xueyao@vt.edu (X.-Y.Z.)
[2] Upper Occoquan Service Authority, 14631 Compton Rd, Centreville, VA 20121, USA; bob.angelotti@uosa.org (B.A.); matt.brooks@uosa.org (M.B.)
* Correspondence: wzw@vt.edu

Abstract: The impact of adding a modified Ludzack–Ettinger (MLE) configuration with Nitrate Recycle (NRCY) on continuous-flow aerobic granulation has yet to be explored. The potential negative effects of MLE on sludge densification include that: (1) bioflocs brought by NRCY could compete with granules in feast zones; and (2) carbon addition to anoxic zones could increase the system organic loading rates and lead to higher feast-to-famine ratios. Two pilot-scale plug flow reactor (PFR) systems fed with real domestic wastewater were set up onsite to test these hypotheses. The results showed that MLE configuration with NRCY could hinder the sludge granulation, but the hindrance could be alleviated by the NRCY location change which to some extent also compensates for the negative effect of higher feast-to-famine ratios due to carbon addition in MLE. This NRCY location change can be advantageous to drive sludge densification without a radical washout of the sludge inventory, and had no effects on the chemical oxygen demand (COD) and nitrogen removal efficiencies. The PFR pilot design for the MLE process with a modified NRCY location tested in this study could be developed as an alternative to hydrocyclones for full-scale, greenfield, continuous sludge densification applications.

Keywords: NRCY; MLE; granular sludge; PFR startup

Citation: Wang, J.-F.; An, Z.-H.; Zhang, X.-Y.; Angelotti, B.; Brooks, M.; Wang, Z.-W. Effects of Nitrate Recycle on the Sludge Densification in Plug-Flow Bioreactors Fed with Real Domestic Wastewater. *Processes* **2023**, *11*, 1876. https://doi.org/10.3390/pr11071876

Academic Editor: Adam Smoliński

Received: 21 May 2023
Revised: 14 June 2023
Accepted: 19 June 2023
Published: 22 June 2023

1. Introduction

Process intensification through the employment of advanced cell immobilization techniques, such as activated sludge densification via aerobic granulation, provides an economical option to expand the treatment capacity of wastewater resource recovery facilities (WRRFs) with a low capital, land, and maintenance investment [1]. To achieve aerobic granulation, hydraulic selection pressure has always been utilized as a driving force to selectively return better settling sludge to the feast zone or the phase of either plug-flow reactors (PFRs) or sequencing batch reactors (SBRs). This mode of operation selectively favors the coaggregation of microbial sludge into aerobic granules by washing out their diffuse flocculent sludge counterpart [1,2]. As a result, the transition from flocculent sludge to densified sludge often came at the cost of substantial sludge washout over the course of the reactor startup. This in turn directly resulted in the loss of treatment capacity until the granular sludge inventory within the reactor was slowly re-established [2,3]. However, in full-scale applications, WRRFs cannot withstand such a treatment interruption, even if it is temporary, and have to comply with discharge permit limits all the time. Therefore, a strategy is needed to drive densified sludge formation by selectively separating granules and bioflocs, while also safeguarding the effluent quality throughout the reactor startup period without treatment degradation or interruption.

Hydrocyclones have been employed in most of the WRRFs to provide such a selection pressure; however, their relatively low hydraulic capacity (e.g., 120–3360 m^3 d^{-1}) limits their application in large systems to sidestream return activated sludge (RAS), where only a small portion, e.g., 5–7%, of the RAS flow can be processed for physical selection [4–6]. This low-strength selection pressure plus the high suspended solids content in RAS, e.g., 5000 mg L^{-1}, compromised the physical selection efficiency of hydrocyclones for sludge densification [7]. As a result, a prolonged startup phase, perhaps as long as a year, was required to observe significant sludge settleability improvements in full-scale continuous-flow reactors in which densified sludge formation is still a challenge [1,7–10]. For this reason, an alternative strategy should be developed to provide sufficient selection strength to promote successful continuous-flow sludge densification, while maintaining the startup treatment performance required for large full-scale applications. To this end, [2] demonstrated a new strategy by continuously returning the sludge in the underflow and overflow of a continuous upflow gravity selector to the feast and famine zones of a plug-flow reactor, respectively, which can effectively drive a smoother transition from flocculent-activated sludge to granular sludge, without dramatic sludge concentration and effluent quality decline during the reactor startup. However, this strategy was only proven successful in conventional-activated sludge pilots without biological nitrogen removal (BNR) [2,11]. Its application in a modified Ludzack–Ettinger (MLE) configuration with Nitrate Recycle (NRCY) has yet to be explored. The potential challenges for having both MLE and continuous-flow aerobic granulation in one system include that: (1) NRCY could bring bioflocs to the feast zone to compete with granules; (2) carbon addition to anoxic zones could lift the system organic loading rates and result in higher feast-to-famine ratios. Our hypotheses are that by changing the NRCY location, bioflocs could be largely separated from densified sludge, which makes the sludge densification achievable. Meanwhile, the higher organic loading rates (OLRs) and feast-to-famine ratios in MLE systems with NRCY caused by carbon addition could hinder the full granulation process. But the hindrance might be compensated to some extent by the NRCY location change. Testing these two hypotheses will not only provide engineering solutions, but also shed light on the continuous-flow aerobic granulation mechanism by crosschecking with the previous understandings of the feast and famine selection roles [2,11,12].

Thus, two PFR pilots fed with real domestic wastewater were set up onsite to test the hypotheses (Figure 1). Compared to the configuration in the study by [2,11], a NRCY system along with an equalization tank was added. The results from this study would shed light on how to drive sludge densification through granulation in an MLE process, while safeguarding effluent quality during reactor startup and providing a novel approach to enable the full-scale application of continuous-flow sludge densification technology with reasonable infrastructure modifications.

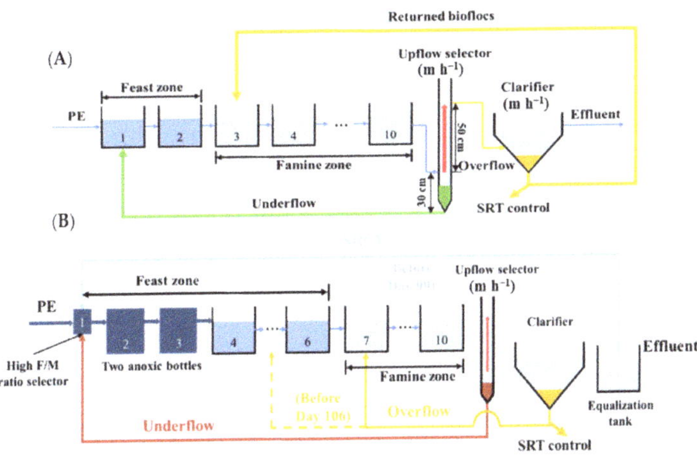

Figure 1. Illustrative design of the PFR systems equipped without NRCY (**A**) and with NRCY (**B**).

2. Materials and Methods

2.1. Reactor Setup and Operation

Two granulation PFR pilot trains were set up as illustrated in Figure 1A,B, and operated in situ by feeding real primary effluent (PE) at the Upper Occoquan Service Authority (UOSA), a WRRF in Centreville, VA, USA. The startup strategy of a conventional-activated sludge process was tested using the PFRs configured in Figure 1A and compared to the MLE PFRs with NRCY configurations in Figure 1B. Each PFR pilot with a 140 L total working volume was made of 10 identical completely mixed chambers connected in a series to create feast/famine conditions along the flow direction. The hydraulic retention time of each pilot was controlled at 6.5 h, similar to that of the full-scale secondary process at UOSA. All chambers in Figure 1A and Chambers 4 to 10 in Figure 1B were aerated at a flow rate of 3 L min^{-1} through aeration stones to ensure dissolved oxygen (DO) > 2 mg L^{-1} and maintain homogenous mixing conditions in each chamber. In contrast, the first three chambers in Figure 1B (MLE system with NRCY) were airtight and kept anoxic by purging nitrogen gas at the beginning and recirculating headspace gas in each chamber afterward. Methanol was added to the first chamber as a carbon source for denitrification.

As shown in Figure 1A,B, a continuous upflow column with an internal diameter of 8 cm and a total working height of 80 cm was installed at the end of the PFR to mimic a high-rate clarifier. The column inflow, i.e., the mixed liquor from the PFR effluent, continuously entered at a height of 30 cm from the column bottom and then flowed up for 50 cm before overflowing from the top of the upflow column (Figure 1A,B). The upflow velocity of this column, which is equivalent to the surface overflow rate (SOR) of a clarifier, was set at 10 m h^{-1} based on the flow rate and column cross-section area, to separate the sludge particles based on their settling velocity. Theoretically, only those particles with a settling velocity greater than 10 m h^{-1} can be retained within the underflow of the velocity selector and returned to the first feast zone chamber of the PFR (Figure 1A,B). The unsettled bioflocs in the overflow were sent to another clarifier designed with a more typical SOR of 1 m h^{-1}. A portion of bioflocs settled under this low SOR was wasted on a daily basis for solids retention time (SRT) control (Figure 1A,B). The remainder of the flocculent sludge was returned to the first famine zone of the PFR. The flow rate ratio of the selector overflow to underflow was set at about 1.6. An airlifting method described in a previous study was used to return sludges in Figure 1A,B [3]. In addition, an equalization tank with an HRT of about 1 h was added to the MLE system after the clarifier (Figure 1B). NRCY in Figure 1B was achieved through a peristaltic pump (Masterflex® L/S, Cole-Parmer, Vernon Hills, IL, USA), and its flow rate to the inflow rate was set at a ratio of 2. The NRCY location was changed from Chamber 10 to the equalization tank on Day 99.

2.2. Analytical Methods

The chemical oxygen demand (COD), nitrate, nitrite and ammonia concentrations were analyzed using TNTplus® 820, 835, 839 and 840 vials in a spectrophotometer (Hach, Loveland, CO, USA). Nitrate, nitrite, ammonia and soluble COD (sCOD) samples were measured with the filtrate from 0.45 μm syringe filters (EZFlow®, Old Saybrook, CT, USA). The total inorganic nitrogen (TIN) concentration was calculated by summing nitrate, nitrite and ammonia nitrogen concentrations. DO was measured using an HQ40D meter (Hach, Loveland, CO, USA). The sludge volume index (SVI), zone settling velocity (V_{zs}), mixed liquor suspended solids (MLSS), mixed liquor volatile suspended solid (MLVSS), as well as the carbonaceous and nitrogenous specific oxygen uptake rates (SOURs) were analyzed according to standard methods [13]. The 5 min and 30 min SVI were measured using a standard 2 L settle meter and recorded as SVI_5 and SVI_{30}, respectively. It is noteworthy that SVI herein was measured after the dilution of mixed liquor samples to total suspended solids (TSS) concentrations of 1000 mg/L to avoid hindered settling [13].

3. Results

3.1. Negative Impacts of NRCY on Granulation

Figure 2 allows for a comparison of the settleability improvement in the last chamber of the two PFRs in Figure 1A,B. One (Figure 1A) is a conventional-activated sludge process without a BNR, and the other (Figure 1B) is an MLE process with NRCY and external carbon (methanol) addition. In general, the former appeared to achieve better granulation than the latter. Figure 2A shows that after a 141-day operation, the SVI_{30} of the systems without and with NRCY dropped to more or less the same level around 74 mL/g, even though the SVI_{30} of the system without NRCY somehow started at a much higher level than that of the system with NRCY (114 mL/g). In addition, SVI_{30} surges in the system without NRCY were observed on Days 36 and 71 due to accidental sludge loss events (Figure 2A). Meanwhile, SVI_5/SVI_{30} evolution in Figure 2B demonstrated that the system without NRCY had much lower values than the system with NRCY. For example, after about a 140-day operation, the SVI_5/SVI_{30} of the system without NRCY plummeted to around 1.5, while the one with NRCY stayed around 2.1. It was reported that the SVI_5/SVI_{30} of successful granulation should approximate 1 [3]. Thus, the system without NRCY likely achieved more granulation than the other with NRCY. This could also be evidenced by the V_{ZS} profiles in Figure 2C. It can be seen that after around a 140-day operation, the V_{ZS} of the system without NRCY rose to 5.4 m/h, which was 42% higher than that of the system with NRCY.

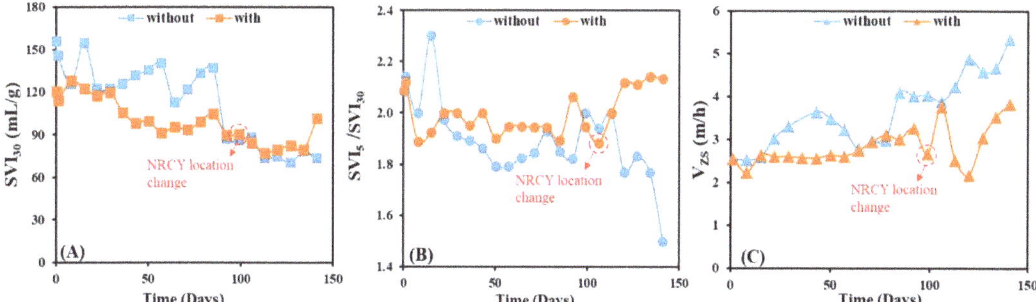

Figure 2. Profiles of (**A**) SVI_{30}, (**B**) SVI_5/SVI_{30} and (**C**) V_{zs} measured in the last chamber of the PFR with and without NRCY.

As discussed above, the system appeared to achieve better granulation and thus, better sludge settleability when NRCY was not employed. In other words, the NRCY configuration could be disadvantageous to the aerobic granulation. The potential reasons for this were twofold. Firstly, NRCY located at the end of the aerobic zone could bring bioflocs to the feast zone competing with granules, which defeated the purposes of the pilot design (Figure 1). The pilot design herein can take advantage of feast/famine conditions in PFRs to provide a biological selection internal to the bioreactors by redirecting the underflow and overflow of the physical velocity selector to the feast and famine zones of the treatment train, respectively [12]. Therefore, the introduction of bioflocs to feast zones might be responsible for the inferior sludge settleability. This challenge was addressed by changing the NRCY location from the last aerobic chamber containing significant MLSS biomass to an equalization tank after biomass separation in the clarifier on Day 99, as shown in Figure 1B. The hypothesis of the recirculated flocculent biomass to the feast zone will be tested and discussed in the following sections. Even so, sludge settleability in the system with NRCY was still inferior to the one without NRCY. This could be ascribed to a second reason, i.e., the higher OLRs and feast-to-famine ratios caused by carbon addition. To denitrify $NO_3{}^-$–N brought by NRCY in the anoxic zone, external carbon (methanol) was added to the first anoxic chamber, as shown in Figure 1B, which led to a higher OLR than the one without NRCY (Figure 1A). According to Zhang et al. [14], a decreased OLR enhances the formation and stability of aerobic granular sludge because granular sludge cultivated under a decreased OLR maintains a good balance between polysaccharides and protein in the extracellular polymeric substance (EPS) content. Similarly, Ghangrekar et al. [15] reported that for a reactor started at a higher OLR, the SVI of the sludge became higher even if the SVI of the inoculum was lower. In addition, average sCOD concentrations in each chamber of the two systems were shown in Figure 3. One can see that without NRCY, a majority of the readily biodegradable sCOD (rbsCOD) was removed in the first two chambers, leaving mostly a refractory sCOD in downstream chambers. A feast-to-famine duration ratio of 0.25, as estimated from Figure 3, was deemed to be beneficial to aerobic granulation in PFRs [12]. In contrast, with the NRCY configuration and external carbon addition, rbsCOD was not consumed until Chambers 5 (before NRCY relocation) and 7 (after NRCY relocation), resulting in feast-to-famine duration ratios of 0.67 and 1.50, respectively, which were considered less favorable for successful granulation [12]. Therefore, the higher organic loading rates and feast-to-famine ratios in the MLE system with NRCY (Figure 1B) could explain its inferior granulation process to the one without NRCY (Figure 1A).

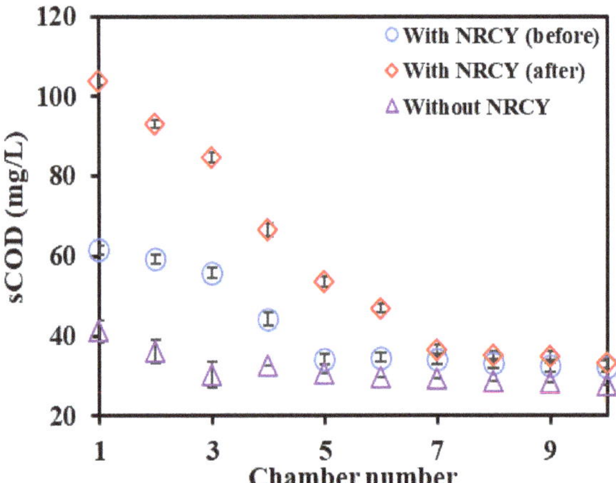

Figure 3. Average sCOD concentration profiles of the PFR pilot without NRCY (using data from Days 120 to 134), with NRCY and before its location change (using data from Days 43 to 57), and after its location change (using data from Days 120 to 134).

3.2. Effects of NRCY Location on Solid Concentrations

As mentioned in the previous section, it is hypothesized that the introduction of bioflocs to the feast zone by NRCY could be avoided by relocating NRCY from the last aerobic chamber to an equalization tank after the clarifier. This hypothesis was tested, and the results were revealed in the following sections.

From Day 1 to 98, NCRY was located at the 10th aerobic chamber, which was similar to the full-scale MLE process, as shown in Figure 1B. Day 1 to 36 was benchmarked as the startup period, as shown in Figure 4. After the startup, the MLVSS profiles of each chamber stabilized during Day 37 to 98 under the combined effects of NRCY and sludge inventory redistribution (Figure 4). Along with $NO_3{}^--N$, a huge amount of flocculent biomass in the NRCY stream was returned to the first feast chamber, namely Chamber 1 (C1). It is well known that bioflocs are very competitive in utilizing readily available substrates for a number of reasons. They can outcompete granules because of their looser structure and higher specific surface area that promotes fast substrate diffusion and utilization [16]. Therefore, returning both flocculent and granular sludges in the NRCY to the same feast zone, as is conventionally practiced is counterproductive to the progression and stabilization of continuous-flow sludge densification. After realizing this, the NRCY location was changed to the equalization tank after the clarifier on Day 99 (Figure 1B). There were very few suspended solids in the equalization tank after the clarifier. Thus, almost no bioflocs were returned to the feast zone after the location change. Consequently, a substantial drop of the MLVSS in all feast chambers was observed, as shown in Figure 4. For example, the average MLVSS concentration in C1 plummeted from 1403 to 498 mg/L after the location change. Conversely, MLVSS concentrations in the famine chambers were not subjected to the NRCY location change, and thus barely changed. This was because a large amount of bioflocs from the clarifier were returned to the first famine chamber, namely Chamber 7, by internal sludge inventory redistribution shown in Figure 1B, regardless of the NRCY location change.

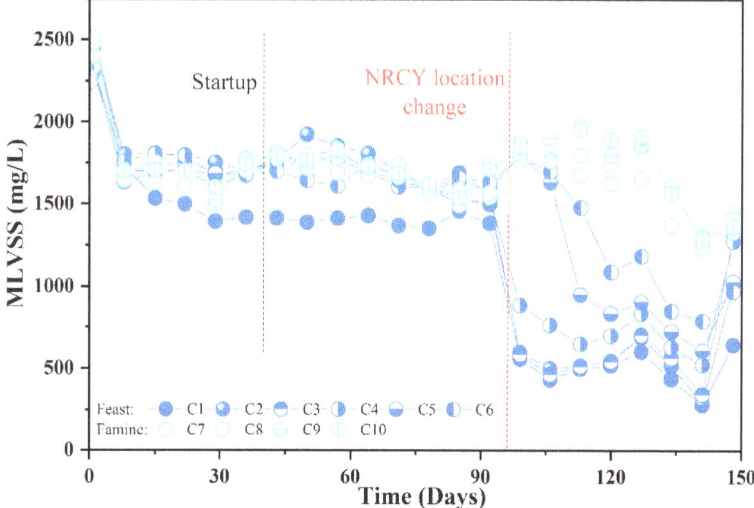

Figure 4. MLVSS profiles in different chambers over the course of the pilot operation with NRCY.

With all this being said, sludge concentrations in the entire PFR actually decreased after the switch to the new NRCY source location. This was ascribed to effluent solid concentration (TSS and VSS) increases (Figure 5A) and an actual SRT decrease (Figure 5B). When the NRCY was relocated from the last aerobic chamber to the equalization tank (Figure 1B), the SOR of the clarifier rose from 1 to 1.7 m/h because the additional recycle flow went through the clarifier. Given this, it was anticipated that effluent TSS concentrations would increase, the degree of which was not well accounted for, from 27 mg/L (Day 106) to 73 mg/L (Day 148), after NRCY location change (Figure 5A). The SRT was calculated based only on the waste-activated sludge (WAS) being purposely discharged, while the actual SRT was impacted by the sludge lost in the clarifier effluent. In retrospect, when the SOR of the clarifier was small ($\leq$1 m/h), the WAS was deemed as the only significant sludge leaving the system, but with the NRCY location change and the increase in SOR, the sludge loss through the effluent ramped up considerably, dropping the actual SRT after Day 99 to as low as 3 days on Day 141 (Figure 5B). Similar to the clarifier, the upflow velocity of the upflow selector also increased from 10 to 17 m/h with the NRCY location change (Figure 1B). Less densified sludge was retained in the underflow and more flocs were washed out, which caused the drop in MLVSS concentrations in the underflow of the upflow selector (Figure 6A). As a result, the dry mass ratios of underflow over overflow also declined (Figure 6B).

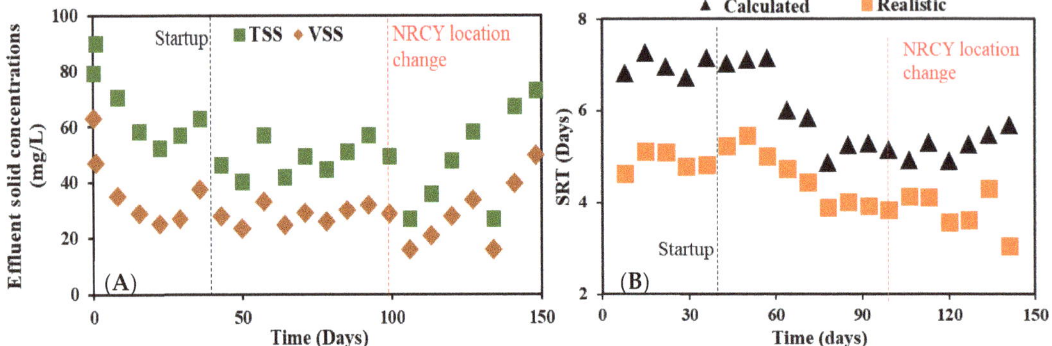

Figure 5. (**A**) Effluent TSS and VSS profiles, and (**B**) the calculated and realistic SRT profiles.

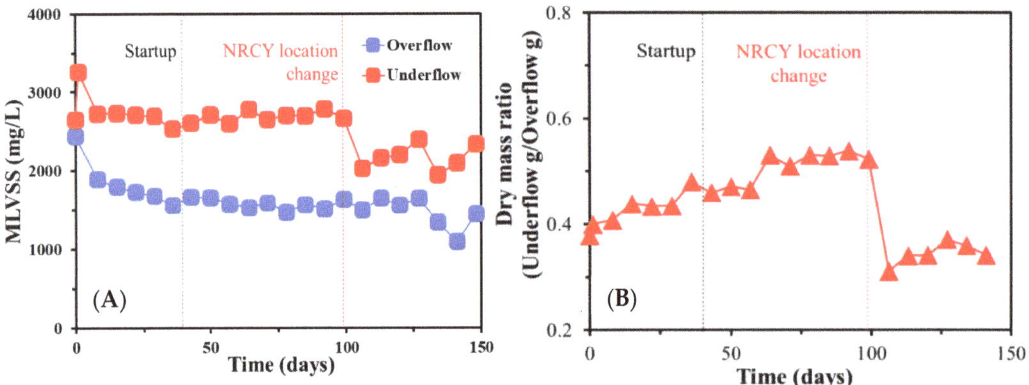

Figure 6. (**A**) MLVSS profiles in overflow and underflow of the upflow selector; (**B**) Effects of NRCY location changes on the biomass ratio in the underflow and overflow of the upflow selector.

3.3. Effects of the NRCY Location on Mixed Liquor Settleability

As mentioned in the previous section, NRCY relocation facilitated the separation and redistribution of heavier densified sludge and lighter bioflocs to the feast and famine zones, respectively. Biofloc out-selection was further boosted by subjecting bioflocs to low substrate availability. As a result, an improvement of sludge settleability, especially in feast chambers, was observed after the NRCY location change.

As shown in Figure 7, the SVI_{30} of all the chambers dropped after the NRCY location change on Day 99. Among those, the SVI_{30} of the feast chambers dropped more than that of the famine chambers. For example, the SVI_{30} values of Chambers 2, 3 and 4 (C2, C3 and C4) dropped from 73, 74 and 90 mL/g to around 51, 49 and 54 mL/g, respectively, around 50 days after the NRCY location change (Figure 7). The research of [1] suggested an upper limit of 60 mL/g SVI_{30} for successful sludge densification or granulation. With that being said, successful sludge densification has been achieved in feast chambers such as C2 and C3, which demonstrated the effectiveness of the upflow selector and sludge redistribution system. Similarly, the drops in SVI_5/SVI_{30} values in feast chambers were also observed (Figure 8). For example, the SVI_5/SVI_{30} of C2, C3 and C4 dropped from 2, 2 and 1.8 to 1.6, 1.3 and 1.5 mL/g, respectively, 50 days after the NRCY location change. It should be pointed out that these SVI_5/SVI_{30} values were still off from 1, which is a theoretical value when successful aerobic granulation was achieved [17], indicating the incomplete sludge granulation process. The improved settleability was also evidenced by V_{ZS} profiles in Figure 9. It can be seen that the V_{ZS} of the sludge in all the feast zones increased substantially. Among those, the values of V_{ZS} in C1, C2, C5 and C6 became greater or equal to the upflow velocity of the selector (17 m/h) 50 days after the NRCY location change, which indicated that sludges in these feast chambers were adapted to the reactor's gravity selection pressure.

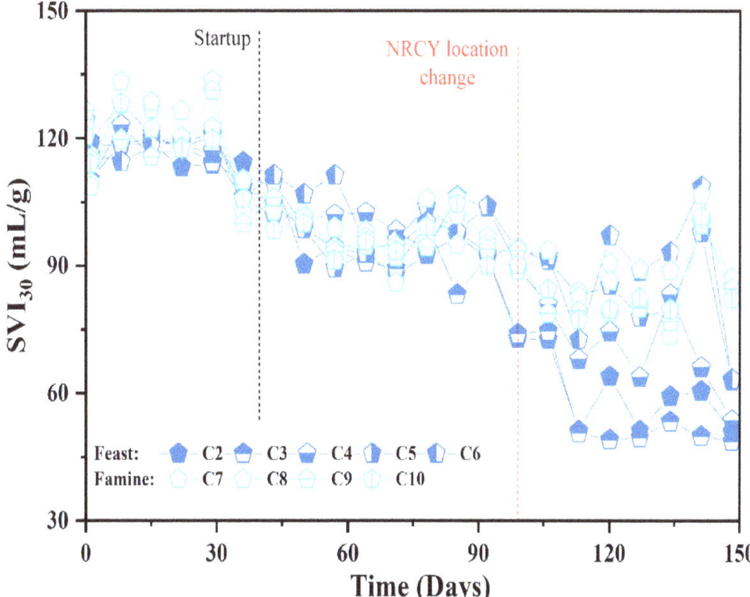

Figure 7. SVI_{30} profiles in different chambers over the course of the pilot operation.

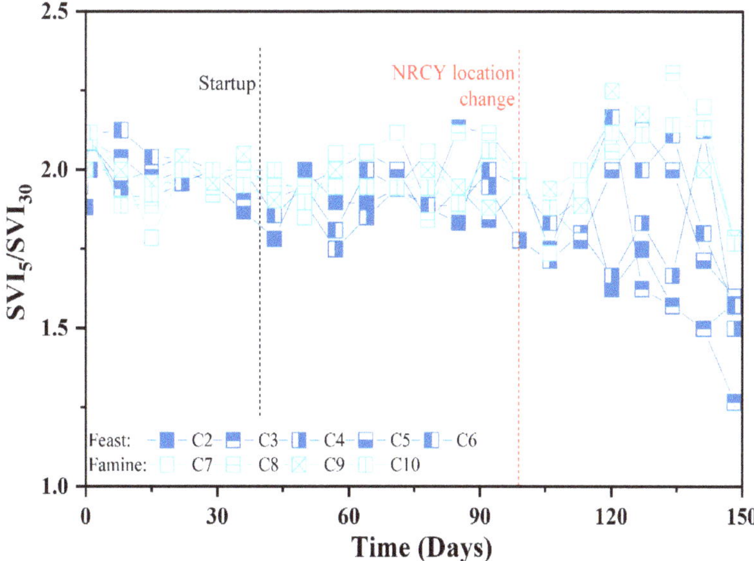

Figure 8. SVI_5/SVI_{30} profiles in different chambers over the course of the pilot operation.

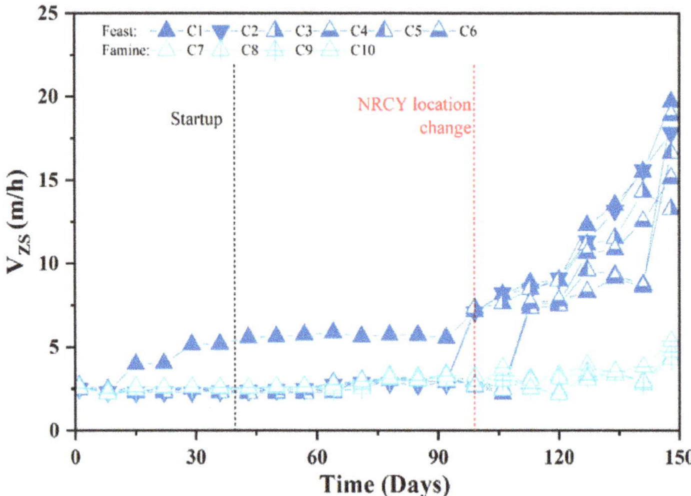

Figure 9. V_{ZS} profiles in different chambers over the course of the pilot operation.

In contrast, the improvement of sludge settleability in the famine chambers after the NRCY location change was very limited in terms of SVI_{30} (Figure 7), SVI_5/SVI_{30} (Figure 8) and V_{ZS} (Figure 9), which indicated that only part of the sludge inventory was densified through the granulation, unlike the system without NRCY mentioned in Section 3.1. It appears that the system was headed in the right direction, but as a result of unfavorable conditions for the entire system granulation, more time would have been needed to nurture the granular sludge formation and weed out the flocculent sludge from the system. One of the bottlenecks is the high feast-to-famine duration ratio due to supplemental carbon addition to the MLE system with NRCY (Figure 3).

3.4. Effects of NRCY Location on COD and Nitrogen Removal

As discussed in the previous section, after the change of NRCY location, better sludge settleability was observed, especially in the feast chambers. However, limited sludge settleability improvement was achieved in the famine zone, which indicated that only part of the sludge inventory was densified through granulation. The potential reason for this could be the relatively higher feast and famine ratios of the system after the NRCY location change (Figure 1B). As shown in Figure 3, before the NRCY relocation, the majority of the rbsCOD was removed in the first four chambers, leaving mostly refractory sCOD in downstream chambers, which resulted in a feast-to-famine duration ratio of 0.67. After NRCY relocation, the MLVSS of the first several chambers declined substantially, as mentioned in Section 3.2. Consequently, under the same OLR, rbsCOD was not used up until the seventh chamber after the change of NRCY location, which led to a feast-to-famine duration ratio of 1.5 (Figure 3). It was reported by [12] that to achieve successful continuous sludge densification in PFRs, a feast-to-famine duration ratio should be equal to or less than 0.6 in addition to an adequate external gravity selection pressure. With this being said, feast-to-famine duration ratios of 0.67 and 1.5 before and after NRCY relocation, respectively, were both considered unfavorable for successful sludge densification in PFRs. However, the change of NRCY location promoted the sludge densification, especially in feast chambers in terms of sludge settleability, despite the unfavorable feast-to-famine duration ratio. One can see that a biological selection internal to the reactor imposed by separately returning heavier densified sludge and lighter bioflocs to the feast and famine zones with the aid of NRCY relocation, and the increased gravity selection pressure in terms of the upflow velocity in the selector could compensate for the increase in feast-to-famine duration ratios to some extent.

Influent tCOD and sCOD removal efficiencies remained around 75% and 65%, respectively, and they were uninterrupted throughout the entire operation (Figure 10A). Similarly, ammonia removal efficiencies also remained stable (Figure 10B), regardless of NRCY location change, excluding some fluctuations from Days 71 to 92 due to SRT alteration (Figure 5B), which indicated the good nitrification performance over the course of the operation. These could be explained by the fact that floc recirculation from the bottom of the clarifier allowed for the retention of an adequate amount of sludge inventory. Figure 11A,B show the nitrogen profiles on Day 57 (before NRCY location change) and Day 127 (after NRCY location change), respectively. The NRCY location change seemed to have no effect on the system TIN removal, as both locations resulted in around 70% TIN removal and met the discharge limit of TIN < 10 mg/L. Specifically, system TIN removals mainly relied on the methanol-driven denitrification in anoxic chambers, especially in C3 (Figure 11A,B). As shown in Figure 12, DO concentrations were above 1 mg/L in C1 and C2, but dropped to around 0.6 mg/L in C3 on both Days 57 and 127. In other words, the chambers did not turn strictly anoxic until C3, although methanol was added in C1, and aeration was not employed in the first three chambers. The DO concentrations in C1 were even higher than the influent. This could be explained by the fact that NRCY and granule recirculation streams to C1 were entrained with large amounts of DO. As mentioned in Section 2, these air-uplifted granule streams from the underflow of the upflow selector contained air. The DO was not consumed by ordinary heterotrophic organisms (OHOs) until the third chamber (C3). As a result, most of the denitrification took place in C3, as shown in Figure 11 A,B. The TIN drop in C1 could be ascribed to the dilution caused by the mix of influent, NRCY and granule recirculation. Nitrification started as aeration began in C4 and basically, all of the NH_4^+–N residuals were oxidized to NO_3^-–N before C9.

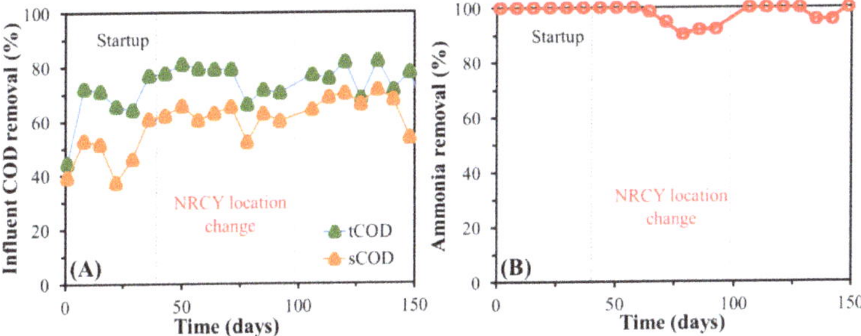

Figure 10. Profiles of (**A**) influent COD and (**B**) ammonia removal efficiencies over the course of the pilot operation.

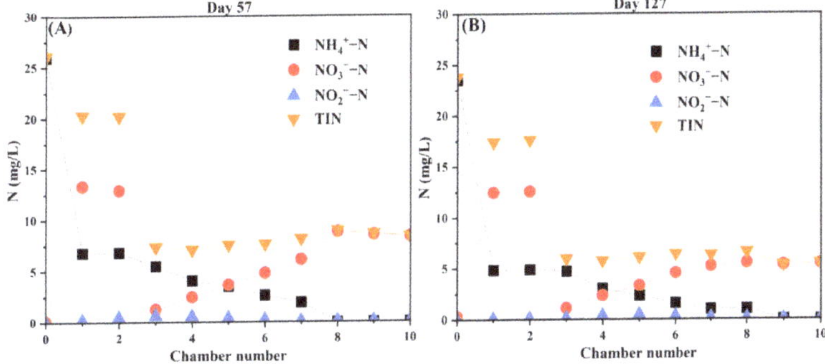

Figure 11. Nitrogen concentration profiles in different chambers on Day 57 before NRCY location change, (**A**) and Day 127 after NRCY location change (**B**). Chamber 0 refers to the influent.

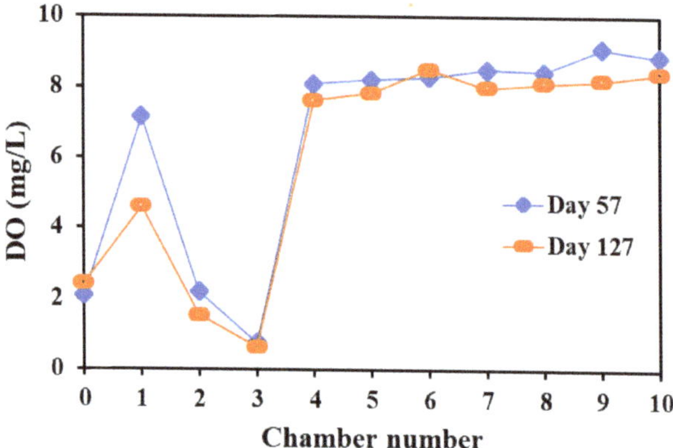

Figure 12. DO concentration profiles in different chambers on Day 57 before NRCY location change, and Day 127 after NRCY location change. Chamber 0 refers to the influent.

4. Discussion

4.1. The Synergy between Physical and Biological Selection Pressures

Granulated cells were not able to compete with dispersed cells (biofocs) for the limited substrate availability because bioflocs have looser structures and a higher specific surface area that promotes fast substrate diffusion and utilization [16]. There is a consensus that sludge densification is normally driven by physical selection pressure that selectively reduces the persistence of bioflocs in a bioreactor [1]. For this reason, extremely low SRTs, e.g., 0.3–0.4 days, had to be employed with such a conventional approach to rid the biomass inventory of bioflocs in a radical way [3,18], because the SRTs of dense granules and bioflocs were not able to be differentiated in previous granulation studies. As an undesirable consequence, the MLVSS in a bioreactor subjected to such a radically short SRT will also have a dramatic reduction in performance, resulting in poor effluent quality [3,18]. This performance loss resulting from the drastic physical out-selection pressure could be resolved by incorporating a strategy that manages to separate flocculent sludge and granule SRTs, while incorporating reactor internal biological selection through controlling feast and famine regimes for the distinctly different sludge morphologies, i.e., selective densified and biofloc returns to feast and famine zones, respectively. As shown in Figure 1B, an upflow selector with a selection velocity of 10–17 m/h offered the possibility for SRT deviation of densified and flocculant sludge. It is noteworthy that other than the loss of solids in the final clarifier effluent, biofloc wasting from the bottom of the clarifier through SRT control is the only outlet for bioflocs to be removed from the PFR system (Figure 1B), and bioflocs were the only type of sludge wasted. The short SRT of bioflocs realized a significant out-selection of this type of biomass, while the retention and recirculation of densified sludge resulted in their persistence and dominance in the sludge bed of feast chambers, as mentioned in Section 3.3. In contrast to the physical selection provided by the upflow selector external to the bioreactors, biological selection internal to the reactor imposed by separately returning a heavier densified sludge and lighter bioflocs to the feast and famine zones further boosted biofloc out-selection by subjecting bioflocs to low substrate availability. To this end, an innovation on the NRCY location of an MLE process in a PFR pilot train was conducted in this study, as mentioned in the previous sections, i.e., returning nitrate-rich mix liquor from the equalization tank after the clarifier instead of the aeration tank to avoid returning bioflocs to the feast chambers. This is a novel application of internal biological selection pressure that can boost a continuous-flow sludge densification and has not been widely accepted in full-scale applications. Consequently, the densified sludge MLSS gradually built up in the PFR, obtaining the advantage from their growth in the feast zone, while bioflocs were steadily starved out.

4.2. Implications of Full-Scale Application

At present, the most popular continuous flow selectors employed in MLE processes in WRRFs are hydrocyclones, which are currently limited to sidestream wasting applications due to technology sizing limitations and associated costs [5,9]. UOSA has also employed hydrocyclones to successfully enhance its full-scale activated sludge settleability. As designed, only 5% of the RAS was processed through the hydrocyclones. Consequently, only moderate sludge settleability improvement was observed over 23 months of operation. In contrast, in this study, it only took 148 days to achieve sludge densification in C2, C3 and C4, as indicated by SVI_{30} profiles in Figure 7. The SVI_{30} of C2, C3 and C4 dropped to below 60 mL/g on Day 148, while the SVI_{30} of the rest of the chambers still kept above 60 mL/g (Figure 7). If we assume that biomass in C2, C3 and C4 with $SVI_{30} < 60$ mL/g was densified sludge and the biomass in other chambers (C5 to C10) was bioflocs, as suggested by [1], then around 32% sludge inventory was densified in terms of the MLVSS shown in Figure 4. According to [10], pursuing full granulation in full-scale application is unnecessary, and only a smaller quantity of aerobic granular sludge proportion, as low as 20%, is sufficient to maintain the SVI below 100 mL/g, and could solve a lot of operational problems for plants suffering acute recurring seasonal bulking or subjected to hydraulic overloading beyond original clarifier design, while experiencing storm-induced peak events.

In addition, the design of the upflow selector and an equalization tank incorporated in Figure 1B is achievable in WRRFs. For example, the upflow selector mimics a conventional clarifier with a very high SOR, so there may be, in some cases, the flexibility to fit upflow physical selector structures within WRRFs. Therefore, the PFR pilot design (Figure 1B) for MLE process with a modified NRCY location tested in this study could be developed as an alternative to hydrocyclones for full-scale, greenfield, continuous sludge densification applications. However, the effluent TSS increase and sludge loss issues mentioned in Section 3.2 after the NRCY location change should be further addressed in future studies.

5. Conclusions

The following concluding remarks can be drawn from this study:

- MLE configuration with NRCY could hinder the sludge granulation, but the hindrance could be alleviated by the NRCY location change from the aerobic chamber to the equalization tank after the clarifier.
- MLE configuration tended to have higher feast-to-famine ratios, the negative effects of which on the sludge densification could be compensated by NRCY location change to some extent.
- The NRCY location change in conjunction with floc recirculation and continuous upflow selection can be advantageous to drive sludge densification without a radical washout of the sludge inventory.
- The NRCY location change had no effects on the COD and nitrogen removal efficiencies.
- Higher feast-to-famine ratios of the MLE system with NRCY could be the bottleneck for full sludge granulation processes.

Author Contributions: Conceptualization, J.-F.W., Z.-H.A. and Z.-W.W.; methodology, Z.-H.A.; writing—original draft preparation, J.-F.W.; writing—review and editing, J.-F.W., X.-Y.Z., Z.-W.W., M.B. and B.A.; supervision, Z.-W.W., M.B. and B.A. All authors have read and agreed to the published version of the manuscript.

Funding: This research received no external funding.

Data Availability Statement: The data presented in this study are available on request from the corresponding author.

Conflicts of Interest: The authors declare no conflict of interest.

References

1. Kent, T.R.; Bott, C.B.; Wang, Z.-W. State of the art of aerobic granulation in continuous flow bioreactors. *Biotechnol. Adv.* **2018**, *36*, 1139–1166. [CrossRef] [PubMed]
2. An, Z.; Wang, J.; Zhang, X.; Bott, C.B.; Angelotti, B.; Brooks, M.; Wang, Z.-W. Coupling physical selection with biological selection for the startup of a pilot-scale, continuous flow, aerobic granular sludge reactor without treatment interruption. *Water Res. X* **2023**, *19*, 100186. [CrossRef]
3. Sun, Y.; Angelotti, B.; Wang, Z.-W. Continuous-flow aerobic granulation in plug-flow bioreactors fed with real domestic wastewater. *Sci. Total Environ.* **2019**, *688*, 762–770. [CrossRef] [PubMed]
4. Avila, I.; Freedman, D.; Johnston, J.; Wisdom, B.; McQuarrie, J. Inducing granulation within a full-scale activated sludge system to improve settling. *Water Sci. Technol.* **2021**, *84*, 302–313. [CrossRef] [PubMed]
5. Partin, A.K. Hydrocyclone Implementation at Two Wastewater Treatment Facilities to Promote Overall Settling Improvement. Master's Thesis, Virginia Tech, Blacksburg, VA, USA, 2019.
6. Regmi, P.; Sturm, B.; Hiripitiyage, D.; Keller, N.; Murthy, S.; Jimenez, J. Combining continuous flow aerobic granulation using an external selector and carbon-efficient nutrient removal with AvN control in a full-scale simultaneous nitrification-denitrification process. *Water Res.* **2022**, *210*, 117991. [CrossRef] [PubMed]
7. Ford, A.; Rutherford, B.; Wett, B.; Bott, C.B. Implementing hydrocyclones in mainstream process for enhancing biological phosphorus removal and increasing settleability through aerobic granulation. *Proc. Water Environ. Fed.* **2016**, *2016*, 2809–2822. [CrossRef]
8. Welling, C.; Kennedy, A.; Wett, B.; Johnson, C.; Rutherford, B.; Baumler, R.; Bott, C.B. Improving settleability and enhancing biological phosphorus removal through the implementation of hydrocyclones. *Proc. Water Environ. Fed.* **2015**, *2015*, 6171–6179. [CrossRef]
9. Wett, B.; Podmirseg, S.M.; Gómez-Brandón, M.; Hell, M.; Nyhuis, G.; Bott, C.B.; Murthy, S. Expanding DEMON sidestream deammonification technology towards mainstream application. *Water Environ. Res.* **2015**, *87*, 2084–2089. [CrossRef] [PubMed]
10. Roche, C.; Donnaz, S.; Murthy, S.; Wett, B. Biological process architecture in continuous-flow activated sludge by gravimetry: Controlling densified biomass form and function in a hybrid granule–floc process at Dijon WRRF, France. *Water Environ. Res.* **2022**, *94*, e1664. [CrossRef] [PubMed]
11. An, Z.; Bott, C.B.; Angelotti, B.; Brooks, M.; Wang, Z.-W. Leveraging feast and famine selection pressure during startup of continuous flow aerobic granulation systems to manage treatment performance. *Environ. Sci. Water Res. Technol.* **2021**, *7*, 1622–1629. [CrossRef]
12. An, Z.; Sun, Y.; Angelotti, B.; Brooks, M.; Wang, Z.-W. Densification dependence in continuous flow and sequential batch granulation systems on reactor feast-to-famine duration ratio. *J. Water Process Eng.* **2021**, *40*, 101800. [CrossRef]
13. Baird, R.B.; Bridgewater, L. *Standard Methzods for the Examination of Water and Wastewater*, 23rd ed.; American Public Health Association: Washington, DC, USA, 2017.
14. Zhang, Z.; Qiu, J.; Xiang, R.; Yu, H.; Xu, X.; Zhu, L. Organic loading rate (OLR) regulation for enhancement of aerobic sludge granulation: Role of key microorganism and their function. *Sci. Total Environ.* **2019**, *653*, 630–637. [CrossRef] [PubMed]
15. Ghangrekar, M.; Asolekar, S.; Joshi, S. Characteristics of sludge developed under different loading conditions during UASB reactor start-up and granulation. *Water Res.* **2005**, *39*, 1123–1133. [CrossRef] [PubMed]
16. Tay, J.H.; Liu, Q.S.; Liu, Y. Characteristics of aerobic granules grown on glucose and acetate in sequential aerobic sludge blanket reactors. *Environ. Technol.* **2002**, *23*, 931–936. [CrossRef] [PubMed]
17. Schwarzenbeck, N.; Erley, R.; Wilderer, P. Aerobic granular sludge in an SBR-system treating wastewater rich in particulate matter. *Water Sci. Technol.* **2004**, *49*, 41–46. [CrossRef] [PubMed]
18. Cofré, C.; Campos, J.L.; Valenzuela-Heredia, D.; Pavissich, J.P.; Camus, N.; Belmonte, M.; Pedrouso, A.; Carrera, P.; Mosquera-Corral, A.; Val del Río, A. Novel system configuration with activated sludge like-geometry to develop aerobic granular biomass under continuous flow. *Bioresour. Technol.* **2018**, *267*, 778–781. [CrossRef] [PubMed]

Article

Feasibility Study of Applying Anaerobic Step-Feeding Mode for the Treatment of High-Strength Wastewater in Granular Sequencing Batch Reactors (GSBRs)

Elsayed Mady [1,2], Jan Oleszkiewicz [1] and Qiuyan Yuan [1,*]

[1] Department of Civil Engineering, Price Faculty of Engineering, University of Manitoba, Winnipeg, MB R3T 5V6, Canada
[2] Public Works Engineering Department, Tanta University, Tanta 31527, Egypt
* Correspondence: qiuyan.yuan@umanitoba.ca

Highlights:

- Granules' EPS content decreased with the disintegration of AGS post anaerobic step-feeding.
- Matured granules stability and reactor performance favored the fast single feeding mode followed by anaerobic mixing than anaerobic step-feeding strategy.
- Granules' EPS protein-to-carbohydrate ratio did not have a significant effect on granulation process either in maturation stage or after applying anaerobic step-feeding strategy.

Abstract: This study investigated the feasibility of applying an anaerobic step-feeding strategy to enhance the performance of granular sequencing batch reactors (GSBRs) in terms of operational stability of the cultivated mature granules and nutrient removal efficiencies. Two identical 5 L reactors were operated with a total cycle time of 8 h. GSBRs were operated with high-strength synthetic wastewater (COD = 1250 ± 43, ammonium (NH_4-N) = 115.2 ± 4.6, and orthophosphate (PO_4-P) = 17.02 ± 0.9 mg/L) for 360 days through three stages: (1) Cultivation, 125 days (>2.1 mm); (2) Maturation, 175 days (>3 mm); (3) alternate feed loading strategy for R2 only for 60 days (anaerobic step-feeding). The granulation process, the physical properties of the granules, the nutrients, and the substrate removal performance were recorded during the entire operational period. For the cultivation and maturation stages, both reactors followed the fast single feeding mode followed by anaerobic mixing, and the results indicated a strong correlation between R1 and R2 due to the same working conditions. During the cultivation stage, adopting high organic loading rate (OLR) at the reactor start-up did not accelerate the formation of granules. Removal efficiency of PO_4-P was less than 76% during the maturation period, while it exceeded 90% for COD, and was higher than 80% for NH_4-N without effect of nitrite or nitrate accumulations due to simultaneous nitrification–denitrification. After changing filling mode for R2 only, there was unexpected deterioration in the performance and a rapid disintegration of the matured granules (poor settleability) accompanied by poor effluent quality due to high content of suspended solids because of applying selection pressure of short settling time. Consequently, GSBRs operation under the effect of fast single feeding mode followed by anaerobic mixing favors stable long-term granule stability.

Keywords: aerobic granular sludge; biological nutrient removal; high-strength organic wastewater; granular size; step-feeding; sequencing batch reactors

Citation: Mady, E.; Oleszkiewicz, J.; Yuan, Q. Feasibility Study of Applying Anaerobic Step-Feeding Mode for the Treatment of High-Strength Wastewater in Granular Sequencing Batch Reactors (GSBRs). *Processes* **2023**, *11*, 75. https://doi.org/10.3390/pr11010075

Academic Editor: Yongqiang Liu

Received: 30 November 2022
Revised: 15 December 2022
Accepted: 20 December 2022
Published: 28 December 2022

1. Introduction

Recently, aerobic granular sludge (AGS) has gained popularity among researchers due to its superior properties as compared to conventional activated sludge (CAS). AGS is an economically and environmentally promising replacement for CAS, for purification of different types of wastewaters (e.g., municipal and heavily loaded industrial wastewater) [1].

Layered structure of AGS is known for high removal performance of different nutrients via multiple biological processes (e.g., nitrification and denitrification) occurring in a single tank [2]. AGS system has biofilms composed of auto-immobilized cells that are generated during treatment in a sequencing batch reactor (SBR), where the clarifier is eliminated, and a settling phase is included in the reactor schedule. Thus, a return activated sludge line is not applicable, depending on the amount of biomass retained in the reactor after decanting. By switching to aerobic granulation system, daily produced biomass is also decreased leading to reduction in the overall cost of up to 25% and energy requirement up to 30% [3]. Since the inception of AGS in the late 1990s as an innovative technology [3], it has been applied successfully for the treatment of different influents; however, there are some unresolved problems such as granule disintegration, long period of granule formation when using sewage with high organic loads, and nitrite accumulation [4].

The literature review findings show the cultivation of AGS in SBR by investigating the impact of various operational parameters that can control the granulation process (i.e., formation and stability), such as alternating feast and famine conditions, reactor configuration (height-to-diameter ratio, hydraulic retention time (HRT), and exchange ratio), hydrodynamic shear forces, influent characteristics (COD:TN:TP ratio), and influent distribution (influence the morphology and microbial community of AGS), selective wasting (in most GSBRs, settling time is the selection pressure for granulation process), organic loading rates, biomass characteristics, and other environmental factors (e.g., dissolved oxygen, pH, and temperature).

Different studies have reported varying conditions for the development of aerobic granulation process. For example, R.A. Hamza et al. (2018) investigated the long-term stability of AGS treating high organic loads through a semi-pilot scale SBR and found that F/M ratio has played a key role in the formation and stability of granules [5]. At F/M ratios between 0.5 and 1.4 gCOD/gVSS·d, stable granules with excellent settleability were maintained, while fluffy granules were caused when the F/M value exceeded 2.2 gCOD/gVSS·d. Another study demonstrated that utilizing longer HRT caused a reduction in the applied OLR, which was found to be more effective for granulation process by providing longer starvation time [6].

Stable AGS was achieved successfully in a previous study by applying low hydrodynamic shear (0.4 cm/s) [7]. In addition, successful removal of the fast-growing organisms (e.g., ordinary heterotrophic organisms (OHO)) from granules' surface has been achieved by applying higher hydrodynamic shear [8,9].

A lot of challenges are related to an efficient and continuous application of the obtained mature granules in the treatment processes. For example, causes of granule instability have also been studied and linked to a range of factors such as: proliferation of filamentous bacteria at high OLR [8,9], as well as decrease and destruction in the structure of extracellular polymeric substances (EPS), with the decrease in protein-to-carbohydrate ratio (proteins play an essential role in granule stability) [10–12]. EPS in the aerobic granulation process, acts similarly to a 'glue' to induce biofilm formation among the microorganisms, where EPS are synthesized as a derivative of organic compounds, such as proteins, polysaccharides, lipids, glycoproteins, nucleic acids [13,14]. It depends on several factors, including aeration time, cycle duration, superficial air velocity, reactor conditions, and type of inoculum sludge that affect the production and composition of EPS.

Feeding mode is one of the key factors in the selection, formation, and stability of AGS. GSBRs are categorized into two types: continuously aerated, and anaerobic influent distribution. Anaerobic filling in GSBR has been widely accepted, in which the duration and influent load are considered efficiency determining factors. A few previous studies have demonstrated that step-feeding can be an efficient strategy to remove organic matter and nutrients simultaneously, especially with great potential to favor granulation in effluents with high OLRs (e.g., sanitary landfill leachate and high-strength industrial discharges). For example, another previous study has evaluated the effect of filling mode on the performance and granules' operational stability to treat high-strength ammonium concentrations through three SBRs: R1, fast feeding in the anaerobic period (20 min); R2, slow feeding in the anaerobic period (40 min); and R3, distribution of feeding throughout the cycle [15]. Results of step-feeding in R3 were the best, where the problems such as granules stability, biomass retention, and nitrite accumulation were minimized while treating leachate, as well as higher nutrients removal efficiencies were achieved compared to R1 and R2. In addition, the results from this study showed the positive impact of applying the strategy of step-feeding in R3 compared to R1 and R2, where the approach of step-feeding distributed throughout the cycle of operation was an excellent method to develop the sedimentability of biomass. Moreover, this strategy enhanced the proliferation of slow-growing organisms through the prevention of growing ordinary and fast-growing heterotrophic bacteria [15].

The effect of different feeding strategies on the proliferation dynamics of polyphosphate-accumulating organism (PAOs) and phosphorus removal has also been investigated [16]. Herein, three feeding strategies (e.g., anaerobic slow feeding, pulse feeding followed by anaerobic mixing, and pulse feeding) were applied through GSBRs. The obtained results show that feeding strategies had no impact on organic and ammonia removal and did not alter the microbial community. However, phosphorus removal performance has been significantly influenced. Reactors that have anaerobic slow feeding and pulse feeding followed by anaerobic mixing demonstrated high phosphorus recovery from wasted sludge; therefore, active PAOs can be accumulated, and GAOs can be significantly inhibited within the bioreactor leading to high specific phosphorus uptake and specific phosphorus release rates [16].

To the best of our knowledge, there is no current information on the effect of applying anaerobic step feeding strategy on matured AGS cultivated and matured under the effect of treatment of high-organic loads in GSBRs systems at the same operational conditions. Thus, the aim of this work is to investigate the potential enhancement in the substrate and nutrients removal performance of large AGS after changing feeding strategy of these GSBRs from fast single feeding mode followed by anaerobic mixing to step-feeding distributed throughout the anaerobic stage, when applied under the same previous operational conditions for treatment of high-strength wastewater.

2. Material and Methods

2.1. Experimental Set-Up

The experiments were conducted through two SBRs named R1 and R2 (cylindrical acrylic reactors) seeded with the same biomass (GSBRs: R1 and R2 in operation are shown in Figure S1—Supplementary File). The reactors had a 6.0 L capacity, with working volume of 5.0 L, internal diameter of 12 cm, and height-to-diameter ratio of 3.7. Volume exchange ratio of 60% was ensured through discharging the top 3 L of supernatant in 10 min following a 10 min of settling period. Reactors were operated with 8 h cycles daily including 10 min feeding (fast filling), 95 min of anaerobic reaction, 350 min of aeration phase, and 20 min of settling and discharging, as well as 5 min of an idle stage before starting a new cycle. Pumping influents and discharging effluents was performed by peristaltic pumps. Treatment cycles were automated using outlet timers (Traceable Digital Outlet Controller, Fisher Scientific, Waltham, MA, USA).

Nitrogen gas was used during the anaerobic phase and fed from the bottom of reactors to create anoxic environment (dissolved oxygen (DO) levels were less than 0.1 mg/L during the total period of anaerobic mixing), and to avoid biomass sedimentation since complete mixing using nitrogen gas enhances the contact between cells and nutrients and improves removal efficiency. Moreover, during the aerobic phase, fine air bubbles were introduced from the bottom of the reactors (DO levels were between 4–5 mg/L). Both nitrogen gas and air were introduced at the same level with a flow rate of 2.92 L/min resulting in superficial air velocity of 0.43 cm/s to induce hydrodynamic shear to achieve compact granules with desirable surface properties [9–12]. The following schematic diagram (Figure 1) describes the GSBRs used in this study.

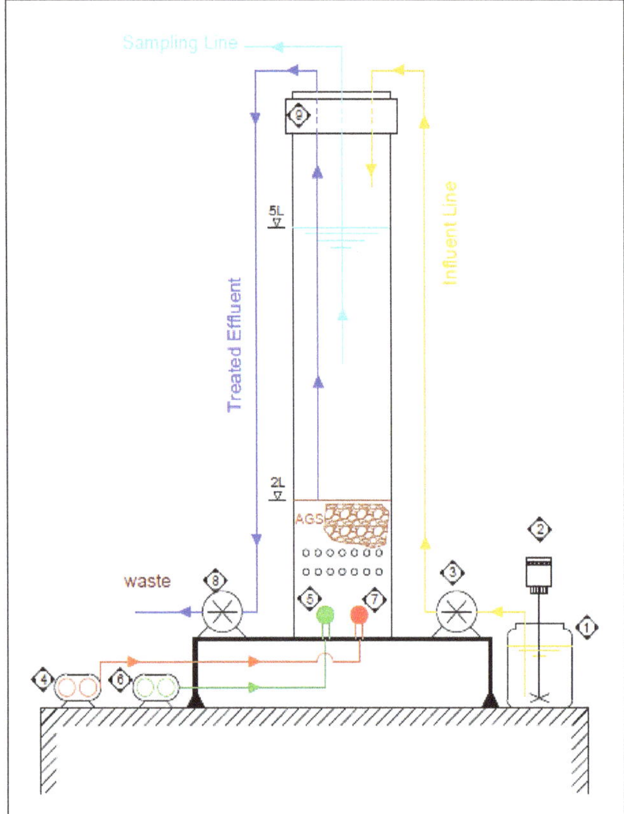

Figure 1. Schematic of the GSBRs in this study; (1) feeding tank; (2) mixer; (3) feeding pump; (4) nitrogen pump; (5) nitrogen diffuser; (6) air pump; (7) air diffuser; (8) decant pump; and (9) reactor's cap.

The total experiment duration for reactors was 360 days to treat high-strength synthetic wastewater (Table 1) under the same operational conditions: (a) 125 days for stage I (Cultivation stage: formation of AGS) after inoculation with the same sludge for R1 and R2, (b) 175 days for stage II (Maturation stage: operation under the same conditions), and (c) 60 days for stage III (R1 continued working under the same operational conditions of stage II, while R2 operated through new anaerobic step-feeding strategy for 60 days). Acclimation period for the first one was two weeks of adaptation after inoculation, while the second period was 10 days after changing filling mode of R2.

The feeding mode of R2 only was changed from fast single feeding mode followed by anaerobic mixing to anaerobic step-feeding strategy, while R1 operated with the same previous filling mode. As mentioned previously, the discharge is 3 L/cycle, and anaerobic stage is performed in 95 min for each cycle. Therefore, a liter is pumped to R2 at the beginning, and another liter is added after 30 min of anaerobic mixing, and the third liter is added after another 30 min. It is important to mention that 10 days after changing the filling mode of R2 was taken as an adaptation period before performing the required measurements of different parameters during this stage.

Table 1. Operation stages for R1 and R2.

	Period (Days) (Start and End Days)	Conditions of GSBRs
Stage I Cultivation	125 days Day 1 to 125	- Two weeks for acclimation of microorganisms after inoculation - 75 days as startup period for R1 and R2 - Cultivation of large AGS were achieved by the day 125 (size $\geq$ 2 mm) - Observation of physical properties (size, sludge volume indexes (SVI_5 and SVI_{30})), and performance (nutrients removal).
Stage II Maturation	175 days Day 126 to 300	- Reactors are operated under the same conditions as stage I - Maturation of the granules were achieved by day 175 (size $\geq$ 3 mm) - Observation of physical properties (size, SVI_5, and SVI_{30}), and performance (nutrients removal), as well as monitoring stability of matured granules.
Stage III Operation under Step-feeding	60 days Day 301 to 360	- R2 only changed to anaerobic step-feeding, then 10 days for adaptation - R1 has been operated under the same conditions of stages I and II - Observation of physical properties (size, SVI_5, and SVI_{30}), and performance (nutrients removal), as well as monitoring stability of matured granules.

The start-up of GSBRs with respect to both physical properties and substrate/nutrient removal is described above. Furthermore, the settling parameters $SVI_{30} < 90$ mL/g and SVI_{30}/SVI_5 ratio > 0.7, the size fraction $d > 0.5$ mm constituting at least 50% of total suspended solids (TSS), granule appearance based on microscopic images, and stable substrate and nutrient removal are provided in supplementary material.

2.2. Sludge Source and Synthetic Wastewater Composition

The same volume (2 L) of return activated sludge (RAS) was used to seed both reactors, which was obtained from the West End Water Pollution Control Centre in Winnipeg, MB, Canada. The mixed liquor suspended solids (MLSS) of the inoculated RAS was measured as 8.1 g/L (average of triplicates samples).

It is important to mention that particulate matter was not used in the substrate of the influent since it has been demonstrated to affect surface properties of granules and would therefore affect the results. During the entire operational period, GSBRs were treating high-strength synthetic wastewater containing an average of 1250 ± 43 mg/L of soluble COD, 144 ± 12 mg/L of TN, and 48 ± 7 mg/L of TP.

The synthetic municipal wastewater feed recipe was based on a previous study [12], which used high-strength synthetic wastewater; therefore, the influent synthetic wastewater comprised 1250 mg/L of yeast extract, 33 mg/L of ammonium chloride (NH_4Cl), and 9 mg/L of dipotassium phosphate ($K_2HPO_4 \cdot 3H_2O$). Other components such as $MgSO_4$ (5.8 mg/L), $CaCl_2$ (12 mg/L) and mineral solution (0.3 mg/L) containing 0.15 g/L H_3BO_3, 0.03 g/L $CuSO_4 \cdot 5H_2O$, 0.03 g/L KI, 0.12 g/L $MnCl_2 \cdot 4H_2O$, 0.06 g/L $Na_2MoO_4 \cdot 2H_2O$, 0.12 g/L $ZnSO_4 \cdot 7H_2O$, 0.15 g/L $CoCl_2 \cdot 2H_2O$, and 1.5g/L $FeCl_3 \cdot 6H_2O$ were also added. Ethylenediaminetetraacetic acid (EDTA: 3 mg/L) is used to prevent the precipitation of salts, such as calcium (Ca_2) and orthophosphate (PO_4^{-3}-P) present in the media. Sodium bicarbonate (12g $NaHCO_3$/g NH_4^+-N) was added to the influent each cycle during the entire experimental duration to achieve sufficient alkalinity (7.14 g $CaCO_3$/g NH_4^+-N) and subsequently achieve complete nitrification [12].

2.3. Analytical Methods

The dissolved oxygen and pH were measured near the center of the GSBRs using a DO probe and a pH probe. The concentrations of MLSS, MLVSS, total suspended solids of the effluent (TSS_{eff}), chemical oxygen demand of the influent and effluent ($sCOD_{inf}$, $sCOD_{eff}$) and Sludge Volume Index after 5 and 30 min (SVI_5 and SVI_{30}) of settling were measured according to standard methods [17]. Concentrations of phosphorous in form of PO_4^{3-}-P and nitrogen in forms of ammonium (NH_4^+-N), nitrite (NO_2^--N), and nitrate (NO_3^--N) for the influent and effluent were measured via a flow injection analyzer (Lachat Quick Chem 8500, HACH, Loveland, CO, USA). Kinetic analysis was performed during operational cycles through the measurements for nitrogen and phosphorus using filtered samples (soluble nitrogen and phosphorus concentration) through medium porosity Q5 filter paper (Fisher Scientific, Waltham, MA, USA).

Granules size was measured to estimate the development of the granulation process and granules' stability through microscopic images in SteREO Discovery (Axio Vert.A1, ZEISS, Gener, Germany). Moreover, surface morphology is investigated using the microscopic analyses on weekly to bi-weekly basis. In this study, we established that the mature aerobic granulation has been achieved (i.e., start of stage II) only when more than 80% of the biomass in R1 and R2 had a diameter greater than 0.2 mm [18], and both of sludge volume indices (SVI_5 and SVI_{30}) are lower than 50 mL/g. It is worth mentioning that the cycle tests were conducted more than three times weekly during the whole period of the experiment (i.e., stage I, II, and III) in order to investigate the performance of GSBRs and understand the mechanisms of SNDPR (simultaneous nitrification, denitrification, and phosphorus removal).

Analysis of the EPS content of the AGS has been performed according to [19]. For the extraction, low frequency ultrasound at 120 W for 5 min in an ice bath (before adding chemicals) extraction was applied for 10 mL of mixed liquor samples. Pre-treatment for the mixed liquor samples using 0.06 mL formaldehyde was conducted prior to storing these samples for 1 h at 4 °C. For the next step, 4 mL of NaOH is added, and stored for 3 h at 4 °C. EPS content was physically extracted at 10,000 rpm for 20 min at the end. Following the extraction, the supernatant for all samples was filtered through 0.2 μm filters to collect soluble fractions. The extracted samples were stored at -20 °C in aliquots until analyses were performed.

3. Results and Discussion

Firstly, there was a strong convergence in the results (as shown in Table 2) such as physical properties (i.e., size, SVI_5, and SVI_{30}), and performance (i.e., removal of organics and nutrients) of R1 and R2 over a period of 300 days of operation after inoculation (through cultivation and maturation of AGS until changing the filling mode of R2 on day 301). This was due to both GSBRs operated under the same conditions (e.g., reactors configuration, design parameters, influent characteristics, seeding sludge, etc.).

Table 2. Removal efficiencies of sCOD, NH_4^+-N, and PO_4^{-3}-P in R1 and R2.

Characteristics	Stage I		Stage II		Stage III	
	R1	R2	R1	R2	R1	R2
$sCOD_{inf}$ (mg/L)			1250 ± 43			
$sCOD_{eff}$ (mg/L)	127.7 ± 24	115.4 ± 25	93 ± 14	91 ± 12	76.82 ± 16	90.74 ± 7.5
sCOD removal (%)	89.5 ± 2.2	90.52 ± 1.6	91.66 ± 2.2	91.06 ± 1.4	94.55 ± 2.3	92.56 ± 2.2
NH_4^+-N_{inf} (mg/L)			115.2 ± 4.6			
NH_4^+-N_{eff} (mg/L)	51.5 ± 18	51.4 ± 13	20.43 ± 7.2	18.83 ± 5.7	12.42 ± 3.6	24.27 ± 6.5
NH_4^+-N removal (%)	55.2 ± 2	55.3 ± 7	82.2 ± 6.2	83.6 ± 5	89.2 ± 1.2	79 ± 1.6
NO_2^+-N_{eff} (mg/L)	27.9 ± 25	25.2 ± 22	14.8 ± 8.1	13.3 ± 9.6	0.67 ± 1.0	10.6 ± 1.3
NO_3^+-N_{eff} (mg/L)	1.6 ± 1.8	1.9 ± 2.55	27.2 ± 12.1	25.43 ± 14.6	0.88 ± 1.3	47.2 ± 2.1
PO_4^{-3}-P $_{inf}$ (mg/L)			17.02 ± 0.9			
PO_4^{-3}-P $_{eff}$ (mg/L)	9.68 ± 2.3	9.8 ± 1.7	4.13 ± 0.8	4.3 ± 0.1	1.39 ± 0.13	9.23 ± 1.4
PO_4^{-3}-P removal (%)	42.8 ± 5.9	42 ± 7.2	75.6 ± 5.9	74.5 ± 9.6	92.1 ± 1.2	46.43 ± 2.7

The mean value of the measured pH through the anaerobic phase for both reactors during cultivation period (first 125 days after inoculation) was about 8.9, and it decreased through the aerobic phase to 8.2. During the maturation stage (day 126 to 300), these values were further decreased to 8.3 and 7.9 through the anaerobic and aerobic phases each cycle. It is worth mentioning that DO concentrations have been measured during the entire period of operation for both reactors in order to continuously monitor the aerobic and anaerobic conditions during operational cycles. DO concentrations during anaerobic phase did not exceed 0.1 mg/L, while it ranged between 4–5 mg/L during the aerobic phase. It is important to mention that the experiment was conducted at room temperature (18 $\pm$ 2 °C).

3.1. Granulation Processes

The seed sludge was dark brown in colour and had MLSS, mixed liquor volatile suspended solids (MLVSS), and SVI_{30} of 8100 mg/L, 6955 mg/L, and 128 mL/g, respectively (average of triplicates samples), as well as a mean particle size of 115 μm. The two SBRs were initially operated with the same volume (2 L). During the acclimation period (15 days), the MLSS concentrations were gradually reduced in GSBRs accompanied with higher TSS_{eff} (974 $\pm$ 122 mg/L). The reason for this reduction is due to the impact of applying short sedimentation phase (10 min) during the cycle operation, which is considered as the selection pressure and a key strategy for developing aerobic granulation in order to wash out the poor settling biomass. However, after adaptation period and even after biomass stabilization, higher amounts of solids existed in the effluents. In other words, constant washout sludge was observed in R1 and R2. This is consistent with previous studies of AGS systems applied to treat high-strength wastewaters [20,21]. Factually, 75 days were estimated as a startup period for both reactors.

Biomass characteristics (MLSS, SVI_5, and SVI_{30}) for R1 and R2 are presented in Figure 2 and Table 3. As mentioned, MLSS concentrations continued to decrease in both reactors from the first day of operation until day 50. During the same stage, in terms of sedimentability and according to previous studies, various researchers demonstrated that the SVI_{30}/SVI_{10} or SVI_{30}/SVI_5 ratios can be considered good indicators of granulation, meaning that a value closer to 1.0 indicates that the sludge consists mainly of mature granules [22–25]. It took about 63 days for flocculent sludge (SVI_{30} more than 125 mL/g and SVI_{30}/SVI_5 ratio lower than 0.4) to convert to small aerobic granules, as shown in Figure 2 (SVI_{30} is 85 mL/g and SVI_{30}/SVI_5 ratio is 0.52 for R1, while SVI_{30} is 93 mL/g and SVI_{30}/SVI_5 ratio is 0.56 for R2). Measurements of the size of AGS that were performed at day 50 were exceeded by 0.5 mm for R1 and R2, meeting the requirement for granular sludge. By the end of cultivation stage (day 125), biomass concentrations (MLSS) increased because of the granules' superior settleability and high biomass density and reached 10 g/L and 11.8 g/L for R1 and R2, respectively, which is consistent with previous studies [22]. Moreover, large AGS have been achieved in both GSBRs SVI_{30} = 48.2 mL/g and SVI_{30}/SVI_5 ratio = 0.78 for R1, and SVI_{30} = 58.5 mL/g and SVI_{30}/SVI_5 ratio = 0.73 for R2. While granules' size reached more than 2.22 mm and 2.13 mm for R1 and R2, respectively, by day 125.

The biomass concentrations for R1 and R2 were 10,164 $\pm$ 111 mg/L and 9156 $\pm$ 157 mg/L, respectively, on day 125 (end of cultivation stage). During the following period (day 126–300), biomass accumulation continued and directly increased biomass concentrations for GSBRs. Table 3 demonstrates the average MLSS concentrations during the whole maturation period. MLSS and MLVSS of both reactors are shown in Figure 3. The rise in biomass concentrations were achieved during the same period on day 294, where the highest MLSS was 14,844 $\pm$ 78 mg/L for R1, and 13,564 $\pm$ 128 mg/L for R2. This is a direct result from continued biomass accumulation. In this context, there were fluctuations in the TSS_{eff} for R1 and R2 which can be attributed to applying shorter settling time (10 min) during the operational cycle (i.e., discharging of poor settling biomass selection pressure). The average concentrations of TSS_{eff} for R1 and R2 during cultivation period (1–125 day) were 180 $\pm$ 18 mg/L and 167 $\pm$ 15 mg/L, respectively (except the first 42 days after inoculation since there were high concentrations of TSS_{eff} and VSS_{eff} of reactors due to discharging the poor settling flocs). In addition, based on the concentration of TSS_{eff} and MLSS, the average solid retention time (SRT) was calculated during the cultivation period (day 15–125) of 18 ± 1 day and 16 ± 2 day for R1 and R2, respectively, while average SRT during the maturation stage (day 126–300) was 28 ± 2 day for R1 and R2 (Table 3).

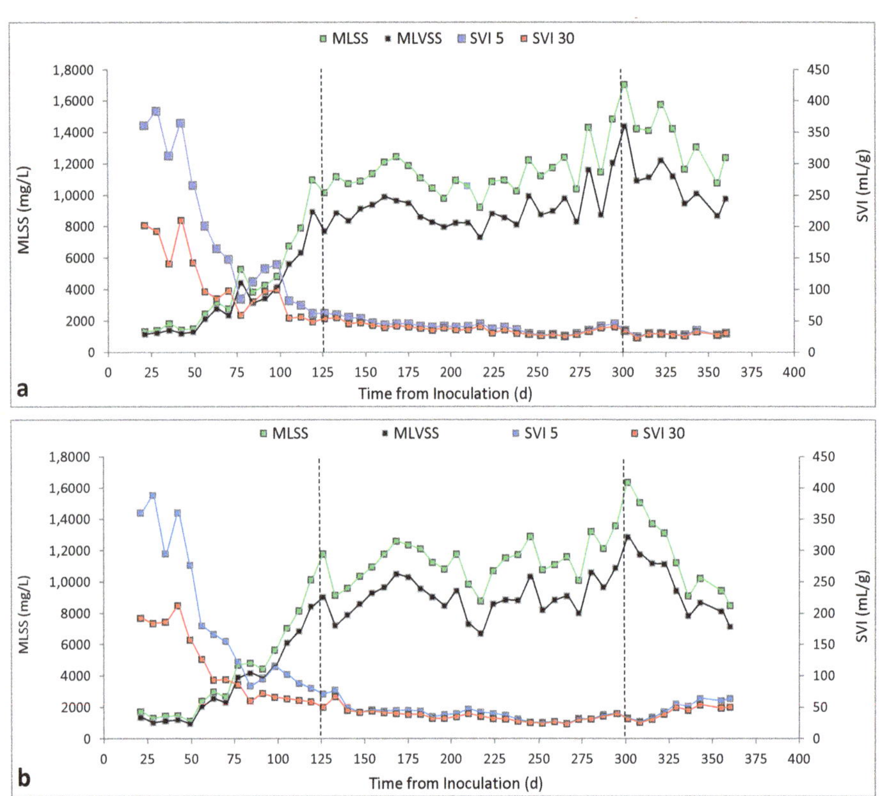

Figure 2. Stability and characteristics of AGS over the entire operational period for (**a**) R1 and (**b**) R2.

Table 3. AGS characteristics (mean values) throughout the operational period for R1 and R2.

Characteristics	Stage I		Stage II		Stage III	
	R1	R2	R1	R2	R1	R2
SVI_5 (mL/g)	62 ± 4.8 at day 120	80 ± 8.8 at day 120	40.85 ± 4.5	39.2 ± 6.2	29.34 ± 1.5	49.91 ± 5.8
SVI_{30} (mL/g)	48.2 ± 12.1 at day 120	58.5 ± 10.8 at day 120	36.62 ± 3.8	35.56 ± 2.1	28.27 ± 1	42.77 ± 6
Mean diameter (mm)	2.22 ± 0.13 day 120	2.02 ± 0.2 day 120	3.02 ± 0.01 at day 300	3.08 ± 0.6 at day 300	3.13 ± 0.08 at day 360	2.94 ± 0.11 at day 360
EPS content (mg/gVSS)	130.78 ± 41.6	141.62 ± 90.1	203.7 ± 23.1	207.9 ± 35.3	228.6 ± 11.4	194.6 ± 15.6
F/M ratio (gCOD/gVSS·d)	0.437 ± 0.77	0.398 ± 0.28	0.24 ± 0.1	0.24 ± 0.06	0.196 ± 0.04	0.22 ± 0.07
OLR (Kg COD·m^{-3}·d^{-1})			2.16 ± 0.25			
MLSS (mg/L)	6265 ± 2523	6415 ± 2136	11348 ± 1260	11295 ± 1219	13695 ± 1964	11860 ± 2808
HRT (h)			13.33			
SRT (d)	18 ± 1	16 ± 2	28 ± 2	28 ± 2	29 ± 1	22 ± 1

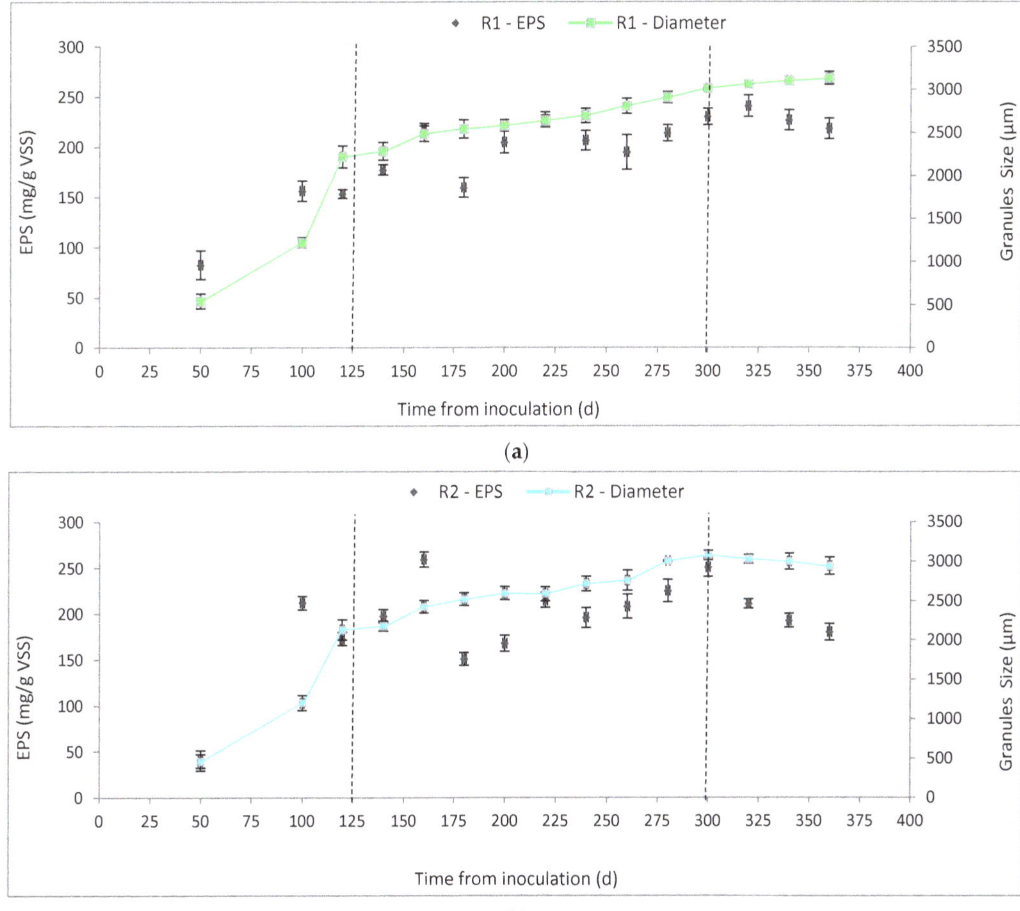

Figure 3. EPS measurements (protein + carbohydrate) over the entire operational period (360 days) of (**a**) R1 and (**b**) R2.

It has been demonstrated through several studies that higher F/M ratios favor formation and development of the AGS because of the enhanced EPS production [12]. Therefore, it should be noted that through the first stage (cultivation stage—day 125) and under continuous treatment of high organic loads (2.25 Kg COD/m^3·d), the gradual increase in the biomass concentrations (MLSS and MLVSS) might be the reason for the delay in the formation of large granules. F/M ratio decreased from 1.71 ± 0.11 gCOD/gVSS·d for R1 and 1.35 ± 0.18 gCOD/gVSS·d for R2 (at the beginning of cultivation period) to 0.29 ± 0.1 gCOD/gVSS·d for R1 and 0.19 ± 0.12 gCOD/gVSS·d for R2 (day 125). Consequently, this led to lower production of EPS compared to the following stage (maturation) as shown in Table 3 and Figure 3.

Regarding EPS, these components are biopolymers consisting of polysaccharides, proteins, and other substances, which play a crucial role in the granules' structure, production, and stability. Several studies have shown that stable AGS have higher protein (PN) structures than polysaccharides carbohydrates (PS), and PN/PS ratio is a way of characterizing its stability [14]. PN/PS ratio higher than 25 [12], as well as lower values of 4.8 [26] and 0.6 [27,28] have been reported. The concentrations of EPS are presented in Figure 3. In the current study, EPS has been estimated as a combination of PN and PS. EPS ranged between 36 ± 1.1 mg/gVSS (PN/PS ratio of 1.11) for R1 and 31 ± 0.8 mg/gVSS (PN/PS ratio of 1.38) for R2 on day 28 and increased to 153.3 ± 4.8 mg/gVSS (PN/PS ratio of 4.3) for R1 and 172.7 ± 2.1 mg/gVSS (PN/PS ratio of 7.2) for R2 on day 120. Thus, the granules' size (Figure 4) increased during the same stage and reached 2.22 ± 0.13 mm for R1 and 2.13 ± 0.13 mm for R2. During the maturation period, there was a strong correlation of the average concentrations of EPS for both reactors. EPS for R1 and R2 were 203.7 mg/gVSS (average PN/PS ratio of 9.17) and 207.9 mg/gVSS (average PN/PS ratio of 9.51). Accordingly, the granules' size (Figure 4) was increased to about 3.02 ± 0.09 mm for R1 and 3.08 ± 0.06 mm for R2, respectively.

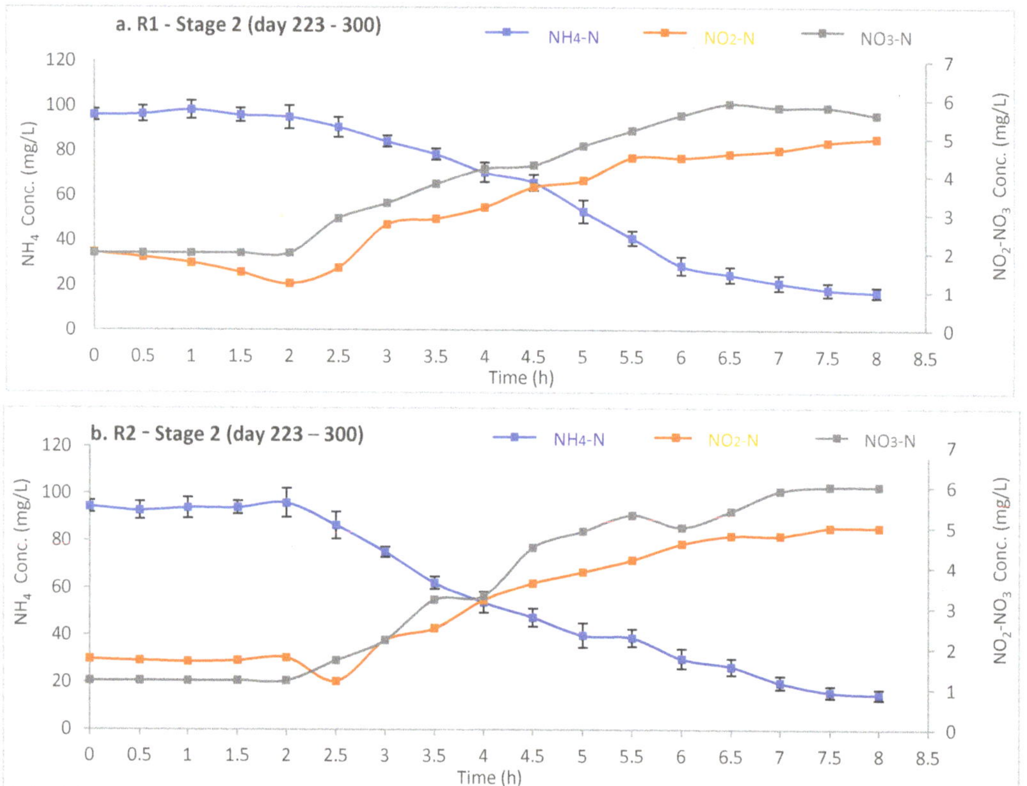

Figure 4. Performance profile of nitrogen distributed over a typical cycle during stage II by (**a**) R1 (**b**) R2.

3.2. Performance of the Reactors

There was a strong correlation of the reactors' performance in terms of sCOD, ammonium (NH$_4$-N), nitrite (NO$_2$-N), nitrate (NO$_3$-N), and phosphorus (PO$_4$-P) concentrations between R1 and R2 through cultivation and maturation stages (day 1–300). These results were expected because of the similar conditions of operation and same influent characteristics. Moreover, both reactors had good COD removal efficiency, but nitrogen and phosphorus removals were different and improved over time as mature granules were formed.

Both reactors exhibited high COD degradation, during the entire operational period (360 days: before and after changing filling mode for R2). R1 and R2 had mean removal efficiencies of 91.5± 2% and 91.2 ± 1%, respectively, during 360 operation days. This suggests that there was no effect of the feeding mode on the removal of organics. The results of COD removal efficiencies showed in this study are consistent with previous AGS systems that used both municipal and industrial wastewaters [12,29]. It should be noted that during the first stage (day 1–125: cultivation), R1 and R2 had mean sCOD concentrations in the effluent of 127.7 ± 24 mg/L and 115.4 ± 25 mg/L, respectively. While during maturation period as well as after changing the feeding mode of R2 (day 126–360), the mean sCOD concentrations in the effluent of R1 and R2 were estimated at 93 ± 14 mg/L and 91 ± 12 mg/L, respectively.

3.2.1. Ammonium Removal Performance

The average influent concentrations of ammonium were about 115.2 ± 4.6 mg/L (Table 2). During the first 27 days and after two weeks of adaptation post inoculation (day 42), there were higher concentrations of ammonium in the effluents in both reactors. The mean value of NH_4-N removal efficiency was about 30% for GSBRs until day 42: NH_4-N concentrations in the effluents of R1 and R2 were found between 73 and 86 mg/L. After this point, an efficient ammonium oxidation was observed with NH_4-N removal efficiency of 51% and 55.9% for R1 and R2 at day 49, respectively. After that, a continuous gradual decrease in the NH_4-N concentrations of the effluents was observed until the end of the cultivation stage, where the mean values of 70.8 ± 1.5% and 69.6 ± 2.1% for R1 and R2 were found, respectively. Only nitritation (NH_4-N converted to NO_2-N) was observed, while full-nitrification ([a] NH_4-N oxidized to NO_2-N then [b] NO_2-N oxidized to NO_3-N) was not observed during this stage. This was indicated by the highest effluent concentrations of NO_2-N accompanied with lowest NO_3-N concentrations. At the end of this stage (day 120), effluent concentrations of NO_2-N were 62.3 mg/L and 58.4 mg/L for R1 and R2, while for NO_3-N concentrations were 4.4 mg/L and 7.3 mg/L for R1 and R2, respectively.

The potential reasons for this behavior might be due to the type of the inoculated sludge that used for both reactors at the beginning of operational period. Moreover, the lower sludge age that estimated for GSBRs during the first stage based on the MLSS and TSS_{eff} concentrations have played a direct role in the difficulty of the proliferation of slow-growing organisms (e.g., ammonia oxidizing bacteria (AOB), and nitrite oxidizing bacteria (NOB)). SRT for R1 and R2 have been estimated by 18 ± 1 and 16 ± 2 days, respectively. In addition, applying higher OLR (2.25 Kg COD/m^3·d) might be considered one of the potential reasons caused these low efficiencies towards the NH_4-N removal through increasing the possibility of fast-growing, heterotrophic bacteria growing on granule's surfaces. Besides, applying higher influent NH_4-N concentrations (120 ± 3.4 mg/L) was one of the reasons of lower performance of ammonium removal. Previous study [12], has demonstrated that an improvement through the nitrification process has been achieved through a reduction in influent NH_4-N from 140 to 80 mg/L. This enhancement in the nitrification process for GSBRs has been confirmed through a gradual decrease in the effluent NO_2-N to less than 1 mg/L accompanied with gradual increase in NO_3-N from 35 to 48 mg-N/L.

During the maturation stage (day 126–300), NH_4-N removal efficiencies for R1 and R2 have gradually increased due to accumulation growth of nitrifiers (AOB and NOB) as well as the average SRT during this stage increased and reached 28 ± 2 days for both R1 and R2. The average NH_4-N concentrations in the effluent of R1 and R2 were estimated at 20.4 ± 1.4 mg/L and 18.8 ± 2.2 mg/L, respectively. Basically, the performance of GSBRs in NH_4-N removal through maturation period can be discussed considering the following three phases:

(a) The commencement of full nitrification, and gradual enhancement in ammonium oxidation was observed between days 130 and 178. Reduction in effluent concentrations of NH_4-N from 24.8 ± 1.7 mg/L and 27.7 ± 1.6 mg/L to 16.7 ± 2.4 mg/L and 15.5 ± 3.1 mg/L for R1 and R2, respectively, was estimated, and these reductions were accompanied with the reduction in NO_2-N concentrations with simultaneous increase in NO_3-N concentrations in the effluent. By day 178, effluent concentrations of NO_2-N decreased and reached 16.7 ± 3 mg/L and 19.1 ± 1.2 mg/L for R1 and R2, respectively, while effluent concentrations of NO_3-N increased to 45 ± 9 mg/L and 49.4 ± 12 mg/L for R1 and R2, respectively.

(b) From day 179 to 222, active and efficient full nitrification of the oxidized ammonium is observed in addition to the continuous reduction in effluent concentrations of NO_2-N, which have been decreased to 7.9 ± 1.2 mg/L and 10.4 ± 0.3 mg/L for R1 and R2, respectively. In addition, there was a reduction in the effluent concentrations of NO_3-N during the same period, and these concentrations reached about 18 ± 0.7 mg/L and 22.4 ± 1.8 mg/L for R1 and R2, respectively. It is worth mentioning that the decreasing in nitrate concentrations accompanied to the reduction in nitrite was due to the action of simultaneous nitrification-denitrification (SND), which will provide an important benefit to the GSBRs system by preventing the accumulation of nitrite and nitrate. The potential reasons for this can be characterized by the compact layer structure of the mature granules and the size of these AGS that can extend beyond 2.6 ± 0.06 mm for R1 and R2 (Figure S2c,d—Supplementary File).

(c) Both reactors exhibited a good tendency to perform SND process for the rest of this stage, between days 223 and 300, (which was confirmed by the results of the kinetic tests as shown in Figure 4a,b) through noticeable reduction in the effluent concentrations of NO_2-N and NO_3-N. However, the effluent concentrations of NO_2-N and NO_3-N for reactors reached more than 6 mg/L and 10 mg/L, respectively. In addition, it should be noted that the removal efficiency of NH_4-N by the end of the maturation stage (day 300) was estimated for R1 and R2 as $85.5 \pm 1.8\%$ and $87.1 \pm 1.6\%$, respectively, (NH_4-N concentrations in the effluents ranged between 15 and 20 mg/L).

Accordingly, and based on the levels of nitrite and nitrate in the effluent, a new operational stage (stage III) was planned to investigate the possibility of the enhancement of denitrification process for reactors and minimizing both of nitrite and nitrate accumulation. Thus, feeding strategy was changed to anaerobic step-feeding for R2 only to investigate its influence and feasibility on the enhancement of the denitrifiers performance, while R1 was operated under the same conditions during the previous 300 days. In addition, and because of the low efficiency of phosphorus removal for GSBRs (about 75% as shown in the following section), studying the impact of changing filling mode on the activity of PAOs will be investigated. Results for GSBRs are discussed in the following section.

3.2.2. Phosphorus Removal Performance

The average influent concentration of PO_4-P was 17.02 ± 0.9 mg/L (Table 2). Until the stage of cultivation (day 125), lower efficiencies of phosphorus removal for GSBRs were observed. The mean phosphorus removal efficiencies for R1 and R2 were $42.8 \pm 5.9\%$ and $42 \pm 7.2\%$, respectively. The potential reasons for these lower efficiencies can be insufficient carbon source (COD = 1250 ± 43 mg/L) for PAOs in the influents (COD:TN:TP ratio is 25:3:1). Thus, there will be a competition of carbon uptake during anaerobic stage (1.5 h) in the operational cycle (8 h) between these PAOs and other denitrifiers. However, higher efficiencies of phosphorus removal have been observed for R1 and R2. For instance, phosphorus removal efficiency for R1 and R2 were estimated at $75 \pm 1.4\%$ and $77.2 \pm 0.9\%$ at day 98.

The seed sludge used in the reactors played a role in the enhancement of phosphorus removal after inoculation in GSBRs and especially during the initial days in cultivation stage (i.e., before formation of large granules) because it was RAS from a BNR wastewater treatment plant. Kinetic tests shown in Figure 5a,b demonstrated relatively lower rates of the anaerobic phosphorus release (4.4 mg-P/gVSS and 4.1 mg-P/gVSS, for R1 and R2, respectively) and slightly higher aerobic phosphorus uptake (6.5 mg-P/gVSS and 6.7 mg-P/gVSS, for R1 and R2, respectively) during the cultivation stage. Phosphorus content in the biomass at the end of an operational cycle was observed at lower concentrations of 72 mg-P/gVSS as a mean value during the first 125 days of operation. The lower phosphorus removal in both reactors implies that the layered structure of the formed AGS (with an aerobic outer layer and an anaerobic core) at this stage (size of granules was about 2 mm) was insufficient to allow anaerobic phosphorus release and efficient phosphorus uptake. Low phosphorus removal was previously reported in GSBRs systems [30]. Comparable results were found by Iorhemen, O.T et al. (2022) [16], where three GSBRs were operated for 84 days to treat an influent with COD:TN:TP ratio of 100:5:1. The results show that the phosphorus removal efficiency was about $55 \pm 11\%$ for the GSBRs that used pulse feeding without anaerobic mixing.

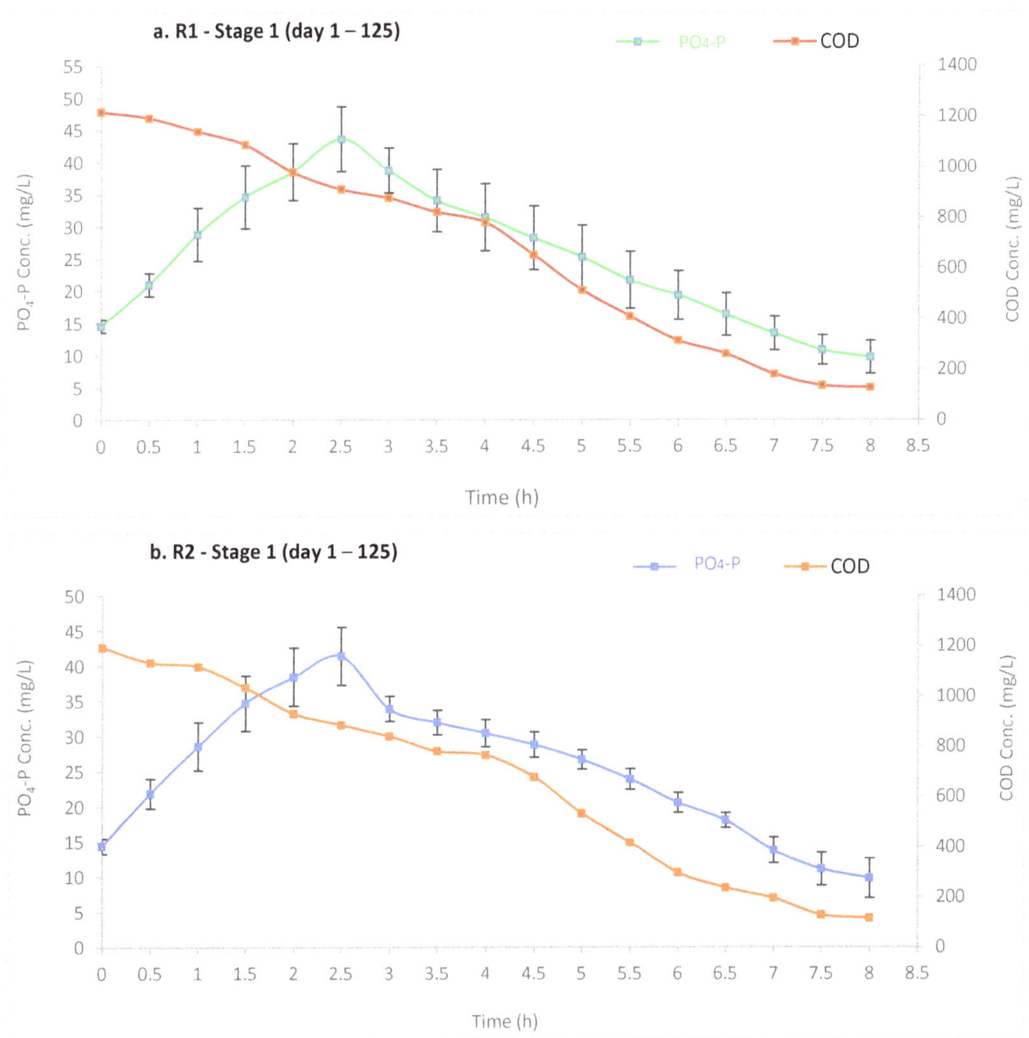

Figure 5. Phosphorus and COD removal through a typical cycle during stage I in (**a**) R1 and (**b**) R2.

Till day 175 of operation (maturation stage), and under the same influent characteristics (high OLR and COD:TN:TP) and same operational design parameters (time of cycle's stages, HRT, superficial air velocity), relatively higher phosphorus removal efficiencies for R1 and R2 were observed compared to the previous cultivation stage. Variation in the phosphorus removal of R1 and R2 was observed, and the mean phosphorus removal efficiencies for R1 and R2 were $75.6 \pm 5.9\%$ and $74.5 \pm 9.6\%$, respectively. In addition, the variance in the removal efficiency of PO_4-P concentrations for the GSBRs was observed and concentrations as high as 7.6 mg/L and 7 mg/L were recorded for R1 and R2, respectively, at the beginning of maturation period (day 140). Moreover, after day 220 until the end of this period (day 300), phosphorus concentrations in the effluents of both reactors were between 2–3 mg/L.

As mentioned previously, the size of the aerobic formed granules in both reactors increased from 2.1 mm (by the end of cultivation stage) to more than 3 mm during this stage. Consequently, the mechanisms for phosphorus removal in these GSBRs systems have improved during this stage due to the impact of enhanced biological phosphorus removal (EBPR), which depends on the enrichment of PAOs in the GSBRs, as well as phosphorus precipitation inside the granular matrix [2,7,31–35]. A slight improvement in PAOs activity has been confirmed during the maturation stage and according to the frequent kinetic analysis (Figures 5 and 6), it was found that influent sCOD concentrations have been degraded during anaerobic phase (95 min) by about 60% for both GSBRs compared to 30% for 125 days of the previous stage (mean values for the initial 175 days of operation). Therefore, this led to higher rates of the anaerobic phosphorus release (6.33 mg-P/gVSS and 6.5 mg-P/gVSS for R1 and R2, respectively) as well as aerobic phosphorus uptake (7.5 mg-P/gVSS and 7.7 mg-P/gVSS for R1 and R2, respectively) during the maturation stage, which was confirmed by the results of the kinetic tests as shown in Figure 6a,b). In addition, the highest phosphorus removal efficiency for reactors was observed, where for R1 it was about 95.1 ± 1.5% at day 238, and 97.6 ± 0.8% for R2. Additionally, high phosphorus content in the biomass at the end of a completed operational cycle was observed (more than 20–50 mg-P/gVSS, which was reported for ordinary heterotrophic bacteria [36]) and reached about 0.21 g-P/gVSS, on day 234 (Figure 5).

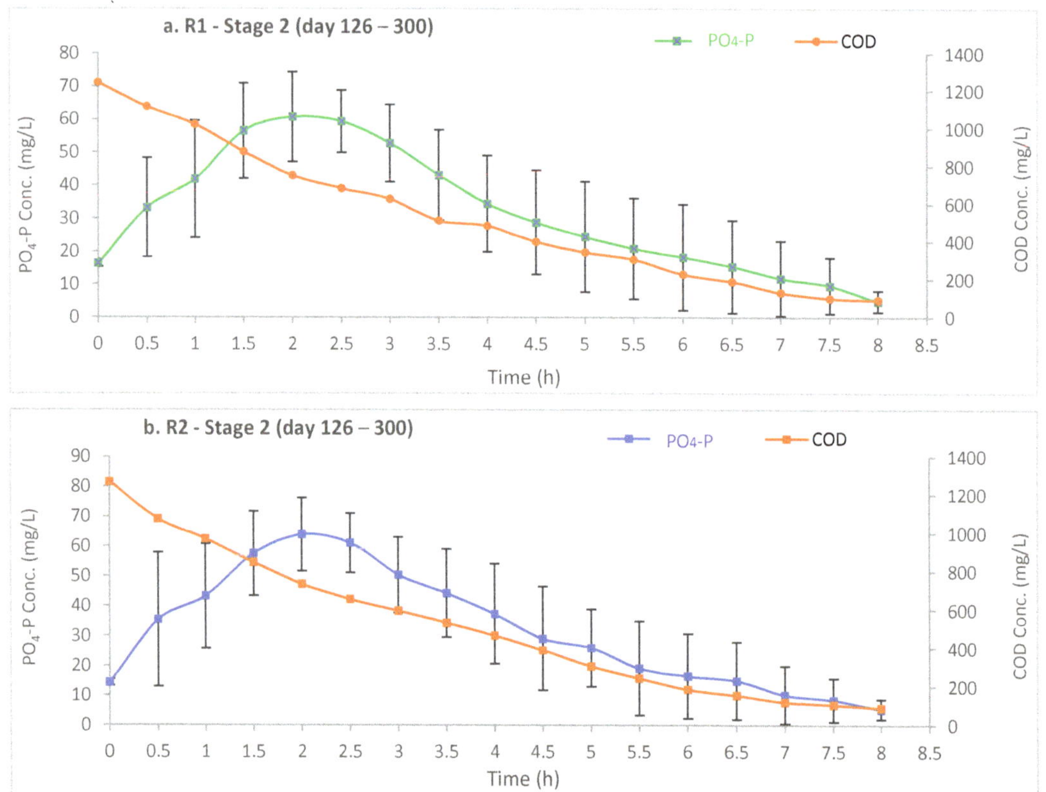

Figure 6. Phosphorus and COD removal through a typical cycle during stage II in (**a**) R1 and (**b**) R2.

3.3. Influence of Anaerobic Step-Feeding

Improving the performance of these GSBRs was the primary objective after finishing 300 days of operation under the effect of high-strength wastewater (microscopic images of the matured AGS through these two stages are shown in the supplementary file, Figures S2 and S3). Therefore, a third operational stage of 60 days was proposed to investigate the influence of changing the filling strategy on the reactors' performance. In addition, the microscopic images of the aerobic granules for both reactors during the last stage are shown in Figure S4 (Supplementary File). The feeding mode of R2 was changed from fast single feeding mode followed by anaerobic mixing to anaerobic step-feeding strategy, while R1 has been operated with the same previous filling mode.

The results demonstrated that changing filling strategy did not influence the removal efficiency of organics. R1 and R2 had similar mean organic removal efficiencies during this stage (60 days) of 94.55$\pm$ 1.4%, and 92.6 $\pm$ 2.3%, respectively. These results are consistent with previous studies of AGS systems that investigated the effect of feeding strategy on the removal efficiencies of organic substrate [15,16].

The measured values of EPS and size of AGS, as demonstrated in Figures 3 and 4, show rapid disintegration of the mature granules in R2. Moreover, at the end of this stage (day 360), high portion of granules had become fluffy and viscous flocs (Figure S6 in the Supplementary File). Consequently, lower physical properties were observed for R2 by the end of the current period (day 360) as shown in Figure 2, where SVI_{30} increased from 31.8 mL/g (end of maturation stage) to 50.2 mL/g and also SVI_{30}/SVI_5 ratio decreased from 0.96 (end of maturation stage) to 0.77 (Figures S4 and S6 in the Supplementary File). Furthermore, granules' size in the same reactor decreased from 3.08 $\pm$ 0.06 mm (end of maturation stage) to 2.94 $\pm$ 0.11 mm (Figure S4 in the Supplementary File). This deterioration of the mature AGS in R2 can be considered a direct result for the notable reduction in the EPS concentrations. EPS concentrations decreased from 250.82 mg/L (end of maturation stage) to 180 mg/L by day 360, as well as the reduction in PN/PS was noted during the third stage (e.g., PN/PS = 9.1 at day 360). In contrast, the performance of R1 as well as the physical properties had the same trend during the initial 175 days of operation during maturation, where the EPS concentrations were higher than those of R2. Thus, larger AGS volume have been achieved in R1, where SVI_{30} = 27.5 mL/g and SVI_{30}/SVI_5 ratio = 0.98, while granules' size became greater than 3.13 $\pm$ 0.07 mm (Figure S3 in Supplementary File).

In addition, biomass concentrations in R2 have decreased in the 60 days of operation after changing feeding mode, where MLSS concentrations were about 8502.3 mg/L. It is important to mention that TSS_{eff} for R2 indicated a deterioration and disintegration of the granules, which was comparable with these concentrations in R1 (Figure S7 in the Supplementary File). The mean value of TSS_{eff} for R2 increased during this period to 251.6 mg/L, while for R1 TSS_{eff} was much lower than R2 at 85.8 mg/L.

Accordingly, and after the evident deterioration of granules in R2 during the third stage of operation, it was expected that the performance of R2 in terms of phosphorus and ammonium removal will be lower than the performance of R1 (as shown in Figures 7 and 8 that show the kinetic analysis during the third stage). For the performance of R1 in terms of phosphorus removal (Figure 7a), the average PO_4-P concentrations in the effluent were about 1.38 $\pm$ 0.5 mg/L. In contrast to the phosphorus removal of R2, the average PO_4-P concentrations were about 9.22 $\pm$ 1.3 mg/L. Deterioration had occurred in PAOs activity for R2, during the last stage after changing the filling mode, and according to the frequent kinetic analysis (Figure 7b), it was found that influent sCOD concentrations have been degraded during 1.5 h anaerobic phase by about 45% for R2 compared to 60% for the previous stage, while for R1 it was about 65% during the same period (stage III). This indicated that the anaerobic mixing time was not sufficient to provide enough degradation of COD during the same phase (i.e., anaerobic mixing). In other words, COD was not readily oxidized during the anaerobic phase and in the first period of the aeration phase, OHO began to develop, mainly in filaments, at expense of the growth of other slow-growing organisms (e.g., PAOs, GAOs, nitrifiers, and denitrifiers). Accordingly, granules stability has been deteriorated, and the performance of R2 for the biological nutrient removal has been dropped. Anaerobic phosphorus release and aerobic phosphorus uptake was about 2.4 mg-P/gVSS and 3.2 mg-P/gVSS, respectively, for R2 during maturation stage.

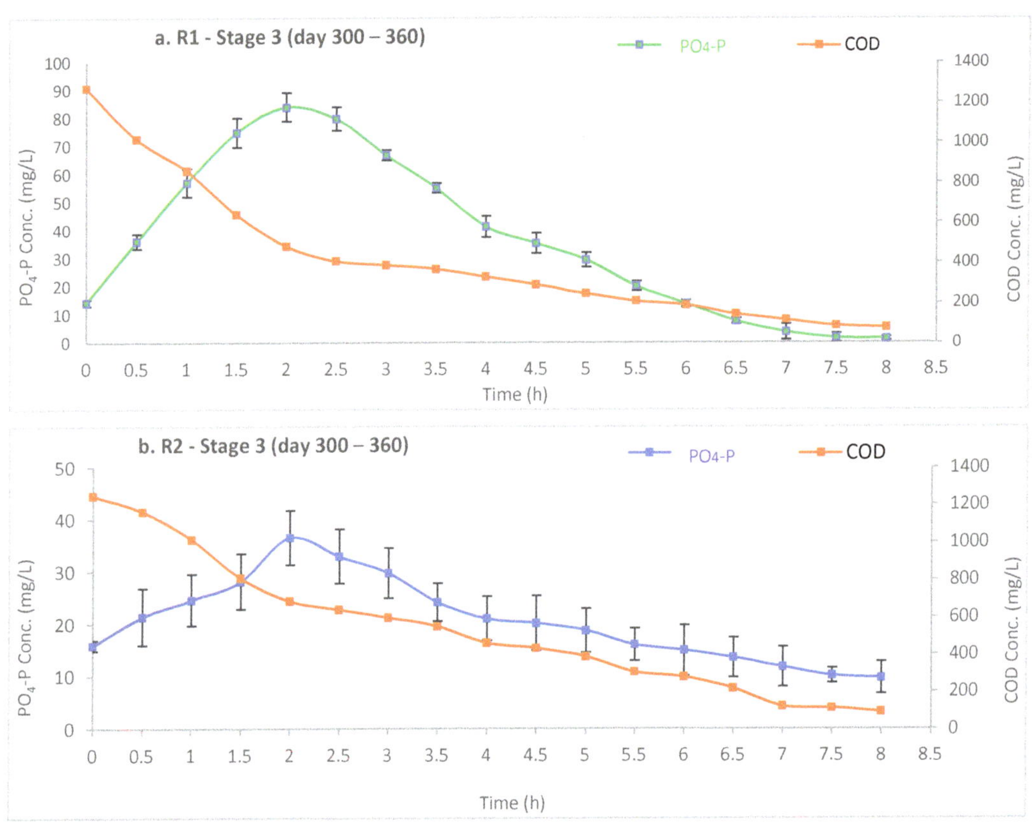

Figure 7. Phosphorus and COD removal through a typical cycle during stage III in (**a**) R1 and (**b**) R2.

Moreover, performance of ammonium removal in R2 based on kinetic tests is demonstrated in Figure 8b for the entire 60 days of the third stage. Due to the similarity and the disintegration of granules occurring in R2, it was expected that performance of R2 in terms of ammonium removal will be minimized compared to R1. The average NH_4-N concentrations in the effluent of R2 have been estimated at 24.3 ± 6.3 mg/L. In addition, a significant rise in the concentrations of nitrate, as well as slight increase in nitrite concentrations during aerobic phase was noticed, which led to nitrite/nitrate accumulation. Effluent concentrations of nitrate were between 46 and 50 mg/L, while it was 10 mg/L for nitrite. For R1 (Figure 8a), the average NH_4-N concentrations in the effluent were about 12.41 ± 3.3 mg/L.

The granules' disintegration in R2 after changing filling mode and recurrent washouts because of applying short settling time (10 min) had resulted in a loss of nitrifying bacteria (AOB and NOB) as well as some PAOs that were present in the broken granules. In addition, when washouts occur at higher frequencies during these 60 days of operation, the sludge age (SRT) is reduced (mean SRT for R2 was estimated at 22 ± 1 days during this period), therefore the performance of ammonium (i.e., SND) and phosphorus removal was negatively affected due to the reduction in sludge age (compared to SRT for R1 which was about 29 ± 1 days). This further substantiates the key role of the STR in granules' stabilization and reactors' performance (nutrients removal) since it is directly related to the accumulation and growth of these slow-growing bacteria, which is also confirmed in other studies [14].

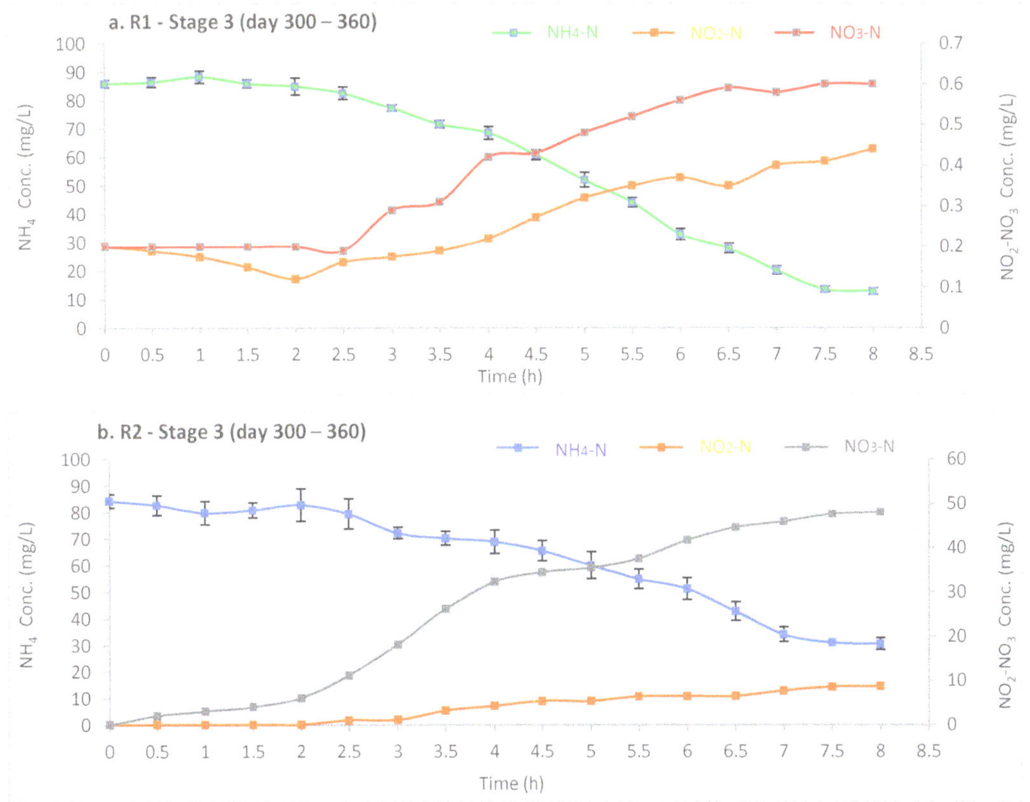

Figure 8. Profile of nitrogen removal distributed over a typical cycle during stage III in (**a**) R1 and (**b**) R2.

Overall, the performance of R1 was continuing in the same trend of enhancement along with physical properties of AGS. However, the slow rate of this improvement of R1 might be due to the high strength of influent wastewater characteristics (high influent concentrations of COD, NH_4-N and PO_4-P) and composition (i.e., COD:TN:TP ratio).

4. Conclusions

Aerobic granules were successfully cultivated (>2.1 mm) and matured (>3 mm) through two SBRs under the effect of high-strength wastewater with insufficient influent substrate (COD:TN:TP ratio of 100:11.5:3.5) through 300 days of operation after inoculation. Aerobic granulation for the purposes of high phosphorus and ammonium removal performance was unsuitable with long start-up periods, and high concentrations of nitrite and nitrate have been observed as well. However, formation of large sized matured granules, there was a low performance of P removal for GSBRs, and this can be attributed to the competition between PAOs and denitrifiers for rbCOD uptake during anoxic mixing phase due to the transition from anaerobic zone to anoxic phase. Therefore, applying anaerobic step-feeding was aiming at decreasing the competition between PAOs and other denitrifiers, by enhancement of the activity for these microorganisms during anoxic phase in order to enhancement of the reactors' performance. Unexpectedly, deterioration in physical properties of AGS has been resulted after changing the filling mode and it was accompanied to a reduction in EPS concentrations, and the protein-to-carbohydrates ratio. Moreover, significant reduction in the phosphorus and ammonium removal efficiencies has been demonstrated, and COD degradation during anaerobic phase has been decreased as well. This implies that the anaerobic mixing duration after applying anaerobic step-feeding was not sufficient, which attributed to lower carbon utilization by PAOs and denitrifiers during the same phase. Therefore, extended anaerobic mixing durations after applying anaerobic step-feeding might provide enhanced activity of PAOs and denitrifiers to ensure higher performance in terms of phosphorus removal and avoid nitrite/nitrate accumulation. Thus, further research is required to identify the influence of this approach on the performance of GSBRs-treated high-strength wastewater with low influent substrate (COD:TN:TP ratio higher than 100:5:1); additionally, further research should be performed to improve SND.

Supplementary Materials: The following supporting information can be downloaded at: https://www.mdpi.com/article/10.3390/pr11010075/s1, Figure S1: GSBRs (R1 and R2) in operation, Figure S2: (a): Microscopic images of the formed granules in both reactors on day 50 (during of the cultivation stage); (b): Microscopic images of the formed granules in both reactors on day 120 (end of the cultivation stage); (c): Microscopic images of the formed granules in both reactors on day 210 (during the maturation stage); (d): Microscopic images of the formed granules in both reactors on day 300 (end of the maturation stage), Figure S3: Microscopic images of the formed granules in R1 during stage III (day 301 until 360), Figure S4: Microscopic images of the formed granules in R2 after applying anaerobic step-feeding mode (during stage III; day 301 until 360), Figure S5: Measurements of SVI5 and SVI30 for both reactors in day 352 (deterioration of the AGS in R2 after applying anaerobic step-feeding strategy), Figure S6: AGS in R2 rapidly lost its structural integrity, after changing the filling mode to anaerobic step-feeding, resulting in loose and fluffy aggregates, Figure S7: Biomass samples for both reactors in day 342 (stage III, after change feeding mode for R2) showing the disintegration of the AGS in R2 which caused higher content of TSS concentrations in its effluents compared to R1.

Author Contributions: Conceptualization, Q.Y.; methodology, E.M. and Q.Y.; software, E.M.; validation, E.M., J.O. and Q.Y.; formal analysis, E.M. and Q.Y.; investigation, E.M. and Q.Y.; resources, Q.Y.; data curation, E.M., J.O. and Q.Y.; writing—original draft preparation, E.M.; writing—review and editing, E.M., J.O. and Q.Y.; visualization, E.M., J.O. and Q.Y.; supervision, J.O. and Q.Y.; project administration, J.O. and Q.Y.; funding acquisition, J.O. and Q.Y. All authors have read and agreed to the published version of the manuscript.

Funding: E.M. is funded by a full scholarship from the Ministry of Higher Education of the Arab Republic of Egypt. In addition, the authors gratefully acknowledge the financial support from the Natural Sciences and Engineering Research Council of Canada (NSERC RGPIN-2014-05510).

Institutional Review Board Statement: Not applicable.

Informed Consent Statement: Not applicable.

Data Availability Statement: The authors confirm that the data supporting the findings of this study are available within the articles listed in the section of reference (and/or its Supplementary Materials).

Acknowledgments: E.M. is funded by a full scholarship from the Ministry of Higher Education of the Arab Republic of Egypt. In addition, the authors gratefully acknowledge the financial support from the Natural Sciences and Engineering Research Council of Canada (NSERC RGPIN-2014-05510).

Conflicts of Interest: The authors declare that they have no known competing financial interests or personal relationships that could have appeared to influence the work reported in this paper.

Abbreviations

AGS, aerobic granular sludge; AOB, ammonia oxidizing bacteria; AS, activated sludge; BNR, Biological nutrient removal; COD, chemical oxygen demand; DO, dissolved oxygen; EBPR, enhanced biological phosphorus removal; EPS, extracellular polymeric substances; GAO, glycogen-accumulating organism; MLSS, mixed liquor suspended solids; MLVSS, mixed liquor volatile suspended solids; NOB, nitrite oxidizing bacteria; ORP, oxidation reduction potential; PAO, polyphosphate-accumulating organism; SBR, sequencing batch reactor; SND, simultaneous nitrification-denitrification; SRT, solids retention time; TN, total nitrogen; TP, total phosphorus, TSS, total suspended solids; VSS, volatile suspended solids; WWTP, wastewater treatment plant.

References

1. Nancharaiah, Y.; Reddy, G.K.K. Aerobic granular sludge technology: Mechanisms of granulation and biotechnological applications. *Bioresour. Technol.* **2018**, *247*, 1128–1143. [CrossRef] [PubMed]
2. Gao, D.W.; Liu, L.; Liang, H.; Wu, W.M. Aerobic granular sludge: Characterization, mechanism of granulation and application to wastewater treatment. *Crit. Rev. Biotechnol.* **2011**, *31*, 137–152. [CrossRef] [PubMed]
3. Sarma, S.J.; Tay, J.H. Aerobic granulation for future wastewater treatment technology: Challenges ahead. *Environ. Sci. Water Res. Technol.* **2018**, *4*, 9–15. [CrossRef]
4. Sarma, S.J.; Tay, J.H.; Chu, A. Finding Knowledge Gaps in Aerobic Granulation Technology. *Trends Biotechnol.* **2017**, *35*, 66–78. [CrossRef]
5. Hamza, R.A.; Sheng, Z.; Iorhemen, O.T.; Zaghloul, M.S.; Tay, J.H. Impact of food-to-microorganisms ratio on the stability of aerobic granular sludge treating high-strength organic wastewater. *Water Res.* **2018**, *147*, 287–298. [CrossRef]
6. Corsino, S.F.; di Biase, A.; Devlin, T.R.; Munz, G.; Torregrossa, M.; Oleszkiewicz, J.A. Effect of extended famine conditions on aerobic granular sludge stability in the treatment of brewery wastewater. *Bioresour. Technol.* **2016**, *226*, 150–157. [CrossRef]
7. Henriet, O.; Meunier, C.; Henry, P.; Mahillon, J. Improving phosphorus removal in aerobic granular sludge processes through selective microbial management. *Bioresour. Technol.* **2016**, *211*, 298–306. [CrossRef]
8. Show, K.-Y.; Lee, D.-J.; Tay, J.-H. Aerobic Granulation: Advances and Challenges. *Appl. Biochem. Biotechnol.* **2012**, *167*, 1622–1640. [CrossRef]
9. Devlin, T.; di Biase, A.; Kowalski, M.; Oleszkiewicz, J. Granulation of activated sludge under low hydrodynamic shear and different wastewater characteristics. *Bioresour. Technol.* **2017**, *224*, 229–235. [CrossRef]
10. Long, B.; Yang, C.-Z.; Pu, W.-H.; Yang, J.-K.; Liu, F.-B.; Zhang, L.; Cheng, K. Rapid cultivation of aerobic granular sludge in a continuous flow reactor. *J. Environ. Chem. Eng.* **2015**, *3*, 2966–2973. [CrossRef]
11. Long, B.; Yang, C.-Z.; Pu, W.-H.; Yang, J.-K.; Liu, F.-B.; Zhang, L.; Zhang, J.; Cheng, K. Tolerance to organic loading rate by aerobic granular sludge in a cyclic aerobic granular reactor. *Bioresour. Technol.* **2015**, *182*, 314–322. [CrossRef] [PubMed]
12. Kang, A.J.; Yuan, Q. Long-term stability and nutrient removal efficiency of aerobic granules at low organic loads. *Bioresour. Technol.* **2017**, *234*, 336–342. [CrossRef]
13. Tay, J.-H.; Liu, Q.-S.; Liu, Y. The effects of shear force on the formation, structure and metabolism of aerobic granules. *Appl. Microbiol. Biotechnol.* **2001**, *57*, 227–233.
14. Rollemberg, S.L.D.S.; Barros, A.R.M.; Firmino, P.I.M.; dos Santos, A.B. Aerobic granular sludge: Cultivation parameters and removal mechanisms. *Bioresour. Technol.* **2018**, *270*, 678–688. [CrossRef]
15. Da Silva, V.E.P.; Rollemberg, S.L.D.S.; dos Santos, A.B. Impact of feeding strategy on the performance and operational stability of aerobic granular sludge treating high-strength ammonium concentrations. *J. Water Process Eng.* **2021**, *44*, 102378. [CrossRef]
16. Iorhemen, O.T.; Ukaigwe, S.; Dang, H.; Liu, Y. Phosphorus Removal from Aerobic Granular Sludge: Proliferation of Polyphosphate-Accumulating Organisms (PAOs) under Different Feeding Strategies. *Processes* **2022**, *10*, 1399. [CrossRef]
17. APHA/AWWA/WEF. *Standard Methods for the Examination of Water and Wastewater*; American Public Health Association/American Water Works Association/Water Environment Federation: Washington, DC, USA, 2012.
18. De Kreuk, M.K.; Kishida, N.; van Loosdrecht, M.C.M. Aerobic granular sludge—State of the art. *Water Sci. Technol.* **2007**, *55*, 75–81. [CrossRef]
19. Adav, S.S.; Lee, D.-J.; Tay, J.-H. Extracellular polymeric substances and structural stability of aerobic granule. *Water Res.* **2008**, *42*, 1644–1650. [CrossRef]
20. Leal, C.; del Río, A.V.; Mesquita, D.P.; Amaral, A.L.; Castro, P.M.; Ferreira, E.C. Sludge volume index and suspended solids estimation of mature aerobic granular sludge by quantitative image analysis and chemometric tools. *Sep. Purif. Technol.* **2020**, *234*, 116049. [CrossRef]
21. Zhang, C.; Zhang, H.; Yang, F. Diameter control and stability maintenance of aerobic granular sludge in an A/O/A SBR. *Sep. Purif. Technol.* **2015**, *149*, 362–369. [CrossRef]
22. De Kreuk, M.; Van Loosdrecht, M. Selection of slow growing organisms as a means for improving aerobic granular sludge stability. *Water Sci. Technol.* **2004**, *49*, 9–17. [CrossRef] [PubMed]
23. Rollemberg, S.L.D.S.; de Oliveira, L.Q.; Barros, A.R.M.; Melo, V.M.M.; Firmino, P.I.M.; dos Santos, A.B. Effects of carbon source on the formation, stability, bioactivity and biodiversity of the aerobic granule sludge. *Bioresour. Technol.* **2019**, *278*, 195–204. [CrossRef] [PubMed]
24. Corsino, S.F.; Di Trapani, D.; Torregrossa, M.; Viviani, G. Aerobic granular sludge treating high strength citrus wastewater: Analysis of pH and organic loading rate effect on kinetics, performance and stability. *J. Environ. Manag.* **2018**, *214*, 23–35. [CrossRef]
25. Liu, X.; Sheng, G.; Luo, H.; Zhang, F.; Yuan, S.; Xu, J.; Zeng, R.; Wu, J.; Yu, H. Contribution of extracellular polymeric substances (EPS) to the sludge aggregation. *Environ. Sci. Technol.* **2010**, *44*, 4355–4360. [CrossRef] [PubMed]

26. Wei, D.; Shi, L.; Yan, T.; Zhang, G.; Wang, Y.; Du, B. Aerobic granules formation and simultaneous nitrogen and phosphorus removal treating high strength ammonia wastewater in sequencing batch reactor. *Bioresour. Technol.* **2012**, *171*, 211–216. [CrossRef]
27. Ren, Y.; Ferraz, F.; Lashkarizadeh, M.; Yuan, Q. Comparing young landfill leachate treatment efficiency and process stability using aerobic granular sludge and suspended growth activated sludge. *J. Water Process. Eng.* **2017**, *17*, 161–167. [CrossRef]
28. Ren, Y.; Ferraz, F.M.; Yuan, Q. Landfill leachate treatment using aerobic granular sludge. *J. Environ. Eng.* **2017**, *143*, 1–7. [CrossRef]
29. Franca, R.D.; Pinheiro, H.M.; van Loosdrecht, M.C.; Lourenço, N.D. Stability of aerobic granules during long-term bioreactor operation. *Biotechnol. Adv.* **2018**, *36*, 228–246. [CrossRef]
30. Iorhemen, O.T.; Hamza, R.A.; Sheng, Z.; Tay, J.H. Submerged aerobic granular sludge membrane bioreactor (AGMBR): Organics and nutrients (nitrogen and phosphorus) removal. *Bioresour. Technol. Rep.* **2019**, *6*, 260–267. [CrossRef]
31. Cassidy, D.; Belia, E. Nitrogen and phosphorus removal from an abattoir wastewater in a SBR with aerobic granular sludge. *Water Res.* **2005**, *39*, 4817–4823. [CrossRef]
32. Mañas, A.; Biscans, B.; Spérandio, M. Biologically induced phosphorus precipitation in aerobic granular sludge process. *Water Res.* **2011**, *45*, 3776–3786. [CrossRef] [PubMed]
33. Filali, A.; Mañas, A.; Mercade, M.; Bessière, Y.; Biscans, B.; Spérandio, M. Stability and performance of two GSBR operated in alternating anoxic/aerobic or anaerobic/aerobic conditions for nutrient removal. *Biochem. Eng. J.* **2012**, *67*, 10–19. [CrossRef]
34. Sarma, S.; Tay, J.H. Carbon, nitrogen and phosphorus removal mechanisms of aerobic granules. *Crit. Rev. Biotechnol.* **2018**, *38*, 1077–1088. [CrossRef] [PubMed]
35. Purba, L.D.A.; Ibiyeye, H.T.; Yuzir, A.; Mohamad, S.E.; Iwamoto, K.; Zamyadi, A.; Abdullah, N. Various applications of aerobic granular sludge: A review. *Environ. Technol. Innov.* **2020**, *20*, 101045. [CrossRef]
36. Metcalf, E.; Eddy, M. *Wastewater Engineering: Treatment and Resource Recovery*; McGraw Hill: New York, NY, USA, 2014; pp. 1530–1533.

Article

The Dynamic Shift of Bacterial Communities in Hybrid Anaerobic Baffled Reactor (ABR)—Aerobic Granules Process for Berberine Pharmaceutical Wastewater Treatment

Yan Wang [1,2], **Yongqiang Liu** [3], **Juan Li** [1,2], **Ruirui Ma** [1,2], **Ping Zeng** [1,2,*], **Choon Aun Ng** [4] and **Fenghua Liu** [5]

[1] State Key Laboratory of Environmental Criteria and Risk Assessment, Chinese Research Academy of Environmental Sciences, Beijing 100012, China
[2] Department of Urban Water Environmental Research, Chinese Research Academy of Environmental Sciences, Beijing 100012, China
[3] Faculty of Engineering and Physical Sciences, University of Southampton, Southampton SO17 1BJ, UK
[4] Department of Environmental Engineering, Faculty of Engineering and Green Technology, Universiti Tunku Abdul Rahman, Kampar 31900, Perak, Malaysia
[5] CECEP Engineering Technology Research Institute Co., Ltd., Beijing 100082, China
* Correspondence: zengping@craes.org.cn

Citation: Wang, Y.; Liu, Y.; Li, J.; Ma, R.; Zeng, P.; Ng, C.A.; Liu, F. The Dynamic Shift of Bacterial Communities in Hybrid Anaerobic Baffled Reactor (ABR)—Aerobic Granules Process for Berberine Pharmaceutical Wastewater Treatment. *Processes* **2022**, *10*, 2506. https://doi.org/10.3390/pr10122506

Academic Editor: Adam Smoliński

Received: 9 October 2022
Accepted: 16 November 2022
Published: 25 November 2022

Publisher's Note: MDPI stays neutral with regard to jurisdictional claims in published maps and institutional affiliations.

Abstract: Because of its anticancer, anti-inflammatory, and antibiotic properties, berberine has been used extensively in medication. The extensive production of berberine results in the generation of wastewater containing concentrated residual berberine. However, to date, limited related studies on the biological treatment of berberine wastewaters have been carried out. A lab-scale anaerobic baffled reactor (ABR)–aerobic granular sludge (AGS) process was developed for berberine removal from synthetic wastewater. The system showed effective removal of the berberine. In order to better understand the roles of the bacterial community, the ABR–aerobic granular sludge system was operated in the state with the highest BBR removal rate in this study. The bacterial community dynamics were studied using the 16S rDNA clone library. The results showed that the hybrid ABR-AGS process achieved 92.2% and 94.8% overall removals of berberine and COD, respectively. *Bacterium* was dominant species in ABR, while the *CFB group bacteria* and *Betaproteobacteria* were dominant species in AGS process. The *uncultured bacterium clone B135, Bacillus endophyticus strain a125, uncultured bacterium mle1-42, uncultured bacterium clone OP10D15,* and *uncultured bacterium clone B21.29F54* in ABR, and *uncultured bacterium clone F54, uncultured bacterium clone ZBAF1-105, uncultured bacterium clone SS-9,* and *uncultured bacterium clone B13* in AGS process were identified as functional species in the biodegradation of berberine and/or its metabolites. Both anaerobic and aerobic bacterial communities could adapt appropriately to different berberine selection pressures because the functional species' identical functions ensured comparable pollutant removal performances. The information provided in this study may help with future research in gaining a better understanding of berberine biodegradation.

Keywords: 16S rDNA clone library; bacterial community structure; berberine wastewater; anaerobic baffled reactor (ABR); aerobic granular reactor

1. Introduction

Berberine (5,6-dihydro-9,10-dimethoxybenzo [g]-1,3-benzodioxolo [5,6-α]) quinolizinium ($C_{20}H_{18}NO_4$, abbreviated BBR) is an isoquinoline quaternary alkaloid. BBR can be extracted from herbal plants, or chemically synthesized, and used as a natural antibiotic against a variety of bacteria [1]. BBR's application has been expanded to antitumor, antioxidation, anti-disease, anti-Alzheimer's, and anti-hyperglycemic due to its anticancer, anti-inflammatory, and antibiotic properties [2,3], resulting in a sharp increase in the demand for BBR. Meanwhile, the widespread production and use of BBR has resulted in the discharge of large amounts of BBR-containing wastewater into the environment (thousands of mg/L). Berberine's IC50, minimum inhibitory concentration (MIC), and minimum micro-bicidal concentration (MMC) values against resistant *Pseudomonas aeruginosa* and *Escherichia coli* were found to be 99.2, 240, and 250 mg/L, and 87.0, 469, and 500 mg/L, respectively [4]. Thus, a significant threat is posed to ecosystems due to its significant inhibitory effects on biological activities [5]. As a result, prior to its discharge into the environment, BBR in wastewater, particularly that from pharmaceutical processes, must be treated.

Physical, chemical, and biological processes are commonly used to treat BBR-containing wastewater [5–7]. Due to their high cost and risk of producing new pollutants, physical and chemical treatments of BBR-containing wastewater have limited application [8,9]. As a result, biological treatment is preferred, due to its lower cost and its potential for complete mineralization. The application of a combined anaerobic–aerobic treatment system is recommended for the treatment of wastewater containing antibiotics and pharmaceutical effluents in the guidelines on the available techniques of pollution prevention and the control of the pharmaceutical industry active pharmaceutical ingredient (fermentation, chemical synthesis, and extraction) and preparation categories [10].

Anaerobic baffled reactor (ABR) showed a better resistance to toxic compounds, which is attractive for the pharmaceutical wastewater treatment. For ABR, every chamber served as a UASB, shielding the vulnerable microorganisms from the toxic substrate [11,12]. Meanwhile, the push-flow pattern is similar to the overall reactor flow pattern. Because of its unique structure, it can process toxic substances and hard-to-degrade inhibitors with a better buffering adaptability. As for the aerobic processes, aerobic granular sludge was found to be a promising alternative in eliminating the recalcitrant and toxic antibiotics [13–15]. Thus, a hybrid ABR–AGS system was setup in order to degrade BBR-containing wastewater. The ABR–AGS system showed a higher endurance with a higher influent COD concentration compared to an up-flow anaerobic sludge blanket (UASB)–membrane bioreactor (MBR) system [16,17]. However, in most industrial pharmaceutical production plants, the anaerobic process showed a low removal efficiency effected by the influent toxic compounds. It is necessary to make clear the roles of the microorganisms during the BBR degradation process in order to understand the mechanisms of BBR degradation.

In this study, the ABR–aerobic granular sludge system was operated in the state with the highest BBR removal rate. The bacterial community dynamics were studied in order to further improve the system degradation efficiency by using the 16S rDNA clone library, which is still widely used for microbial community studies in wastewater treatment processes [18,19]. Theoretical guidance is provided in order to further improve the treatment capacity and the stability of aerobic granular sludge. The goal of this study was to provide some basic information on the bacterial community composition in the biological berberine treatment process. It is anticipated to serve as a resource for further pilot-scale or industrial wastewater treatment.

2. Materials and Methods

2.1. Hybrid ABR–AGS System

The ABR (Figure S1) was made in the shape of a rectangle, with 610 mm length, 300 mm width, and 430 mm height, which provided the effective volume of 30 L [20–22]. The reactor consisted of 4 chambers. The widths of the upper flow chamber and the lower flow chamber were 90 mm and 30 mm, respectively [23,24]. The angle of the baffle plate set at 45° to get a higher lower flow speed, which could push the sludge in the bottom of reactor up to the floating level so that the sludge and liquid could be mixed completely. The sampling ports were set at the top and bottom of each chamber and were used to take supernatant and sludge, respectively. A gas collection port was arranged at the top of each chamber. The temperature of the reactor was kept at 32 ± 1 °C by binding the reactor walls with a tubular heater.

A sequencing batch reactor (SBR) was a cylinder with 600 mm internal diameter and 1000 mm height, with H/D of 16:1, providing an effective volume of 2.8 L (Figure S1).

Reactors were operated periodically at 4 h as one cycle. Every cycle included influent time of 4 min, aeration time of 225 min, settling time of 5 min, and effluent time of 5 min.

The effluent was discharged from a port with 50 cm distance from the bottom of the reactor, thus, the volumetric exchange ratio was 50%. A total of 4.0 L min^{-1} of aeration was provided by an air dispenser located at the bottom of the reactor. The hydraulic retention time (HRT) was 8 h.

No excess sludge was discharged from the hybrid ABR–AGS system.

2.2. Inoculum

Inoculums, with an initial concentration of 13,570 mg MLSS/L for the operation of ABR, were obtained from a chemical synthetic pharmacy company's wastewater treatment plant (hydrolysis/acidification tank) in Shenyang, China. As shown in Table S1, the inoculum of SBR, with an initial concentration of 2350 mg MLSS/L, was obtained from the aeration tank of the same pharmaceutical wastewater treatment plant.

The hybrid ABR–AGS system's operation conditions are summarized in Table S2.

2.3. Medium

The hybrid ABR–AGS system was fed with synthetic wastewater with the components of glucose and the industrial berberine wastewater, providing COD of 4253 ± 102 mg/L and berberine of 121.6 ± 2.4 mg/L. The effluent from ABR was provided to SBR as an influent. The high-concentration berberine mother liquid was discharged from the separation process during berberine production as the main component of the industrial berberine wastewater. The concentration of the berberine mother liquid was quite high, with COD of 4166 ± 102 mg/L and berberine of 900 ± 100 mg/L [25]. The compositions of the influent wastewater are summarized in Table S3.

2.4. Analytical Methods

The samples of the ABR and SBR influent and effluent were taken every day. Standard methods were used to determine the COD and NH_4^+-N contents of the wastewater samples. The limit of quantitation (LOD) for COD and NH_4^+-N were 5.0 mg/L and 0.025 mg/L, respectively [26]. High performance liquid chromatography (HPLC) was used to determine the concentration of berberine (Agilent 1100, Santa Clara, CA, USA) at 345 nm (LOQ is 0.05 mg/L), with a 0.05 M KH_2PO_4/Acetonitrile (30:70 v/v) solution as the mobile phase. The flow rate was 1.0 mL/min. A 0.4 μm polytetrafluoroethene microfiltration membrane was used to filter 20 μL of the wastewater sample, which was injected with an auto-sampler into HPLC for analysis. A column of Agilent HB-C8 (150 mm 4.6 mm, 5 μm) was used to separate analytes at 30 °C [25].

2.5. Sampling and DNA Extraction

The clone libraries were constructed using biomass samples that were collected at different operation times [27]. The floc sludge levels in ABR and SBR were sampled separately, and the aerobic granules were dispersed using sonication and suspended in sterile water. After centrifuging all of the samples for 5 min at $1000 \times g$, the supernatant was decanted and the pellet was re-suspended in Tris-EDTA buffer (10.0 mM Tris-HCl, 1.0 mM EDTA, pH 8.0). After resuspension, all samples were immediately frozen and stored at -80 °C until DNA extraction.

According to the manufacturer's instructions, DNA was extracted from the samples using a QIAamp DNA Mini Kit (Qiagen, Valencia, CA, USA). The DNA that was extracted in triplicate for each sample was mixed to create the templates used for PCR amplification in order to reduce variations in DNA extraction.

2.6. PCR Amplification and 16S rDNA Cloning

The universal primers 27F and 1492R were used to amplify the 16S rDNA genes from the DNA extracts [11]. A DNA thermocycler (Bio-Rad, Richmond, CA, USA) was used to amplify PCR in a total volume of 50 L in 200 L tubes. Every sample for PCR contained 40 ng of template DNA, 200 μM of each deoxynucleoside triphosphate, 0.5 μmol of each primer, 1.25 U of Taq polymerase (Pro- mega, Madison, WI, USA), 1 × PCR buffer, and 2 mM $MgCl_2$. The temperature cycling conditions were as follows: pre-incubation at 95 °C for 2 min, 25 cycles of 95 °C for 1 min, 62 °C for 1.5 min, and 72 °C for 1 min, followed by 10 min at 72 °C.

Following the manufacturer's instructions, a Qiaquick PCR cleanup kit (Qiagen, Valencia, CA, USA) was used to purify the PCR products, which was ligated into a PCR 2.1-TOPO vector, then transformed into TOP 10 *E. coli* competent cells (Invitrogen, Carlsbad, CA, USA). For each sample, approximately 100 clones were randomly selected for analysis using ampicillin and x-gal.PCR amplification. The PCR amplification with the primer pair M13 was used to identify positive clones, which used the same program as for 16S rDNA amplification. Sequencing was performed using an ABI 3730 automated sequencer for all positive clones (Invitrogen, Carlsbad, CA, USA).

3. Results

3.1. The Performance of Hybrid ABR–AGS System

The hybrid ABR–AGS system was run for 125 days with berberine concentrations of 121.62.4 mg/L as the influent. The berberine and COD removal rates were high and stable overall. Table 1 summarizes the system's treatment performance. A significant proportion of the berberine (57.0 ± 0.2%) and COD (71.9 ± 1.0%) removal rates was achieved in ABR. The SBR containing aerobic granules removed COD and berberine at rates of about 81.8 ± 0.3% and 81.6 ± 0.5%, respectively, resulting in effluent COD and berberine levels of less than 219.7 mg/L and 9.5 mg/L, respectively. These findings suggest that the hybrid ABR–AGS system was effective at both berberine reduction and COD removal. Furthermore, these findings indicate that a functional stable bacterial community for berberine degradation had already been established. The biomass samples in the ABR and SBR reactors were subjected to 16S rDNA clone library analysis in order to better understand the bacterial community composition and to identify the key functional bacterial species/groups in the berberine biodegradation process.

Table 1. The results of the berberine wastewater treatment in the hybrid ABR–AGS system.

	Influent (mg/L)	ABR Effluent (mg/L)	AGS Effluent (mg/L)	ABR Removal Efficiency (%)	AGS Removal Efficiency (%)	Overall Removal Efficiency (%)
Berberine	121.6 ± 2.4	52.3 ± 1.3	9.5 ± 0.4	57.0 ± 0.2	81.8 ± 0.3	92.2 ± 0.7
COD	4253 ± 104	1193.1 ± 46	219.7 ± 2.4	71.9 ± 1.0	81.6 ± 0.5	94.8 ± 1.3

3.2. Differences in the Structure of Total Bacterial Communities in ABR Reactor Chambers

At the steady operation condition, with the feeding influent berberine concentration of 120 mg/L, the 16S rDNA clone libraries of the total bacteria were produced from the sludge samples in each chamber of the ABR reactor. After the comparison study, the total bacterial clone libraries in each chamber of the ABR reactor were 31, 32, 35, and 38 operational taxonomic units (OTUs) during stable operation. The results were shown in Figure 1.

Under the stable operating conditions, there were 31 OTUs of total bacteria in the A1 chamber of the ABR reactor, with a coverage of 84%, and a Shannon–Wiener diversity index of 2.84. These clones had a 99% maximum similarity and an 85% minimum similarity to the known bacteria in the Gen-bank. According to an analysis of the 31 OTU sequence data, the 31 OTUs belonged to six different taxa within the bacterial domain. The *uncultured bacterium partial 16S rRNA gene* from *clone 053B03_B_DI_P58* was the most dominant group, accounting for 24% of the total bacterial flora, and its clone was 98% similar to the known bacteria in the Gen-bank; The *uncultured Bacteroides* sp. *Bacteroides* sp. *clone J3* (*Bacteroides* sp.) and the *uncultured bacterium 054B06_B_DI_P58*, were the second abundant group, accounting for 11% and 9% of the total bacterial community, respectively. The *uncultured bacterium clone B135* and *Bacillus endophyticus strain a125* with antibiotic resistance propertiesaccounted for 6% and 2% of the total bacterial flora, respectively; Each of the *uncultured bacterium clone DC75* and the *uncultured Clostridia bacterium clone L24 (Clostridium perfringens)* accounted for 3% of the total bacterial flora and both of them were hydrolytic acidifying functional colonies.

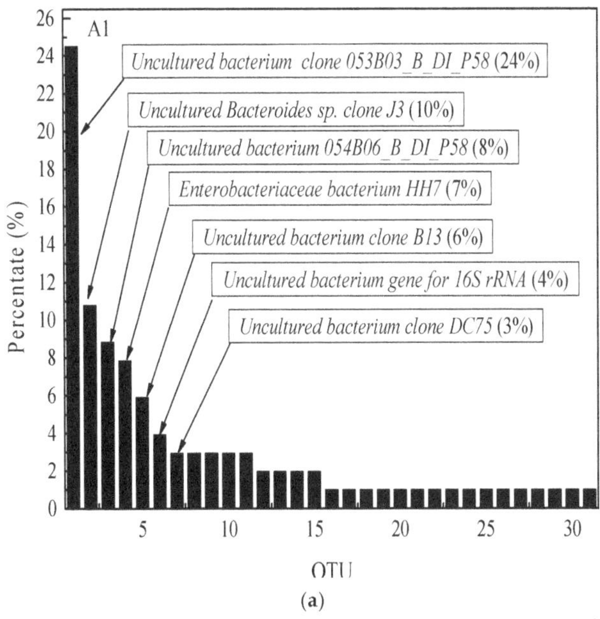

(a)

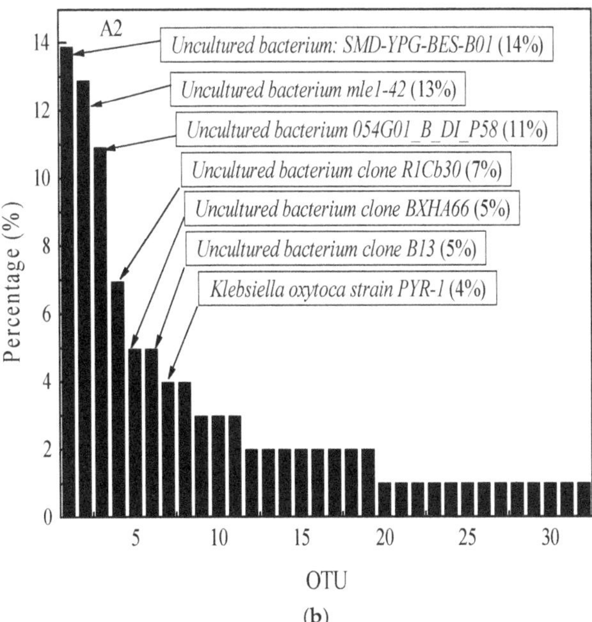

(b)

Figure 1. *Cont.*

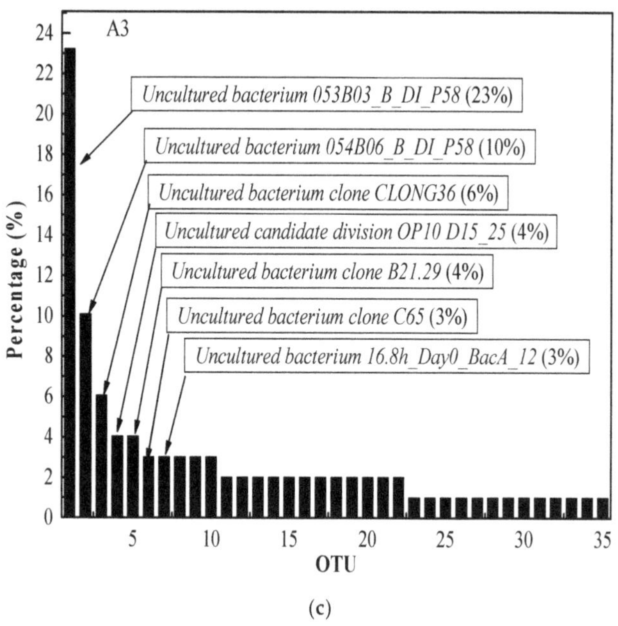

(c)

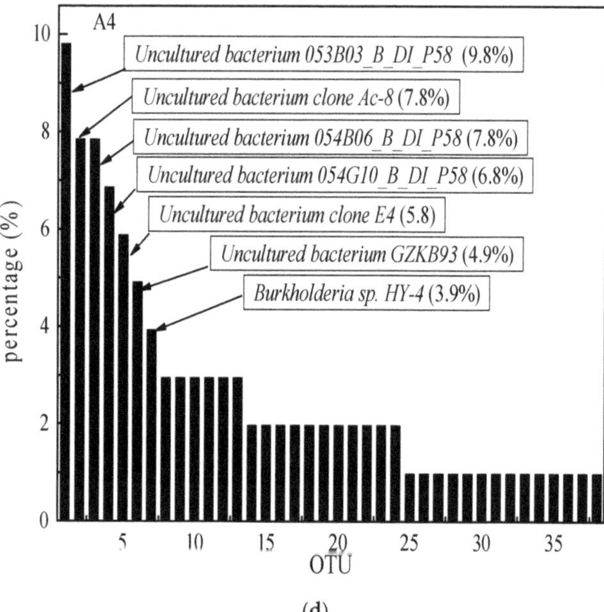

(d)

Figure 1. The total bacterial community in compartments of ABR (**a**) chamber A1, (**b**) chamber A2, (**c**) chamber A3, and (**d**) chamber A4.

For chamber A2, the OTU number of total bacteria was 32, with a coverage of 87%, a Shannon–Wiener diversity index of 3.04, and a minimum similarity of 86% to the known sequences, belonging to five taxa. The top three dominant groups were the *uncultured bacterium SMD-YPG-BES-B01*, the *uncultured bacterium mle1-42*, and the *uncultured bacterium 054G01_B_DI_P58*, with 14%, 13%, and 11% of the total bacterial community of the system, respectively. The *uncultured bacterium mle1-42* was the dominant colony for the degradation of the pharmaceutical wastewater pollutants. The bacterial communities with antibiotic resistance increased to 16% of the total bacterial community. The *uncultured bacterium gene for 16S rRNA* accounted for 1% of the total bacterial community and was a volatile fatty-acid-producing community.

For chamber A3, the OTU number of total bacteria was 35, with 86% coverage, and the Shannon–Wiener diversity index was 3.07, with a minimum similarity of 90% to the known sequence comparisons, belonging to five taxa. Among them, the *uncultured bacterium 053B03_B_DI_P58* was the most dominant group, accounting for 23% of the total bacteria, and was an anaerobic archaeal community. The next dominant colonies were the *uncultured bacterium 054B06_B_DI_P58* and the *uncultured bacterium clone CLONG36*, accounting for 10% and 6% of the total bacterial community, respectively, with the *uncultured bacterium clone CLONG36* belonging to the anaerobic granular sludge community. The bacterial communities with antibiotic resistance accounted for 7% of the total bacterial community. *Bacterium K-4b6*, which are acidophilic alkane-producing bacteria, accounted for 1% of the bacterial community.

For chamber A4, the OTU number of total bacteria was 38, with 86% coverage, and the Shannon–Wiener diversity score was 3.35, with a minimum similarity of 83% to the known sequence matches, belonging to four taxa. The *uncultured bacterium 053B03 B DI P58* was the most prevalent, accounting for 9.8% of the total microorganisms. *Acidophilic alkane-producing bacteria* made up 3% of the total bacterial population.

A phylogenetic tree was constructed for the bacteria in each chamber of the ABR reactor, and the results are shown in Figure 2.

(1)   The total bacteria in chamber A1 of the ABR reactor:

In A1, 80 clones were retrieved and were grouped into 23 operational taxonomic units (OTUs). *Bacteria* was the dominant species, accounting for 68% of the total bacterial community; *CFB group bacteria* and *firmicutes* were the next two dominant species, each accounting for 11% of the total bacterial community;

(2)   The total bacteria in the A2 chamber of the ABR reactor:

In A2, 87 clones were retrieved and were grouped into 25 OTUs. *Bacteria*, as the dominant species, accounted for 86% of the total bacterial community;

(3)   The total bacteria in chamber A3 of the ABR reactor:

In A3, 91 clones were retrieved and were grouped into 30 OTUs. *Bacteria* were the dominant species, accounting for 92% of the total bacterial community;

(4)   The total bacteria in chamber A4 of the ABR reactor:

In A4, 90 clones were retrieved and were grouped into 33 OTUs. *Bacteria* were the dominant species, accounting for 92% of the total bacterial community. *Characteria* served as the dominant species, accounting for 88% of the total bacterial community.

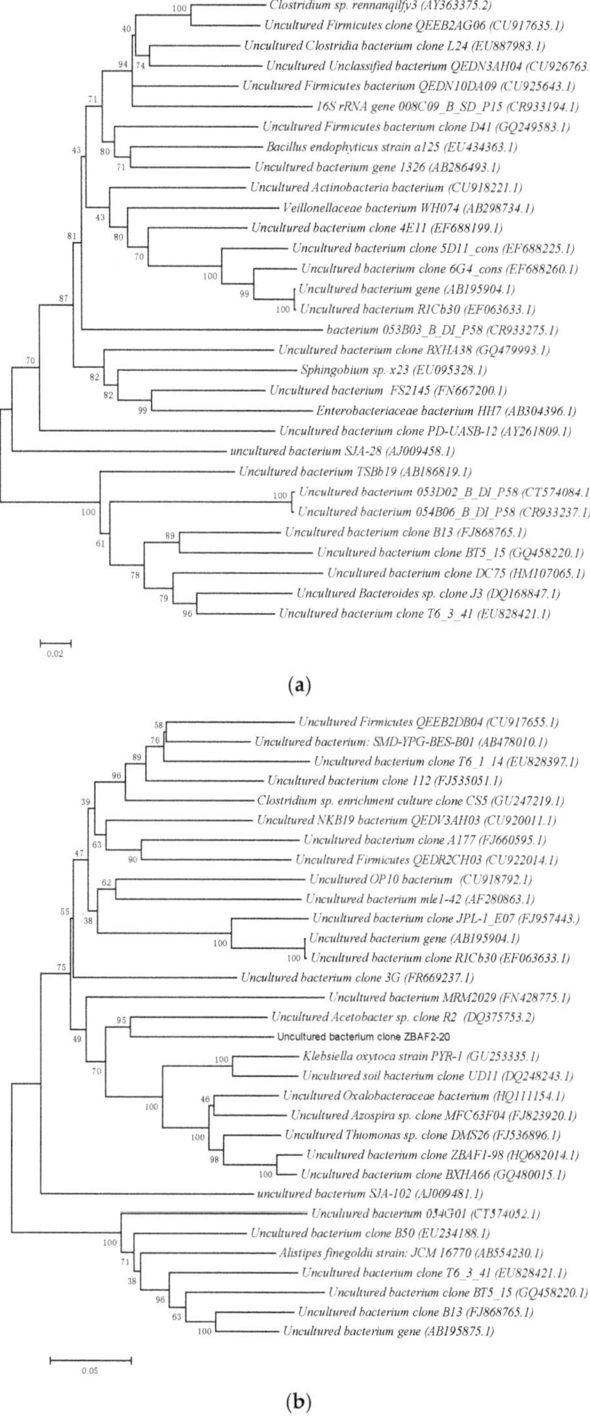

(**a**)

(**b**)

Figure 2. *Cont.*

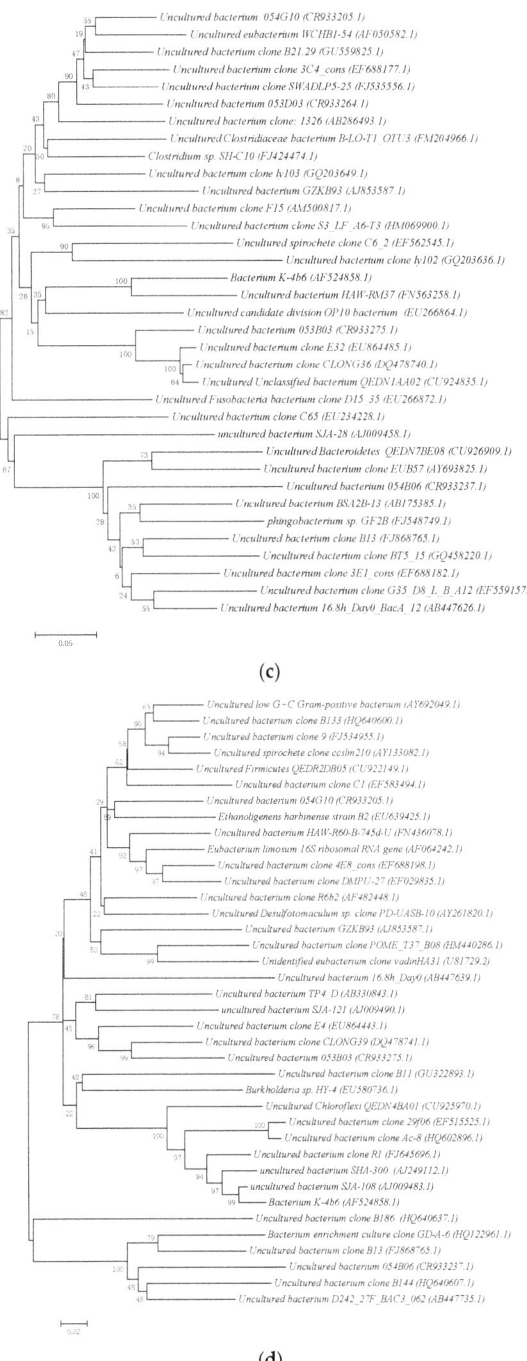

(c)

(d)

Figure 2. Phylogenetic tree for the total bacterial community in ABR chambers (**a**) chamber A1, (**b**) chamber A2, (**c**) chamber A3, and (**d**) chamber A4.

3.3. Differences in the R3 Reactor Sludge Granulation Process's Overall Bacterial Community Structure

The total bacterial 16S rDNA clone libraries were constructed for the sludge samples at different operation times of the aerobic granulation process. Each library contained about 100 clones (with a sequence length of about 1400 bp), and the sequencing results were compared by BLAST in the Gen-bank. The total bacterial clone library in the aerobic granular sludge (AGS) reactor after 15, 40, and 80 reactor starts was 38, 34, and 35 OTUs, respectively. The results are shown in Figure 3.

When the AGS reactor started up, the aerobic granular sludge appeared after 15 days of operation. The OTU number of total bacteria in the reactor was 38, with 77% coverage, and the Shannon–Wiener diversity index was 3.01. The maximum similarity between these clones and the known bacteria in the Gen-bank was 100%, while the minimum similarity was 90%. Among them, OTU1 was the uncultured bacterial clone. In addition, the most dominant species was the *Uncultured bacterium clone A156*, which accounted for 28% of the total bacterial community, and its clones were 93% similar to the known bacteria in the Gen-bank. It was followed by the *Uncultured bacterium clone AS-19* and *Hydrogenophaga sp. EMB 7* as the dominant species, which accounted for 8% and 6% of the total bacterial community, respectively. The *uncultured bacterium clone F54*, *Uncultured bacterium clone ZBAF1-105*, and the other antibiotic-resistant bacterial communities accounted for 11% of the total bacterial community; the *Uncultured bacterium clone M01* accounted for 3% of the total bacterial community by enhancing the sludge settling ability, and the *Bacterium clone M0111 48*, which represented for 3% of the overall bacterial community, helped with the generation of the aerobic granular sludge.

After 40 days of operation, the aerobic granular sludge was the dominant form in AGS reactor. The OTU number of the total bacteria in the AGS was found to be 34, with 89% coverage, a Shannon–Wiener diversity index of 3.21, and a minimum similarity of 90% to the known sequence comparisons, belonging to six taxa. The *uncultured Bacteroidetes bacterium clone Skagenf54 bacteria* was the most dominant species, belonging to *CFB group bacteria*, accounting for 19% of the entire bacterial community. The second most dominant species was the *Uncultured bacterium clone SS-9*, accounting for 4% of the overall bacterial community, and belonging to the pyridine, quinoline, and derivative degradation group. The dominant species for the granulation in the aerobic granular sludge was the *Uncultured bacterium clone EUB72-2*, which consisted of uncultured clones with the same percentage as the *Uncultured bacterium clone SS-9*. At the same time, the microbial community of the *uncultured bacterium clone 77*, which was fitted for the nutrients cyclical changing operation mode, was generated, accounting for 4% of the total bacterial community. The antibiotic-resistant bacterial populations then accounted for 16% of the entire bacterial community.

After 80 days' operation, the AGS reactor was filled with mature aerobic granular sludge. The OTU number of total bacteria in the AGS was found to be 35, with 82% coverage, and the Shannon–Wiener diversity index was 3.11, with a minimum similarity of 95% to the known sequences from eight taxa. *Comamonas* sp. *PP3-1*, which is a *Betaproteobacteria* of the phylum Amoeba, was the most abundant group, accounting for 20% of the total bacteria. The following dominant species were *Comamonas* sp. *XJ-L67*, the *uncultured bacterium clone A_SBR_1*, and the *uncultured bacterium clone B13*, which accounted for 8%, 7%, and 5% of the total bacterial community, respectively. As one of these, the *uncultured bacterium clone B13* made up the majority of the bacterial communities with antibiotic wastewater treatment capability. The antibiotic-resistant bacterial populations made up 15% of the entire bacterial community.

A phylogenetic tree was constructed for the bacteria in the AGS reactor at different operation times, and the results are shown in Figure 4.

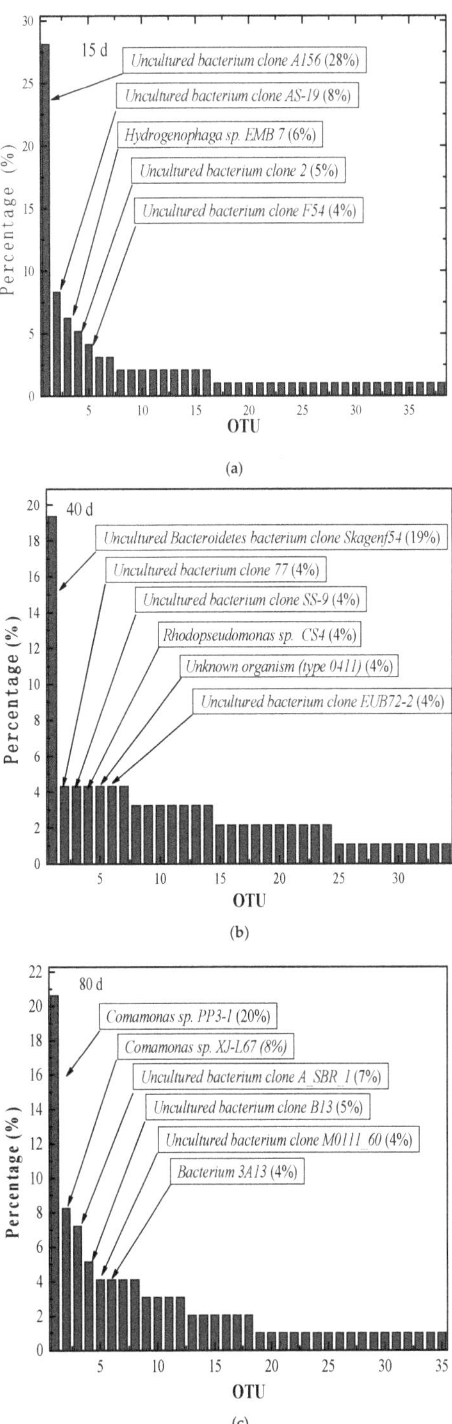

Figure 3. The total bacterial community in AGS at different stages (**a**) 15 d, (**b**) 40 d, and (**c**) 80 d.

(1)   The AGS reactor with 15 days of operation:

In the early stage of aerobic granular sludge formation, the *bacteria* and the *CFB group bacteria* dominated the reactor's bacterial ecology, accounting for 84% and 6% of the overall bacterial community, respectively;

(2)   The AGS reactor with 40 days of operation:

The aerobic granular sludge was the dominant form after 40 days of AGS reactor startup. The *Bacteria* and the *CFB group bacteria* were the main bacterial species in the reactor, accounting for 60% and 19% of the total bacterial community, respectively. The *a-proteobacteria* were close following them, accounting for 13% of all bacteria;

(3)   The AGS reactor with 80 days of operation:

When the AGS reactor ran for 80 days, the *Bacteria* and the *Betaproteobacteria* dominated the bacterial community, accounting for 43% and 34% of the total bacterial community, respectively.

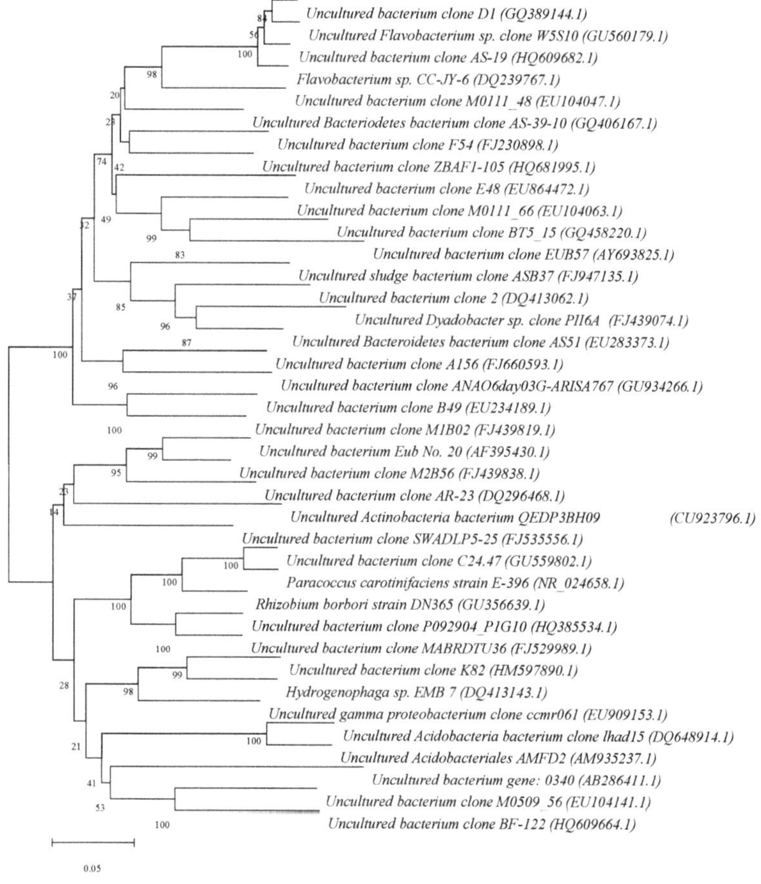

(a)

Figure 4. *Cont.*

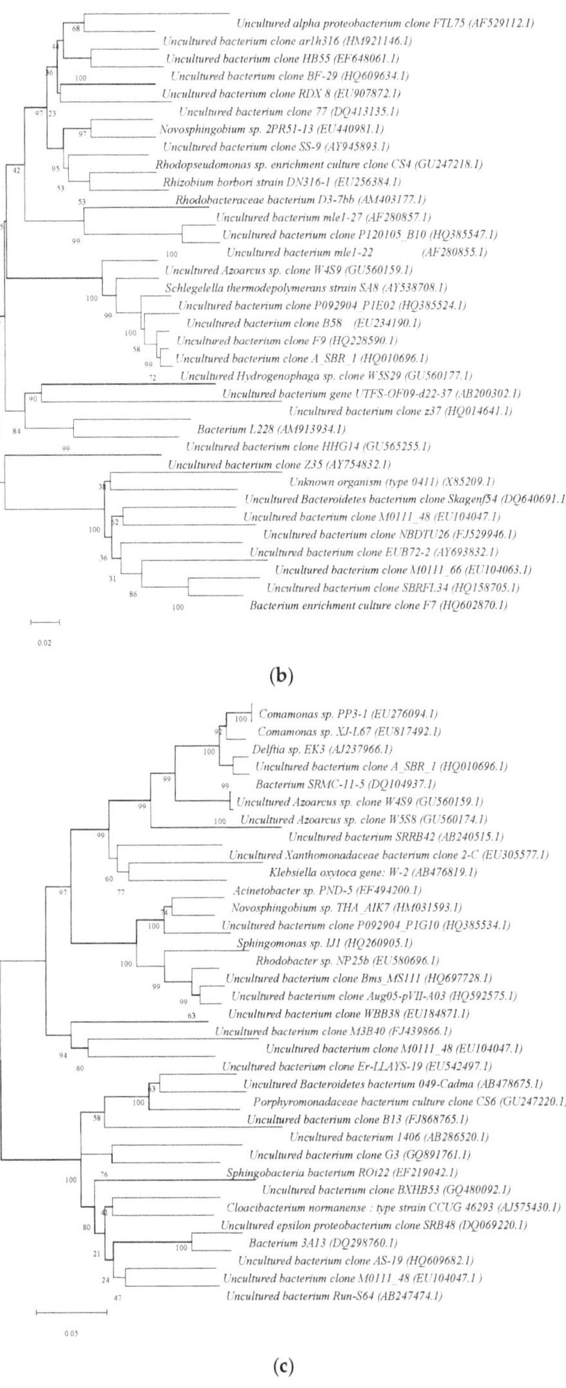

Figure 4. Phylogenetic tree for the total bacterial community in AGS at different stages (**a**) 15 d, (**b**) 40 d, and (**c**) 80 d.

4. Discussion

Berberine is an anti-inflammatory quaternary ammonium salt [2,3]. A high concentration of berberine is toxic to microbial organisms because it damages their cytoplasmic membranes and deactivates their enzymes. Berberine removal from synthetic wastewater was proposed using a hybrid ABR–AGS process. The promising results indicated that the hybrid ABR–AGS process is feasible and efficient in the degradation of berberine. The dynamic shift of bacterial communities in the ABR–AGS system is shown in Figure 5.

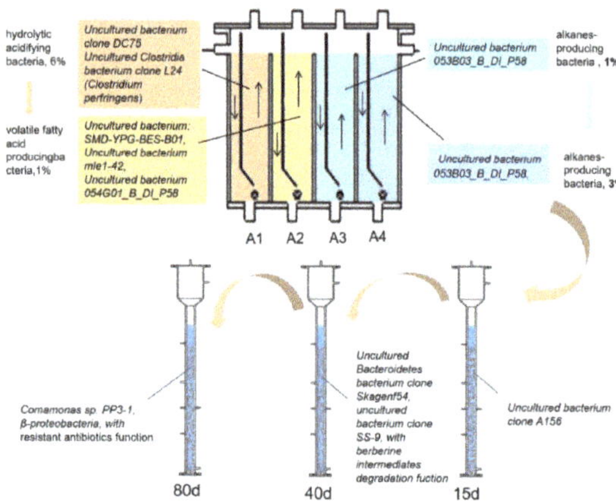

Figure 5. The schematic diagram of dynamic shift of bacterial communities in the hybrid ABR–AGS system.

When the influent berberine concentration was 120 mg/L, the dominant community in the ABR shifted as follows: *clone 053B03_B_DI_P58* (chamber A1) → *uncultured bacterium SMD-YPG-BES-B01, uncultured bacterium mle1-42, uncultured bacterium 054G01_B_DI_P58* (chamber A2) → *uncultured bacterium 053B03_B_DI_P58* (chamber A3) → *uncultured bacterium 053B03_B_DI_P58* (chamber A4). The results were consistent with the results that were provided by the phylogenetic tree that was constructed by the 16S rDNA clone library methods. The function of the chambers changed from hydrolytic acidification to methane production, with the influent flowing from A1 to A4. Thus, the colonies of the *uncultured bacterium clone DC75* and the *uncultured Clostridia bacterium clone L24* (*Clostridium perfringens*) with hydrolytic acidifying function occupied 6% of the total bacterial community in the chamber. *Clostridium* is a *Firmicutes* genus that has a hard cell wall and can produce endospores [28,29]. In addition, *Clostridium* was observed in a UASB reactor feed, with berberine as a carbon source [26]. There was a high proportion of Gram-positive low GC bacteria in the other anaerobic bioreactors as well [30]. Some *Clostridium* strains were found to be capable of using the methyl group of aromatic methyl ethers as a carbon source via O-demethylation [29]. It has also been reported that another stain can cleave aromatic rings [28]. Thus, there may be a functional species that is involved in the cleavage of the aromatic rings and/or the degradation of the methoxyl groups in berberine molecules.

In chamber A2, the *uncultured bacterium gene* of 16S rRNA with the function of volatile fatty acid production decreased to 1% of the total bacterial community. While *Bacteria*, as the dominant species, accounted for 86% of the total bacterial community.

In chamber A3, the acidophilic alkane-producing bacteria consisted of 1% of the bacterial community, indicating that the function of the microorganism changed from acid production to alkane production.

In chamber 4, the rate of alkane-producing bacteria in the total bacterial community increased to 3%. Since berberine is a drug with an antibiotic function, the rate of microorganisms with antibiotic resistance in the total bacterial community increased from 8% in chamber A1 to 16% in A2 and decreased to 7% in A3. Thus, most of the berberine was degraded in chamber A1 and A2. The results were consistent with the result of berberine degradation in the ABR (Figure S2).

For the AGS operation, the aerobic granules appeared on day 15, dominated on day 40, and matured on day 80. The dominant community in the AGS shifted as follows: *uncultured bacterium clone A156* → *uncultured Bacteroidetes bacterium clone Skagenf54* → *Comamonas* sp. *PP3-1*. The rate of antibiotic-resistant microorganisms in the total bacterial community was found to change with operation time, increasing from 11% on day 15 to 16% on day 40, and remaining at 15% until day 80, indicating that the AGS's antibiotic resistance capability increased with the operation time. After day 40 of operation, an *uncultured bacterium clone SS-9* appeared, which was responsible for the degradation of pyridine, quinoline, and derivatives, perhaps because the degradation intermediates of berberine, pyridine, quinoline, and derivatives were aromatic compounds. The communities that were responsible for aerobic granules' formation were important. The *Uncultured bacterium clone M0111_48* accounted for 3% of the total bacterial community after 15 days of operation and the *Uncultured bacterium clone EUB72-2* accounted for 4% of the total bacterial community after 40 days of operation.

The *Proteobacteria* species was observed after 40 days of operation. The *α-Proteobacteria* appeared after 40 days of operation in the AGS reactor, while *β-Proteobacteria* appeared after 80 days of operation in the AGS reactor. According to Xia et al. (2012) and Zhang et. al. (2004), the bacterial strains that have antibacterial properties are important in wastewater treatment. [31,32]. Thus, *β-proteobacteria* may be the functional group to resistant antibiotics.

5. Conclusions

For the synthetic wastewater containing 120 mg/L berberine, the hybrid ABR–AGS process achieved 92.2% and 94.8% overall removals of berberine and COD, respectively. *Bacteria, CFB group bacteria*, and *Betaproteobacteria* dominated the AGS system, whereas the *Bacterium* dominated the ABR. The *Uncultured bacterium clone B135, Bacillus endophyticus strain a125, Uncultured bacterium mle1-42, Uncultured bacterium clone OP10D15,* and *Uncultured bacterium clone B21.29F54* in the ABR, and the *Uncultured bacterium clone F54, Uncultured bacterium clone ZBAF1-105, Uncultured bacterium clone SS-9,* and *Uncultured bacterium clone B13* in the AGS were identified as functional species in biodegradation of berberine and/or its metabolites. Both anaerobic and aerobic bacterial communities could adapt appropriately to different berberine selection pressures, since the functional species' identical functions ensured comparable pollutant removal performances. The information that has been provided in this study may help with future research in gaining a better understanding of the berberine biodegradation process.

Supplementary Materials: The following supporting information can be downloaded at: https://www.mdpi.com/article/10.3390/pr10122506/s1, Figure S1: The image of the hybrid ABR–AGS system; Figure S2: The change in the berberine concentration from A1 to A4; Table S1: Physical and chemical properties of inoculum; Table S2: The operation conditions of the hybrid ABR–AGS system; Table S3: The composition of influent wastewater.

Author Contributions: Conceptualization, P.Z.; methodology, J.L. and F.L.; investigation, Y.W. and R.M.; data curation, Y.W.; writing—original draft preparation, P.Z. and Y.W.; writing—review and editing, Y.L. and C.A.N.; project administration, P.Z.; funding acquisition, P.Z. All authors have read and agreed to the published version of the manuscript.

Funding: This research was funded by the National Key Research and Development Program of China (No. 2022YFC3203300), the Chinese Research Academy of Environmental Sciences Central Public Welfare Scientific Research Project (2022YSKY-63), and the National Major Scientific and Technological Projects for Water Pollution Control and Management (2017ZX07402003).

Institutional Review Board Statement: Not applicable.

Informed Consent Statement: Not applicable.

Conflicts of Interest: The authors declare no conflict of interest.

References

1. Wojtyczka, R.D.; Arkadiusz, D.; MalGorzata, K.; Robert, K.; Agata, K.-D.; Tomasz, M.; Danuta, I. Berberine enhances the antibacterial activity of selected antibiotics against coagulase-negative *Staphylococcus* strains in vitro. *Molecules* **2014**, *19*, 6583–6596. [CrossRef] [PubMed]
2. Li, Y.; Yin, Y.M.; Wang, X.Y.; Wu, H.; Ge, X.Z. Evaluation of berberine as a natural fungicide: Biodegradation and antimicrobial mechanism. *J. Asian Nat. Prod. Res.* **2017**, *20*, 1. [CrossRef] [PubMed]
3. Othman, M.S.; Safwat, G.; Aboulkhair, M.; Moneim, A.E.A. The potential effect of berberine in mercury-induced hepatorenal toxicity in albino rats. *Food Chem. Toxicol.* **2014**, *69*, 175–181. [CrossRef] [PubMed]
4. Čerňakova, M.; Košťálová, D. Antimicrobial activity of berberine—A constituent of mahonia aquifolium. *Folia Microbiol.* **2002**, *47*, 375–378. [CrossRef] [PubMed]
5. Li, J.; Wang, L.J.; Liu, Y.Q.; Zeng, P.; Wang, Y.; Zhang, Y.Z. Removal of berberine from wastewater by MIL-101(Fe): Performance and mechanism. *ACS Omega* **2020**, *5*, 27962–27971. [CrossRef]
6. Liu, S.Y.; Zhang, Y.; Zeng, P.; Wang, H.L.; Song, Y.H.; Li, J. Isolation and Characterization of a Bacterial Strain Capable of Efficient Berberine Degradation. *Int. J. Environ. Res. Public Health* **2019**, *16*, 646. [CrossRef]
7. Shan, Y.P.; Song, Y.H.; Liu, Y.Q.; Liu, R.X.; Du, J.J.; Zeng, P. Adsorption of berberine by polymeric resin H103: Kinetics and thermodynamics. *Environ. Earth Sci.* **2015**, *73*, 4989–4994. [CrossRef]
8. Qin, W.; Song, Y.; Dai, Y.; Qiu, G.; Ren, M.; Zeng, P. Treatment of berberine hydrochloride pharmaceutical wastewater by $O_3/UV/H_2O_2$ advanced oxidation process. *Environ. Earth Sci.* **2015**, *73*, 4939–4946. [CrossRef]
9. Ren, M.; Song, Y.; Xiao, S.; Ping, Z.; Peng, J. Treatment of berberine hydrochloride wastewater by using pulse electro-coagulation process with Fe electrode. *Chem. Eng. J.* **2011**, *169*, 84–90. [CrossRef]
10. Ministry of Ecology and Environment of People's Republic of China. *Guideline on Available Techniques of Pollution Prevention and Control for Pharmaceutical Industry Active Pharmaceutical Ingredient (Fermentation, Chemical Synthesis, Extraction) and Preparations Categories*; Ministry of Ecology and Environment of People's Republic of China: Beijing, China, 2022.
11. Krishna, G.G.; Kumar, P.; Kumar, P. Treatment of low-strength soluble wastewater using an anaerobic baffled reactor (ABR). *J. Environ. Manag.* **2009**, *90*, 166–176. [CrossRef]
12. Wang, J.L.; Huang, Y.H.; Zhao, X. Performance and Characteristics of an anaerobic baffled reactor. *Bioresour. Technol.* **2004**, *93*, 205–208. [CrossRef]
13. Zeng, P.; Moy, B.; Song, Y.; Tay, J. Biodegradation of dimethyl phthalate by *Sphingomonas* sp. isolated from phthalic-acid-degrading aerobic granules. *Appl. Microbiol. Biotechnol.* **2008**, *80*, 899–905. [CrossRef]
14. Zhang, Y.; Tay, J. Toxic and inhibitory effects of trichloroethylene aerobic co-metabolism on phenol-grown aerobic granules. *J. Hazard. Mater.* **2015**, *286*, 204–210. [CrossRef]
15. Jiang, H.L.; Tay, J.H.; Tay, S.T.L. Changes in structure, activity and metabolism of aerobic granules as a microbial response to high phenol loading. *Appl. Microbiol. Biotechnol.* **2004**, *63*, 602–608. [CrossRef]
16. Qiu, G.; Song, Y.H.; Zeng, P.; Duan, L.; Xiao, S. Characterization of bacterial communities in hybrid upflow anaerobic sludge blanket (UASB)-membrane bioreactor (MBR) process for berberine antibiotic wastewater treatment. *Bioresour. Technol.* **2013**, *142*, 52–62. [CrossRef]
17. Liu, F.H. The Characteristics of Microbial Molecular Ecology of Hybrid ABR-Aerobic Granulation Reactor for Pharmaceutical Wastewater Treatment. Ph.D. Thesis, University of Science and Technology, Beijing, China, 2011.
18. Liu, Y.Q.; Kong, Y.H.; Zhang, R.; Zhang, X.; Wong, F.S.; Tay, J.H.; Zhu, J.R.; Jiang, W.J. Microbial population dynamics of granular aerobic sequencing batch reactors during start-up and steady state periods. *Water Sci. Technol.* **2010**, *62*, 1281–1287. [CrossRef]
19. Deng, Y.; Zhang, Y.; Gao, Y.; Li, D.; Liu, R.; Liu, M.; Zhang, H.; Hu, B.; Yu, T.; Yang, M. Microbial community compositional analysis for series reactors treating high level antibiotic wastewater. *Environ. Sci. Technol.* **2012**, *46*, 795–801. [CrossRef]
20. Ma, L.L.; Tank, Y.K.; Zhang, H.B. Development and design anaerobic baffled reactor for wastewater treatment. *J. Guangxi Univ.* **2008**, *33*, 378–380.
21. Nachaiyasit, S.S.; Stuckey, D.C. The effect of shock loads on the performance of an anaerobic fabled reactor (ABR):2.step and transient hydraulic shocks at constant feed strength. *Water Res.* **1997**, *31*, 2747–2754. [CrossRef]
22. Zhou, H. Analysis of ABR's engineering design consideration. *Water Wastewater Eng.* **2009**, *35*, 226–229.
23. Hu, X.Q.; Liu, D.Y.; Cai, H.S. Influence of Structure on Hydrodynamic Characteristics of Anaerobic Baffled Reactor. *Earth Sci.* **2004**, *29*, 369–374.
24. Geng, Y.G.; Zhang, X.; Zhang, H.Q.; Liu, J.D. Discussion on Engineering design of anaerobic baffled reactor. *Technol. Water Treat.* **2009**, *35*, 103–107.
25. Qiu, G.; Song, Y.H.; Zeng, P.; Duan, L.; Xiao, S. Combination of upflow anaerobic sludge blanket (UASB) and membrane bioreactor (MBR) for berberine reduction from wastewater and the effects of berberine on bacterial community dynamics. *J. Hazard. Mater.* **2013**, *246*, 34–43. [CrossRef] [PubMed]
26. State Environmental Protection Administration Monitoring (Ed.) *Analytical Methods of Water and Wastewater*, 4th ed.; China Environmental Science Press: Beijing, China, 2002.
27. Qiu, G. Hybrid UASB-MBR Process for Berberine Pharmaceutical Wastewater Treatment and the Process Mechanisms. Ph.D. Thesis, Beijing Normal University, Beijing, China, 2011.

28. Winter, J.; Popoff, M.R.; Grimont, P.; Bokkenheuser, V.D. *Clostridium orbiscindens* sp. nov., a human intestinal bacterium capable of cleaving the flavonoid C-ring. *Int. J. Syst. Evol. Microbiol.* **1991**, *41*, 355–357. [CrossRef] [PubMed]
29. Kasmi, A.E.; Rajasekharan, S.; Ragsdale, S.W. Anaerobic pathway for conversion of the methyl group of aromatic methyl ethers to acetic acid by Clostridium thermoaceticum. *Biochemistry* **1994**, *33*, 11217–11224. [CrossRef]
30. Narihiro, T.; Terada, T.; Kikuchi, K.; Iguchi, A.; Ikeda, M.; Yamauchi, T.; Shiraishi, K.; Kamagata, Y.; Nakamura, K.; Sekiguchi, Y. Comparative analysis of bacterial and archaeal communities in methanogenic sludge granules from upflow anaerobic sludge blanket reactors treating various food-processing, high strength organic wastewaters. *Microbes Environ.* **2009**, *24*, 88–96. [CrossRef] [PubMed]
31. Xia, S.; Jia, R.; Feng, F.; Xie, K.; Li, H.; Jing, D.; Xu, X. Effect of solids retention time on antibiotics removal performance and microbial communities in an A/OMBR process. *Bioresour. Technol.* **2012**, *106*, 36–43. [CrossRef] [PubMed]
32. Zheng, H. Experimental and Theoretical Studies of Protoberberine Alkaloids on the Mechanism of Multiple Activities. Ph.D. Thesis, Tianjin Medical University, Tianjin, China, 2004.

Article

Cultivation of Nitrifying and Nitrifying-Denitrifying Aerobic Granular Sludge for Sidestream Treatment of Anaerobically Digested Sludge Centrate

Guillian Morgan and Rania Ahmed Hamza *

Department of Civil Engineering, Ryerson University [†], 350 Victoria Street, Toronto, ON M5B 2K3, Canada
* Correspondence: rhamza@ryerson.ca
† In April 2022, the university announced the new name of Toronto Metropolitan University, which will be implemented in a phased approach.

Abstract: In this study, three 1.2-L aerobic granular sludge sequencing batch reactors (AGS-SBRs) were used to cultivate nitrifying and nitrifying-denitrifying granules (w/supplemental carbon) and investigate sidestream treatment of synthetic-centrate and real-centrate samples from Ashbridges Bay Treatment Plant (ABTP) in Toronto, Ontario, Canada. Results showed that although the cultivation of distinct granules was not observed in the nitrifying reactors, sludge volume index (SVI_{30}) values achieved while treating real and synthetic centrate were 72 ± 12 mL/g and 59 ± 11 mL/g (after day 14), respectively. Ammonia-nitrogen (NH_3-N) removal in the nitrifying SBRs were $93 \pm 19\%$ and $94 \pm 16\%$ for real and synthetic centrate, respectively. Granules with a distinct round structure were successfully formed in the nitrifying-denitrifying SBR, resulting in an SVI_{30} of 52 ± 23 mL/g. NH_3-N, chemical oxygen demand (COD) and phosphorus (P) removal in the nitrifying-denitrifying SBR were $92 \pm 9\%$, $94 \pm 5\%$, and $81 \pm 14\%$ (7th to 114th day), respectively with a low nitrite (NO_2-N) and nitrate (NO_3-N) concentration in the effluent indicating simultaneous nitrification-denitrification (SND) activity. High nutrient removal efficiencies via the nitrification and SND pathways shows that AGS technology is a viable process for treating sidestreams generated in a WWTP.

Keywords: aerobic granular sludge; ammonia oxidation; centrate; nitrification; denitrification; sidestream treatment

Citation: Morgan, G.; Hamza, R.A. Cultivation of Nitrifying and Nitrifying-Denitrifying Aerobic Granular Sludge for Sidestream Treatment of Anaerobically Digested Sludge Centrate. *Processes* **2022**, *10*, 1687. https://doi.org/10.3390/pr10091687

Academic Editor: Adam Smoliński

Received: 28 June 2022
Accepted: 22 August 2022
Published: 25 August 2022

Publisher's Note: MDPI stays neutral with regard to jurisdictional claims in published maps and institutional affiliations.

1. Introduction

Wastewater treatment plants (WWTPs) face the challenge of implementing sustainable unit processes that enhance nutrient removal and contribute to the recovery of energy and value-added products [1]. In recent years, technologies that permit sustainable wastewater management and solids handling are being increasingly employed worldwide [2]. Anaerobic digestion (AD) is a widely used technology for solids handling that utilizes biochemical processes to facilitate the degradation of organic matter (OM) by a consortium of bacteria under anaerobic conditions. The process of anaerobic digestion also leaves behind digestate (or anaerobically digested sludge) which is a liquid slurry of microorganisms, digested substrates, and the carrier medium [3]. Dewatering anaerobically digested sludge generates a liquid sidestream–commonly referred to as centrate or reject water–with elevated concentrations of N (predominantly NH_3-N) and P. In some installations, AD is preceded by thermal pre-treatment which increases the concentration of NH_3-N up to 2500 mg/L in centrate [4,5]. Centrate can consists of 0.5–1% of mainstream flow, 10–30% of the plant's influent nitrogen load and 20–40% of the plant's phosphorus load [6]. WWTPs predominantly return centrate to the mainstream process for treatment, but the additional nutrient loading negatively impacts removal efficiency. Returned centrate can result in shock loads, process instability and increased operation costs [7]. The inclusion of separate sidestream treatment can improve the performance and reliability of a WWTP [8].

Centrate can be adequately treated in a separate sidestream process using various wastewater technologies that belong to mainly two categories, (1) biological, and (2) physical/chemical [9]. Ammonia stripping is a physical/chemical treatment technology which has been used to remove 80–90% of nitrogen from digester supernatant using air at full scale [10]. Alternatively, ammonia can be recovered as an ammonia salt solution through acid stripping. NH_3-N concentration in a feed stream was drastically reduced from 21,006 to 10.6 mg/L (99.95% NH_3-N removal) at a normal temperature of 25 °C using acid stripping [11]. Ion exchange is another physical/chemical treatment technology that can achieve high ammonia removal from a liquid waste stream [12]. Target ions present in the liquid/aqueous phase such as ammonium ion (NH_4^+) are capable of displacing similar cations on the exchange resin; resulting in nutrient removal [7]. Digested sludge liquor with an ammonia concentration of >600 mg/L was successfully treated using a clay-based ion exchange medium (MesoLite), resulting in >95% ammonia removal [13].

Biological treatment is an inexpensive alternative compared to physical/chemical processes. Conventional activated sludge (CAS) is the most widely applied biological treatment technology. In CAS systems, nitrogen is removed from wastewater via biological nitrification and denitrification [14]. The first step, nitrification, occurs under aerobic conditions, where NH_3-N is oxidized to NO_2-N by ammonia oxidizing bacteria (AOB) and further to NO_3-N by nitrite oxidizing bacteria (NOB) [15]. In the denitrification step, organic loading in the influent or supplemental organic carbon (e.g., methanol) acts as an electron donor for the conversion of NO_3-N to dinitrogen gas (N_2) by heterotrophic denitrifying bacteria [9,16].

A submerged activated growth bioreactor (SAGB) that was fed centrate and operated with a nitrogen loading between 0.54 to 1.51 kg-N/m^3·d, achieved stable NH_3-N oxidation and a nitrification efficiency of 98% according to Pedros et al. [17]. Furthermore, treatment of co-digested piggery/poultry manure and agro-waste sludge liquor from a full-scale anaerobic digester was investigated by Scaglione et al. [18] in a pilot-scale SBR for partial-nitritation (PARNIT). In conjunction with a bench-scale SBR that was inoculated with anaerobic ammonium oxidation (anammox) granular sludge and fed with a combination of PARNIT effluent and synthetic wastewater, an average nitrogen removal of 91 ± 10% was achieved Scaglione et al. [18]. The anammox step allows for the conversion of NH_3-N to N_2 using NO_2-N as the electron acceptor instead of supplemental carbon source, leading to considerable cost savings [19]. Typically, processes that utilize a shortcut route to nitrogen removal require less oxygen and biodegradable carbon [8,20]. Nutrient-rich centrate typically comprises of low biodegradable COD and insufficient alkalinity-both of which are required for biological treatment [9].

Similar to the above-mentioned technologies, aerobic granular is a fitting technology when attempting to treat a sidestream such as centrate with fluctuations in the NH_3-N concentration and limited biodegradable COD [21]. An earlier study on formation of nitrifying granules was investigated by Liu et al. [22] using acclimated sludge and synthetic inorganic wastewater in an SBR. Nitrifying granules were reportedly formed within 21 days with a size of 240 μm and the biomass displayed a sludge volume index (SVI) of 40 mL/g [22]. Both nitrification and partial-nitrification was reported, with the authors stating that the system was negatively impacted by changing influent NH_3-N concentrations (100 to 450 mg/L) [22]. López-Palau et al. [19] also worked with anaerobic digestion supernatant to form nitrifying (NIT) granules in an SBR. A conventional SBR with suspended growth biomass was operated for comparison with the characteristics of aerobic granules. Granules with a distinct and slightly spherical shape were formed with the diameter ranging from 2 to 5 mm [19]. In terms of performance, partial nitrification was achieved in both systems with the NO_2-N/NH_4^+ ratio equating to approximately 1 in the effluent–mainly due to insufficient alkalinity in the system which induced pH suppression, thus resulting in partial nitrification [19,23].

Aerobic granular sludge (AGS) has also gained interest for its robustness and ability to maintain performance in adverse conditions [24]. The impact of COD/N ratio is a topic that is being continuously investigated by researchers worldwide. After investigating different COD/N ratios using ammonia-rich synthetic wastewater, Wei et al. [25] reportedly achieved the highest nitrogen removal at a ratio of 9. Kocaturk and Erguder [26] recommended a COD/N ratio of 7.5–30 for stable granules that favor the growth of heterotrophs and 2–5 for slow growing nitrifying granules. A recent investigation by Figdore et al. [27] pursued the growth of nitrification-denitrification (NDN) and phosphorus accumulating organisms (PAOs) granules from sidestream centrate. Growth of NDN-PAO granules were successful as reported by the author and removal efficiencies were 95% and 88% for NH_3-N and total nitrogen (TN), respectively [27]. The application of AGS to centrate—containing high nutrient load and low biodegradable COD—is practical through the well-known nitrification-denitrification reaction pathway. Additionally, the microbial orientation of AGS offers the potential for multiple redox reactions across the layers of a granules in a single system [28].

This research aims to evaluate the application of AGS–to separate sidestream treatment of centrate. The experimental work explored formation of nitrifying granules with and without an organic carbon source and assessed the treatment of centrate at a low COD/N ratio. The objectives of this study were to: (1) Compare the cultivation aerobic granular sludge and treatment of diluted centrate via nitrification only and nitrification-denitrification reaction pathways, (2) determine the impact of increasing nitrogen (N) loading and a low COD/N (i.e., <5) ratio on nitrification and/or nitrification-denitrification, and (3) determine the rate of nitrification with respect to aeration time and volatile suspended solids (VSS) concentration and the efficiency of nitrification-denitrification. First, the cultivation of nitrifying granules without the addition of external carbon was investigated using real centrate and synthetic wastewater. Secondly, the cultivation nitrifying-denitrifying granules were investigated using synthetic wastewater with the addition of an external carbon source follow by the treatment of real centrate with supplemental carbon. Pollutant degradation was monitored during a designated cycle to determine nitrification rate (NR) and specific nitrification rate (SNR). Simultaneous nitrification-denitrification efficiency was computed using influent and effluent data. The effect of a low COD/N was also investigated during the treatment of centrate (i.e., after granule cultivation using synthetic organic wastewater).

2. Materials and Methods

2.1. Reaction Configuration and Operation

The experimental study was performed in three column-type SCH 40 PVC reactors—AGS-SBR-1, AGS-SBR-2, and AGS-SBR-3—with a 5-cm internal diameter, height of 70 cm and a total volume of 1.38-L. The reactors were operated with a working volume of 1.2-L, equivalent to a working height of 61 cm and height-to-diameter (H/D) ratio of 12.2. Aeration was provided by fine air bubble diffusers located at the bottom of the reactors and flow meters were used to control the airflow at 3 L/min, resulting in a superficial air velocity (SAV) of 2.55 cm/s. Air flow in AGS-SBR-1 was reduced to 1 L/min (SAV = 0.8 cm/s) during the final period (i.e., days 162–183) of the reactor's operation. The AGS-SBRs were designed to accommodate static feeding from the bottom of each reactor and upwards through the bed of settled biomass. The decant ports were conveniently located at the midpoint of each reactor column (i.e., at the 0.6-L mark), providing a volume exchange ratio (VER) of 50%. Peristaltic pumps were used to feed the reactors and discharge treated effluent. The reactors were operated sequentially on a 6-h cycle followed by a 4-h cycle to allow for a reduction in hydraulic retention time (HRT) as displayed in Table 1. The duration of influent filling, effluent discharge, and idle remained constant throughout the experimental study. AGS-SBR-3 was operated with a 60-min anoxic filling phase to stimulate denitrification. Settling times were decreased during the granule cultivation stage with the excess time added to the aeration phase.

Table 1. Cycle time and sequential phases for the investigated AGS-SBRs.

	AGS-SBR-1		AGS-SBR-2		AGS-SBR-3	
Phase	**6-h**	**4-h**	**6-h**	**4-h**	**6-h**	**4-h**
Influent Filling (min)	15	15	15	15	60	60
Aeration (min)	265–315	160–195	265–315	190–195	220–270	145–150
Settling (min)	10–60	10–45	10–60	10–15	10–60	10–15
Effluent discharge (min)	15	15	15	15	15	15
Idle (min)	5	5	5	5	5	5

The operating parameters for each reactor are shown in Table 2. The experiments were conducted at ambient temperature (22–25 °C). A common peristaltic pump was used to dose an 80 g/L solution of sodium bicarbonate ($NaHCO_3$) to AGS-SBR-1 and AGS-SBR-2 for pH control. In AGS-SBR-3, pH was controlled using a 1 M hydrochloric acid (HCl) solution and an Etatron DLX series metering pump attached to a probe. Dissolved oxygen (DO) was considered >2 mg/L or non-limiting since air flow was controlled by a flow meter.

Table 2. Summary of operating conditions for aerobic granular sludge SBR systems.

Parameter	Unit	AGS-SBR-1	AGS-SBR-2	AGS-SBR-3
Operating Duration	days	183	72	212
pH	-	7.8 ± 0.5	8 ± 0.5	7.8 ± 0.3
Temperature	°C	22–25	22–25	22–25
VER	%	50	50	50
DO	mg/L	>2 (non-limiting)	>2 (non-limiting)	>2 (non-limiting)

Operational periods for each reactor are displayed in Tables 3 and 4. The experimental study consisted of several changes that were carried out in different periods. The experimental schematic is shown in Figure 1.

Table 3. Summary of reactor operational parameters for AGS-SBR-1 & 2.

Reactor	Period	Operation Days	Influent NH_3-N Concentration (mg/L)	HRT (h)
AGS-SBR-1	I	Days 1–61	123 ± 5	12
	II	Days 71–99	180 ± 17	12
	III	Days 106–120	182 ± 9	8
	IV	Days 134–155	255 ± 7	8
	V	Days 162–183	292 ± 11	8
AGS-SBR-2	I	Days 1–14	84 ± 2	12
	II	Days 24–51	187 ± 7	12
	III	Days 58–72	192 ± 13	8

Table 4. Summary of reactor operational parameters for AGS-SBR-3.

	Period	Operational Days	Influent NH_3-N Concentration (mg/L)	Supplemental COD Concentration (g/L)	HRT (h)	COD:N:P
	I	Days 1–38	77 ± 12	2.09 ± 0.15	12	105:4:1
		Days 38–72	88 ± 3	2.20 ± 0.07	8	110:4:1
	II	Days 79–93	84 ± 2	1.52 ± 0.01	8	76:4:1
AGS-SBR-3	III	Days 100–114	74 ± 11	0.95 ± 0.06	8	47:4:0.4
	IV	Days 121–135	149 ± 7	0.72 ± 0.05	8	36:8:2
		Days 149–156	131 ± 0	0.54 ± 0.06	8	27:7:1
	V	Days 163–212	121 ± 5	1.39 ± 0.05	8	70:6:2

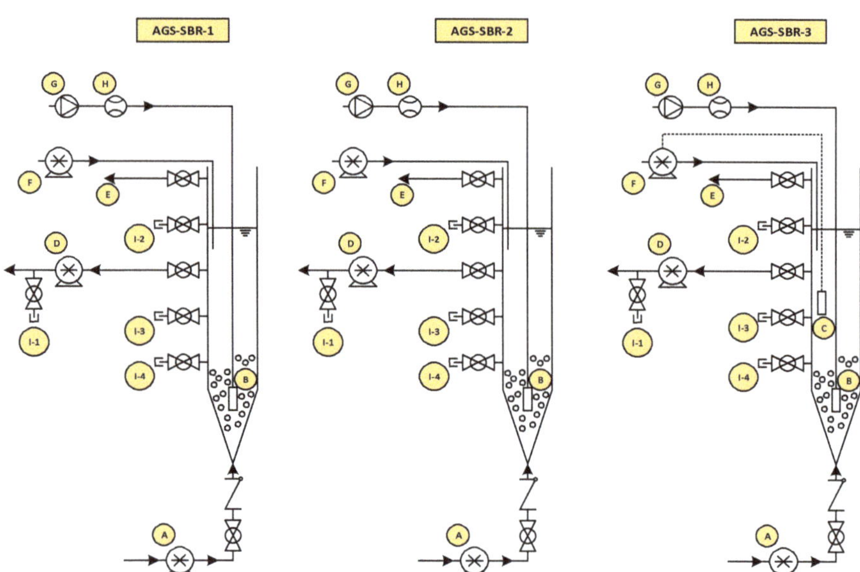

Figure 1. Schematic of AGS-SBR-1, AGS-SBR-2, and AGS-SBR-3: (A) feed pump, (B) air diffuser, (C) pH probe, (D) decant pump, (E) overflow, (F) NaHCO$_3$/HCl pump, (G) air pump, (H) flow meter, (I-1) effluent sample port, (I-2-3) spare sample ports, and (I-4) sludge sample port.

2.2. Seed Sludge

The AGS-SBRs were inoculated with return activated sludge (RAS) from the Ashbridges Bay Treatment plant (ABTP) in Toronto, Ontario, Canada, resulting in an initial sludge concentration of 4.3–5.2 g/L (65–80% volatile). Biomass in AGS-SBR-1 and AGS-SBR-2 was augmented with RAS during periods of low sludge concentration or following a process upset. Additionally, biomass augmentation with RAS was performed to maintain mixed liquor suspended solids (MLSS) and mixed liquor volatile suspended solids (MLVSS) concentrations above 2 g/L and 1.5 g/L, respectively. Biomass augmentation points are indicated on Figure 2a,b,d,e. Seed sludge sampled from ABTP had an average suspended solids (SS) concentration 10 ± 1 g/L (70% volatile) and sludge volume index (SVI$_{30}$) of 115 ± 9 mL/g.

2.3. Wastewater Media Characteristics

Several feed streams were used to provide substrate to the AGS-SBRs in this experimental study. The characteristics of centrate collected from ABTP and composition of synthetic wastewater is described in Table 5. ABTP utilizes the conventional activated sludge process for biological treatment of wastewater and anaerobic digestion for sludge stabilization followed by the addition of a polymer compound prior to dewatering.

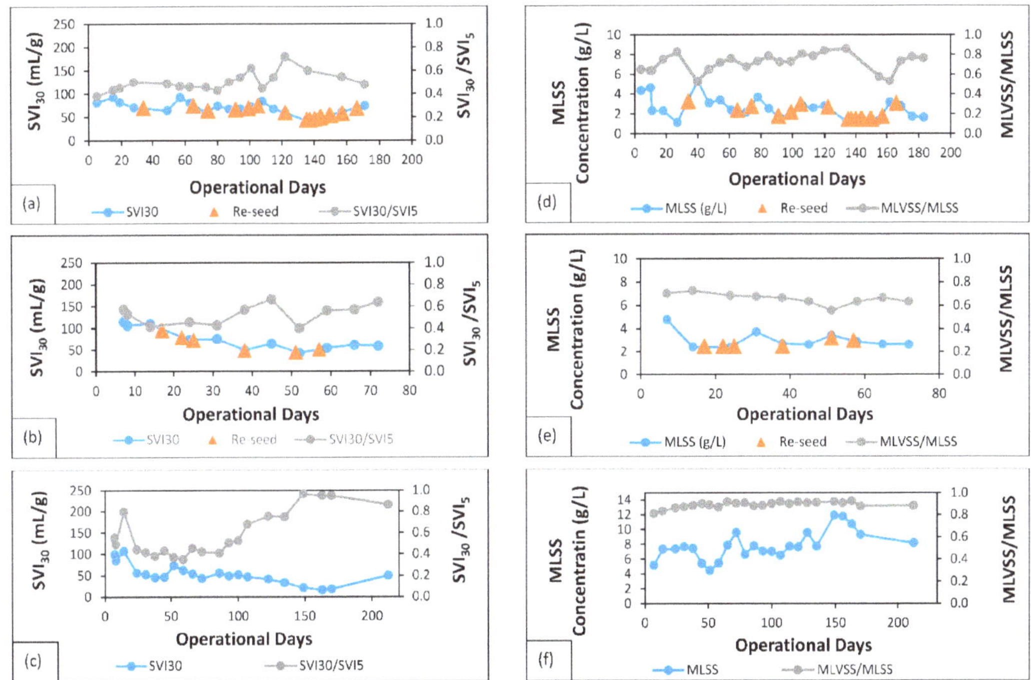

Figure 2. Sludge Volume Index (SVI_{30}) for (**a**) AGS-SBR-1, (**b**) AGS-SBR-2, and (**c**) AGS-SBR-3; MLSS concentration and MLVSS/MLSS ratio for (**d**) AGS-SBR-1, (**e**) AGS-SBR-2, and (**f**) AGS-SBR-3.

Table 5. Characteristics of centrate and synthetic centrate used for AGS-SBR-1, 2 & 3.

	AGS-SBR-1	AGS-SBR-2		AGS-SBR-3 [§]	
Parameter	**Average Concentration (Per Litre)**	**Component**	**Per Litre**	**Component**	**Per Litre**
NH_3-N	484.83 ± 46.96 mg	$NaHCO_3$	1.2 g	CH_3COONa	2.56 g
NO_2-N	0.074 ± 0.05 mg	NH_4Cl	0.35 g	NH_4Cl	0.35 g
NO_3-N	0.91 ± 0.13 mg	K_2HPO_4	0.05 g	K_2HPO_4	0.03 g
TSS	0.089 ± 0.01 mg	KH_2PO_4	0.045 g	KH_2PO_4	0.025 g
VSS	0.075 ± 0.01 mg	$CaCl_2 \cdot 2H_2O$	0.03 g	$CaCl_2 \cdot 2H_2O$	0.03 g
tCOD	674.90 ± 139.62 mg	$MgSO_4 \cdot 7H_2O$	0.025 g	$MgSO_4 \cdot 7H_2O$	0.025 g
sCOD	547.68 ± 136.34 mg	$FeSO_4 \cdot 7H_2O$	0.02 g	$FeSO_4 \cdot 7H_2O$	0.02 g
$CaCO_3$	1922.58 ± 263.84 mg	Micronutrients *	1 mL	Micronutrients *	1 mL
TN	524.75 ± 29.29 mg				
PO_4^{3-}	103.10 ± 11.27 mg				

[§] AGS-SBR-3 feed solution changed to centrate with CH_3COONa as supplemental carbon source after day 79.
* The micronutrient solution consisted of (in g/L): H_3BO_3, 0.05; $ZnCl_2$, 0.05; $CuCl_2$, 0.30; $MnSO_4 \cdot H_2O$, 0.05; $(NH_4)_6Mo_7O_{24} \cdot 4H_2O$, 0.05; $AlCl_3$, 0.05; $CoCl_2 \cdot 6H_2O$, 0.05 and $CoCl_2 \cdot 6H_2O$, 0.05.

- AGS-SBR-1: Centrate collected from ABTP was diluted to provide NH_3-N feed concentrations between 80–300 mg/L. $NaHCO_3$ was added as inorganic carbon source and pH control.
- AGS-SBR-2: Synthetic centrate with $NaHCO_3$ as inorganic carbon source and pH control, NH_4Cl as nitrogen source and K_2HPO_4 & KH_2PO_4 as phosphorus sources.
- AGS-SBR-3: Synthetic centrate with acetate as organic carbon source, NH_4Cl as nitrogen source and K_2HPO_4 & KH_2PO_4 as phosphorus sources. The feed stream

was switched to diluted centrate after day 79 with sodium acetate (CH_3COONa) as supplemental carbon source.

2.4. Analytical Methods

2.4.1. Biomass (Solid Phase) Analysis

MLSS and MLVSS tests were performed according to Standard Methods 2540 D, 2540 E, and 2710 D, respectively [29]. Sludge was allowed to settle in the reactor and timed for 30 min to determine SVI_{30}; similarly, to standard method 2710 D [29]. SVI_{30} tests were carried out in the reactor to prevent process disruptions associated with dismantling the system to access mixed liquor.

2.4.2. Wastewater (Liquid Phase) Analysis

Nutrient concentrations for influent and effluent were measured using a HACH DR 3900 (HACH company) portable spectrophotometer for the entire study period. NH_3-N was analysed using the Salicylate method (10031). NO_2-N, NO_3-N and alkalinity analysis were performed using HCH test kits TNT plusTM839 & 840, TNT plusTM835 and TNT plusTM870, respectively. TN was determined by the persulfate digestion method (10071). Reactive phosphorus (PO_4^{3-}) was determined using the Molybdovanadate method (8114). Total chemical oxygen demand (tCOD) and soluble COD (sCOD) (0.45 μm-filtered) were determined by using the reactor digestion method. Digestion was performed using a Labnet Accublock Digital dry bath (Labnet International Inc., Mayfield, NJ, USA) followed by spectrophotometric measurement. Total suspended solids (TSS) and Volatile suspended solids (VSS) were analysed in accordance with standard methods 2540 D, and 2540 E, respectively [29]. Fisher Scientific Accumet basic pH meter was used to measure pH in AGS-SBR-1 and AGS-SBR-2.

2.4.3. Reactor Performance and Removal Analysis

Removal efficiencies for NH_3-N, sCOD, and PO_4^{3-} were determined by Equation (1) [30]. The accumulation of nitrite in the reactors was determined by Equation (2) [31]. In AGS-SBR-3, the efficiency of simultaneous nitrification-denitrification was determined by Equation (3) [31].

$$\text{Efficiency } (\%) = \frac{C_{influent} - C_{effluent}}{C_{influent}} \times 100 \tag{1}$$

Where, C = Concentration in mg/L

$$\text{Nitrite Accumulation Rate (NAR) } (\%) = \frac{NO_2 - N}{NO_2 - N + NO_3 - N} \times 100 \tag{2}$$

$$\text{Simultaneous Nitrification} - \text{Denitrification (SND)Efficiency } (\%) = \left(1 - \frac{[NO_x - N]_{accumulated}}{[NH_3 - N]_{removed}}\right) \times 100 \tag{3}$$

Sludge production based on wastewater characteristics is determined by Equation (4). Parts A, B & C were computed to determine sludge production in AGS-SBR-3. Sludge production in AGS-SB-1 and AGS-SBR-2 were determined with only Part C. The concentration of nitrogen oxidized as NH_3-N was determined with Equation (5). Nitrification rate was calculated using Equation (6) and specific nitrification rate was determined by taking into the consideration the VSS concentration as shown in Equation (7) [30].

$$P_{x, bio} = \text{Part A (Heterotrophic Biomass)} + \text{Part B (Cell debris)} + \text{Part C (Nitrifying bacteria biomass)}$$

$$P_{x, bio} = \frac{QY_H(S_0-S)\left(\frac{1kg}{10^3 g}\right)}{1+b_H(SRT)} + \frac{(f_d)(b_H)QY_H(S_0-S)SRT\left(\frac{1kg}{10^3 g}\right)}{1+b_H(SRT)} + \frac{QY_n(NO_x)\left(\frac{1kg}{10^3 g}\right)}{1+b_n(SRT)} \tag{4}$$

$$NO_x = \text{concentration of } NH_3 - N \text{ in the influent flow that is nitrified, mg/L}$$
$$Y_H = \text{heterotrophic bacteria synthesis yield coefficient, } 0.4 \text{ g VSS/g COD}$$
$$b_H = \text{endogenous decay coefficient for heterotrophic organisms, } 0.16 \text{ g VSS/g VSS d}$$
$$b_n = \text{endogenous decay coefficient for nitrifying organisms, } 0.2 \text{ g VSS/g VSS d}$$
$$Y_n = \text{nitrifying bacteria synthesis yield coefficient, } 0.12 \text{ g VSS/g } NH_4 - N$$
$$f_d = \text{fraction of cell mass that remains as cell debris, } 0.10 - 0.15 \text{ g VSS/g biomass VSS depleted by decay}$$

$$\text{Nitrogen oxidized} = \text{nitrogen in influent} - \text{nitrogen in effluent} - \text{nitrogen in cell mass}$$
$$(NO_x) = (TKN_o) * -N_e - 0.12\left(\frac{P_{x,\ bio}}{Q}\right) \tag{5}$$

$$*\text{Influent } NH_3 - N \text{ used in place of } TKN_o$$

$$\text{Nitrification Rate} = \frac{NO_x}{t_{aeration}} \tag{6}$$

$$\text{Specific nitrification rate (SNR)} = \frac{\text{Nitrification rate}}{MLVSS_{reactor}} \tag{7}$$

2.4.4. Granule Structure and Morphology

Sludge was routinely collected from the lowest sample port of each reactor and poured into a Petri dish. Photographic images were captured with a Samsung Galaxy A50 smart phone camera and microscopic images were taken with a Leica DM1000 LED microscope. Following installation and start-up of the microscope, the Leica representative returned to calibrate the device using precise measurement bars.

3. Results and Discussion

3.1. Characteristics of Granular Sludge

3.1.1. Centrate Nitrifying Sludge (AGS-SBR-1)

In AGS-SBR-1, SVI_{30} averaged 72 ± 12 mL/g; displaying good sludge settleability. The SVI_{30}/SVI_5 ratio averaged 0.5 ± 0.08, equating to 50% granulation as shown in Figure 2a. Settleability of the sludge was observed to be influenced by the reactor design and operation: (1) H/D ratio of 12.2 and, (2) air flow rate of 3 L/min. A high H/D ratio represents a slender column, which influences the settling time [24]. Non-limiting DO air flow during the aeration phase resulted in adequate oxidation of ammonia.

Sludge concentration for reactor start-up was approximately 4.37 g MLSS/L and 2.85 g MLVSS/L. After 11 days of operation, sludge concentration decreased to approx. 2.34 g MLSS/L and 1.50 g MLVSS/L. To maintain the sludge concentration, RAS samples from ABTP were added to AGS-SBR-1 throughout the operation. On average, a sludge concentration of 2.68 ± 1 g MLSS/L and 1.87 ± 0.63 g MLVSS/L was achieved for the entire duration of the study. This includes fluctuations in the MLSS and MLVSS concentration between 1.16–5.33 g/L and 0.79–2.97 g/L, respectively. In Figure 2d, instances in which RAS was used to augment the sludge concentration are indicated on the MLSS (g/L) concentration curve. The ratio of MLVSS/MLSS fluctuated in response to the sludge concentration, but an average vale of 0.72 ± 0.09 was determined for AGS-SBR-1. According to Czarnota et al. [32], MLVSS/MLSS ratio signifies sludge activity or organic content, and it is typically in the range 0.7–0.8 [30]. In Figure 2d, reactor re-seed on day 34 in response to a substantial decrease in sludge concentration resulted in a lower than usual MLVSS/MLSS ratio. Evidently, a lower MLVSS/MLSS concentration ratio indicated a change in the biomass component such as a decrease in the concentration of viable sludge [33].

3.1.2. Synthetic Centrate Nitrifying Sludge (AGS-SBR-2)

Sludge settleability for AGS-SBR-2 in terms of SVI_{30} averaged 111 ± 4 mL/g for days 4–14 and 59 ± 11 mL/g for the remainder of the reactor operation. The average granulation percentage for the entire research study based on the ratio SVI_{30}/SVI_5 was $53 \pm 9\%$. Visual inspection of the sludge in AGS-SBR-2 did not show obvious evidence of distinct granule formation, but the settling behaviour was like that of a typical granular sludge reactor.

Activated sludge systems are known to have an average SVI_{30} between 110 and 160 [34], but AGS-SBR-2 averages for settling was substantially lower, especially after day 14 of operation as shown in Figure 2b.

Sludge concentration for AGS-SBR-2 at start-up was approx. 4.78 g MLSS/L and 3.35 g MLVSS/L. The sludge concentration decreased to 2.43 g MLSS/L and 1.76 g MLVSS/L by day 14 as shown in Figure 2e. RAS from ABTP was added to AGS-SBR-2 to maintain sludge concentration and the MLVSS/MLSS ratio. In comparison to AGS-SBR-1, MLSS and MLVSS concentration experienced higher stability in AGS-SBR-2. The average MLSS and MLVSS concentration for AGS-SBR-2 were 3.03 ± 0.75 g/L and 1.99 ± 0.54 g/L, respectively. The average MLVSS/MLSS ratio for AGS-SBR-2 was 0.66 ± 0.05, which was lower than that of AGS-SBR-1 at 0.72 ± 0.09. The average MLVSS/MLSS ratio of 0.66 ± 0.05 was also lower than the typical ratio of 0.7–0.8 [30]. The ratio between MLVSS and MLSS highlighted sludge inactivity or subpar organic content [32].

3.1.3. Nitrifying-Denitrifying Sludge (AGS-SBR-3)

The transition from activated to predominantly granular sludge occurred in parallel with decreasing values for SVI_{30} as shown in Figure 2c, highlighting the progression of sludge settleability. The curve for SVI_{30} undergoes a downward trend, which begins with SVI testing performed on day 7 and onwards. In Figure 2c, peaks on SVI_{30} curve on days 14 and 52 were associated with excess sludge production and washout, respectively. On day 14, increased sludge production in AGS-SBR-3 resulted in a MLSS concentration of 7.37 g/L which increased settling time due to excess biomass present in the system. A longer sludge settling time on day 52 was recorded after a reactor upset two days prior. The system is estimated to have completed 2 cycles without receiving influent. Aeration of settled sludge with one-half of the reactor working volume led to disruption of biological aggregates that previously enhanced the settling time. The average SVI_{30} was 52 ± 23 mL/g for the entire duration of the study and 45 ± 15 mL/g for all tests performed after day 14. The SVI_{30}/SVI_5 ratio for the entire duration of the study was 0.59 ± 0.21, which quantifies the percent granulation in the system. In Figure 2c, SVI_{30}/SVI_5 ratio is shown to have gradually increased after day 59 and reached a maximum of 0.96 on day 149. Granules were occasionally collected to monitor sludge morphology through visual inspection and microscopic imaging. As the research objectives were completed by day 163, the reactor operation was maintained to accommodate sludge sampling and storage. On day 212, final sampling and testing were completed and the SVI_{30} narrowly increased in response to a power outage and a routine campus lockout.

The potential for aerobic granular sludge reactors to operate at high sludge concentrations compared to activated sludge was witnessed in AGS-SBR-3. Irrespective of the fluctuations in MLSS concentration displayed in Figure 2f, consistent biomass growth allowed for concentrations of >7 g/L in AGS-SBR-3. Furthermore, MLSS concentration appeared to trend upwards throughout the experimental study. Sludge concentration increased from 5.20 to 7.37 g MLSS/L and 4.21 to 6.14 g MLVSS/L between days 7 and 14. A high sludge concentration was maintained until day 45 when MLSS and MLVSS decreased to 5.41 and 4.86 g/L, respectively. On day 51, the lowest MLSS (4.49 g/L) and MLVSS (3.99 g/L) concentrations for the entire experimental were recorded. A reduction in sludge concentration was caused by an improper timer setting for the decant pump on day 43, which led to a discharge of approximately 50% of the reactor's volume as mixed liquor. The robustness of the AGS-SBR-3 was confirmed as the sludge concentration increased to 9.62 g MLSS/L and 8.66 g MLVSS/L by the 72nd day of the experiment study. Numerous process modifications were applied to AGS-SBR-3 between days 72 and 156. Process modifications included altering the concentration of COD and NH_3-N in the feed solution to lower the COD/N ratio. On day 159, both COD and NH_3-N were adjusted to 1500 mg/L and 120 mg/L, respectively. Sludge analysis on day 163 confirmed a reduction in the sludge concentration to 10.76 g MLSS/L and 9.95 g MLVSS/L. Similarly, a reduction in sludge concentration was also recorded on day 170. AGS-SBR-3 rebounded prior to the end of the experiment with the sludge concentration on day 212 remaining at >8 g MLSS/L and >7 g MLVSS/L. The MLVSS/MLSS ratio curve in Figure 2f, displays a slightly horizontal curve that gradually trend upwards. The average MLVSS/MLSS ratio of 0.89 ± 0.03 proves that MLVSS concentration was also high and continued to increase with the progression of the research study. A high MLVSS concentration signifies good microbial activity within AGS-SBR-3 [32]. The presence of various organisms was captured with the assistance of a microscope.

3.2. Morphology and Structure of Aerobic Granules

Images are organized according to the timeline of the study, which shows the following: (1) Transition from activated to granular sludge, and (2) Shape of granules. In Figure 3a,b,e,f, distinct nitrifying granules were not found even though the system displayed good settleability. In Figure 3e, several small aggregates in the sample were observed to be >100 µm. Similarly, the formation of small aggregates in AGS-SBR-2 was also observed in Figure 3f. In a study by Liu et al. [22], nitrifying granules with a size of 240 µm were formed using synthetic wastewater with an inorganic carbon source as feed. Figure 3c,g provide adequate evidence of granule formation as early as day 32 in AGS-SBR-3. In Figure 3h a granule with diameter 1130 µm (1.13 mm) was captured using the microscope. The diameter could have been slightly less since the cover slide crushed the granules during microscopic analysis. Nonetheless, large-sized granules were sampled as shown in Figure 3d.

Figure 3. Images of sludge morphology in AGS-SBR-1 (**a,e**), AGS-SBR-2 (**b,f**), and AGS-SBR-3 (**c,d,g,h**).

3.3. Pollutants Removal Efficiencies

3.3.1. Centrate Nitrifying SBR (AGS-SBR-1)

AGS-SBR-1 displayed evidence of nitrification through the production of NO_x-N [NO_2-N + NO_3-N]. During start-up, RAS was acclimated with undiluted centrate for three days to ensure adequate oxidation of NH_3-N could be achieved during the subsequent 6-h (days 0–99) and 4-h (days 99–183) cycles. Following three days of acclimation, the reactor was fed with diluted centrate. Experimental analysis of samples collected on day 4 confirmed a 96% NH_3-N removal and the production of NO_3-N which signifies the occurrence of nitrification. NH_3-N removal >96% was maintained throughout days 4–27 in AGS-SBR-1. Sludge sample collected and analyzed on the 27th day resulted in concentrations of 1.16 g MLSS/L and 0.96 g MLVSS/L. Though NH_3-N removal was approx. 100%, a weekend shutdown was completed between days 34 to 36. With a reactor working volume of 1.2 L, half volume was re-seeded with RAS and the remainder was fed with undiluted centrate. On day 36, AGS-SBR-1 regained operation with a 6-h cycle time and a diluted centrate feed targeting 100 mg/L of NH_3-N. Between days 36–61, it can be seen in Figure 4a that NH_3-N removal was low on day 40, but it increased with continued operation. During this period, the production of NO_3-N was hampered and NO_2-N in the effluent was 109.8 mg/L. Secondly, on day 40 the pH reached 8.7 compared to a typical value within the range 7.3–7.9. The presence of AOBs in the mixed liquor allowed for efficient oxidation of NH_3-N to NO_2-N, resulting in partial nitrification. Further oxidation of NO_2-N to NO_3-N was impacted due to the absence of NOBs coupled with the increased pH resulting from the addition of $NaHCO_3$ to supplement the alkalinity in centrate. A high pH would have impacted the growth of NOBs, which are considered to be more sensitive to increasing pH [35]. By day 61, experimental data obtained from analyzed sample confirmed complete recovery, i.e., low NO_2-N and high NO_3-N concentration in the effluent.

Increasing the concentration of NH_3-Ncentrate feed to approx. 200 mg/L after day 61 did not impact performance. Between days 71 and 120, influent NH_3-N averaged 180 $\pm$ 14 mg/L, equating to an NH_3-N loading of 0.36 $\pm$ 0.03 g/L.d and the removal efficiency was 98 $\pm$ 3 %. A second increase in NH_3-N influent concentration was interrupted due to a malfunction with the ancillary equipment. Prior to operational disruption, NH_3-N removal was 88% on day 134, effluent NO_3-N and NO_2-N were 200 mg/L and 15 mg/L, respectively. The malfunctioning feed pump failure was discerned on day 137 and with unsuccessful attempts to guarantee adequate operation overnight, the remaining sludge in the reactor was augmented with 200 mL of RAS from ABTP on day 140. Reactor performance was severely impacted as shown by the low removal of NH_3-N on day 155 at 11% in Figure 4a. The concentration of NO_3-N slightly decreased to 198 mg/L, but NO_2-N increased to a value of 60 mg/L. In addition to the effluent NH_3-N of 232 mg/L, the nitrogen balance for day 155 was affected by the operational disruption–effluent concentrations of NH_3-N, NO_3-N and NO_2-N exceeded the influent NH_3-N concentration. The sludge concentration also decreased during this period, therefore, to prevent further sludge loss through resuspension while allowing for ammonia oxidation, SAV was lowered to 1 L/min after day 155. The impact of operating at a lower SAV was observed with samples collected between days 162 and 183. Excellent NH_3-N removal was achieved, averaging 97 $\pm$ 2% but partial nitrification was dominant with NO_2-N concentration remaining consistently high between 208–268 mg/L. Over the course of the entire research study (i.e., 183 days), average NH_3-N removal was determined to be 93 $\pm$ 19%; inclusive of the reactor disruption periods. In Figure 4a, the downward peaks are indicative of the reactor disruptions periods. Additional information on the reactor's performance can be found in the electronic Supplementary Material (Table S1).

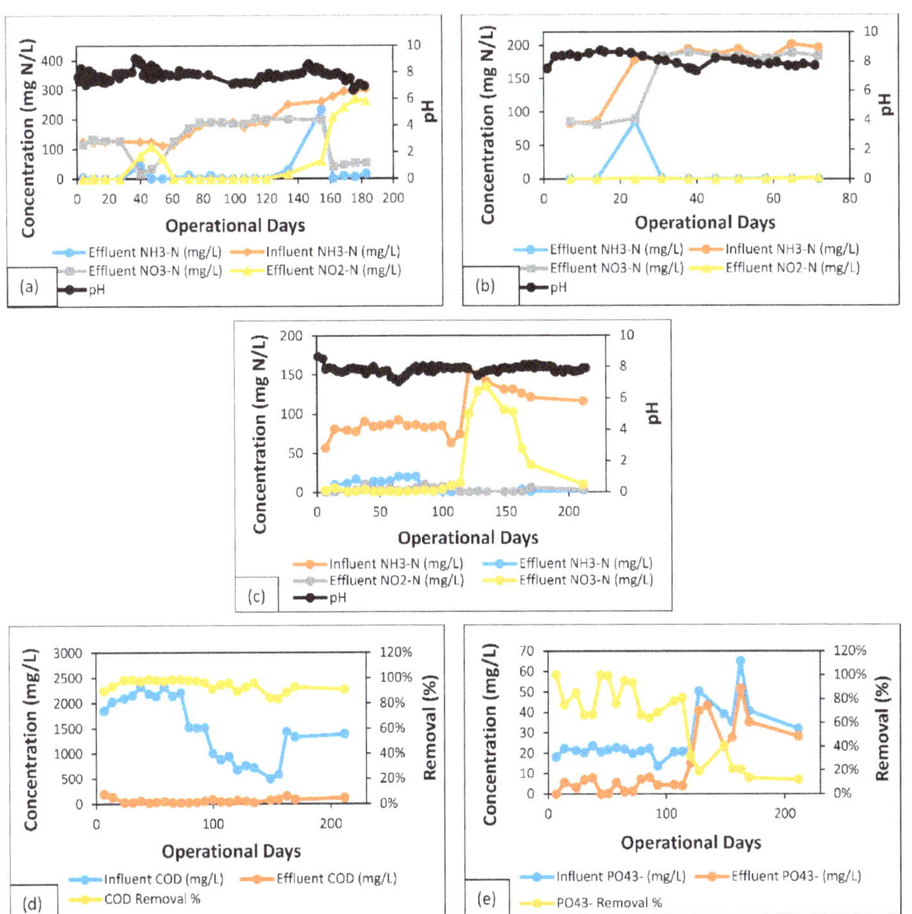

Figure 4. Nitrification profile for (**a**) AGS-SBR-1, (**b**) AGS-SBR-2, and (**c**) AGS-SBR-3; (**d**) COD profile for AGS-SBR-3, and (**e**) Phosphorus (as PO_4^{3-}) profile for AGS-SBR-3.

3.3.2. Synthetic Centrate Nitrifying SBR (AGS-SBR-2)

The occurrence of nitrification in AGS-SBR-2 was confirmed after 7 days of operation, inclusive of an acclimation phase. An average NH_3-N concentration of 84 ± 2 mg/L was fed to AGS-SBR-2 during acclimatization until day 14. Experimental analyses for days 7 and 14 confirmed 100% NH_3-N removal, conversion of NO_2-N during nitrification and an average effluent concentration of 83 ± 3 mg/L for NO_3-N. Increasing the concentration of influent NH_3-N after day 14 negatively impacted the performance of the system as shown in Figure 4b; NH_3-N removal decreased to 52% and NO_3-N concentration was 90 mg/L on day 24. On day 31, NH_3-N returned to approximately 100% removal and NO_3-N present in the effluent reflected the influent NH_3-N concentration. Between days 31 and 72, influent NH_3-N of 190 ± 8 mg/L was completely removed while average effluent concentration of NO_3-N was 185 ± 3 mg/L. NO_2-N effluent concentration was undetectable for majority of the period (i.e., <0.2 mg/L) until day 72 where a concentration of 2.19 mg/L was detected. Overall, NO_2-N was consistently converted to NO_3-N. The growth of distinct nitrifying granules was not detected in AGS-SBR-2, but the reactor's performance in oxidizing NH_3-N is an indication of the technology's capability. The difficulty in cultivating nitrifying granules is related to the slow growth rate of autotrophic bacteria, which are responsible for nitrification [36]. Previously, researchers have confirmed the presence of nitrifying granules [22,37,38], but the cultivation process is understood to be time-consuming. Recent research have shifted to shortening the cultivation time through inoculation with mature granules or supplementing the carbon source [36,39].

The relationship between influent NH_3-N and effluent NO_3-N in AGS-SBR-2 between days 31 and 72, proved that the use of synthetic wastewater positively impacted nitrogen balance in the system. Synthetic centrate prepared in the laboratory contained considerably less impurities and was void of polymer compounds in comparison to real centrate collected from ABTP. In comparison to AGS-SBR-1, NO_3-N in the effluent of AGS-SBR-2 was lower than the influent NH_3-N concentration expect for days 31 and 58. The difference in concentration for days 31 and 58 were 2% and 3%, respectively. Similarly, day 7 shows NO_3-N slightly higher than influent NH_3-N with the difference being 3%. The average pH in AGS-SBR-2 for the duration of the research study was 8 ± 0.5. Minor fluctuation in pH was observed as depicted in Figure 4b, but the system displayed adequate stability. Over the course of the entire research study (i.e., 72 days), average NH_3-N removal was determined to be 94 ± 16%. This takes into consideration a disruption in performance captured on day 24 in Figure 4b. Additional information on the reactor's performance can be found in the electronic Supplementary Material (Table S2).

3.3.3. Nitrifying-Denitrifying SBR (AGS-SBR-3)

In AGS-SBR-3, the marginal difference between influent and effluent NH_3-N represented an average removal of 92 ± 9% for the entire duration of the research study. Complete nitrogen removal was achieved in AGS-SBR-3 prior to day 100 with low NO_x-N concentration in the effluent. Therefore, it was evident that simultaneous nitrification and denitrification (SND) is possibly the main removal mechanism involved. Aerobic granules are capable of SND due to its layered structure which can provide anoxic/anaerobic conditions deep within the granule for conversion of NO_x-N to dinitrogen gas (N_2) [40]. An increase in NH_3-N concentration in the feed stock with reduced COD resulted in excess NO_3-N in the effluent as shown in Figure 4c. NO_3-N accumulation is shown to moderately increase between days 100 and 114 with an average COD/N ratio of 13 ± 1. A tremendous increase in effluent NO_3-N is seen between days 114 and 135; the latter representing the peak value of 134.8 mg/L. During this period, the average COD/N ratio was 5 ± 0.4. Effluent NO_3-N concentrations of 105.2 mg/L and 102.4 mg/L were recorded on days 145 and 156, respectively. NO_3-N concentration > 100 mg/L is considered high in comparison to recorded values prior to the first substantial spike on day 114. The COD/N ratio averaged 4 ± 0.5 for day 145 and 156. Overall, the average COD/N ratio for days 114 to 156 was 5 ± 0.5. NO_x-N accumulation at a low COD/N highlights the dependence on readily biodegradable in the feed for adequate SND. In a study by Ren et al. [31], NO_x-N accumulation and low SND efficiencies occurred in an aerobic granular sludge system when biodegradable COD in the feed decreased. NO_3-N continued to decrease after day 156 through to the 212th day with an average COD/N ratio of 12 ± 0.5 recorded during this period. The intermediate product, NO_2-N, generated during nitrification was continuously converted to NO_3-N. The maximum concentration of NO_2-N recorded in the effluent was 10.28 mg/L on day 38, while 42% of samples had concentration of <0.2 mg/L. SND efficiency for the period with low NO_x-N accumulation was 87 ± 11%. pH readings were recorded at various instances during a cycle and automatic adjustments was performed with the help of a pump used to dose HCl. An average pH of 7.86 ± 0.29 was tabulated for the duration of the experimental study. Additional information on the reactor's performance can be found in the electronic Supplementary Material (Table S3).

COD concentration between days 7 and 72 averaged 2143 ± 139 mg/L as shown in Figure 4d. After confirming the presence of granules in AGS-SBR-3, COD concentration in the feed was gradually lowered in stages until day 159. COD concentration in the feed was increased to 1445 mg/L on day 159 and an average of 1390 ± 54 was maintained for the remainder of the study as shown in Figure 4d. CH_3COONa was included in both the synthetic wastewater and diluted centrate feed solution as the main source of COD; with a switch in feed streams occurring after day 79. Supplemental acetate-COD played two important roles in AGS-SBR-3: (1) Nitrogen removal—Heterotrophic organisms utilize acetate-COD as electron donor for energy generation [30,41], and (2) Phosphorus removal—Acetate-COD is hydrolyzed into organic acids to aid in phosphorus release and uptake [42]. Excellent COD removal is supported by the fact that it is the limiting substrate in nitrogen and phosphorus removal [43].

Effluent monitoring of AGS-SBR-3 captured the removal of phosphorus as shown in Figure 4e, where an average removal of $81 \pm 14\%$ was achieved between days 1–114 of operation. During this period, COD in the feed was gradually decreased after day 72 and the feed stream was switched to centrate instead of synthetic wastewater after day 79. Synthetic wastewater consisted of K_2HPO_4, KH_2PO_4 as a phosphorus source. Phosphorus removal drastically decreased on day 121, at which point the COD was <800 mg/L and NH_3-N was increased to >120 mg/L in the feed. Average phosphorus removal determined for the remainder of the study was $23 \pm 10\%$. Sequential operation of AGS-SBR-3 did not change; therefore the presence of anoxic-aerobic conditions would have influenced the growth of phosphate-accumulating organisms (PAOs)—widely known their role in biological phosphorus removal [43]. A reduction in phosphorus removal on day 121 indicates that a low COD concentration negatively impacts the function of PAOs. According to Zeng et al. [43], COD is a limiting substrate in nitrogen and phosphorus removal processes. Furthermore, the connection between nitrogen and phosphorus removal (i.e., the denitrification step) led researchers to recognizing the function of denitrifying PAOs [44]. Pronk et al. [45] expressed that dPAOs can operate with double functionality in an AGS system: (1) an electron donor in the denitrification process, and (2) store phosphate as poly-P. In this study, NH_3-N was removed without NO_2-N and NO_3-N accumulation, indicating simultaneous nitrification, denitrification, and phosphorus removal. In a study by Figdore et al. [27], a cycle test using NDN-PAO granules resulted in concurrent phosphate uptake and NH_3-N removal without accumulation of NO_2-N and NO_3-N. Furthermore, it was concluded that NO_2-N and NO_3-N occurred during the aeration phase—thus confirming SND mechanism in a granular system [27].

3.4. Pollutants Degradation in SBR Cycle

3.4.1. Cycle Test: AGS-SBR-1 (Day 120)

The cycle test completed for AGS-SBR-1 captured the concentrations of NH_3-N, NO_3-N and NO_2-N during a 4-h cycle period in Figure 5a. The pH curve depicts the effect of nitrification reaction on alkalinity during the cycle. It is evident in Figure 5a that the pH decreases from 7.68 to 5.99 within the first hour of the cycle. $NaHCO_3$ was dosed to the reactor at two different points during the 4-h cycle; the 1-h and 1.5-h mark. Nitrification reduces alkalinity and subsequently the pH of the system through oxidation of NH_3-N [46]. The H^+ ions are freed from the nitrogen bond as NH_3-N is converted to NO_2-N and inevitably NO_3-N, thus creating acidic conditions and altering the pH buffer capability.

Ammonia removal in parallel with the production of NO_3-N via conversion of NO_2-N occurred with 1.5 h of the cycle as shown in Figure 5a. Concentrations of NH_3-N, NO_3-N and NO_2-N measured at time zero were 72.5 mg/L, 109.5 mg/L and 7.85 mg/L, respectively. At the 1.5-h mark, both NH_3-N and NO_2-N were depleted, and the concentration of NO_3-N was 197.0 mg/L, approximately 7 mg/L more than the sum of concentrations for all three nitrogenous compounds at time zero. While this confirmed ammonia oxidation and complete nitrification within 1.5 h, the additional production of NO_3-N is presumably linked to the presence of organic nitrogen in the centrate. Organic nitrogen can form NH_3-N and undergo nitrification. The rate of NH_3-N uptake was determined by performing linear regression on NH_3-N depletion over time. In Figure 5b, NH_3-N is shown to have been depleted within 1.5 h. Linear regression of the curve resulted in a nitrification rate (NR) of 45.3 mg NH_3-N/L·h. Furthermore, the average MLVSS concentration was 2.36 ± 0.02 g/L, resulting in a specific nitrification rate (SNR) of 19.19 mg NH_3-N/g VSS·h. Calculated values for NR and SNR using Equations (4)–(7) were 64.16 mg NH$_3$-N/L·h and 26.96 mg NH$_3$-N/g VSS·h, respectively.

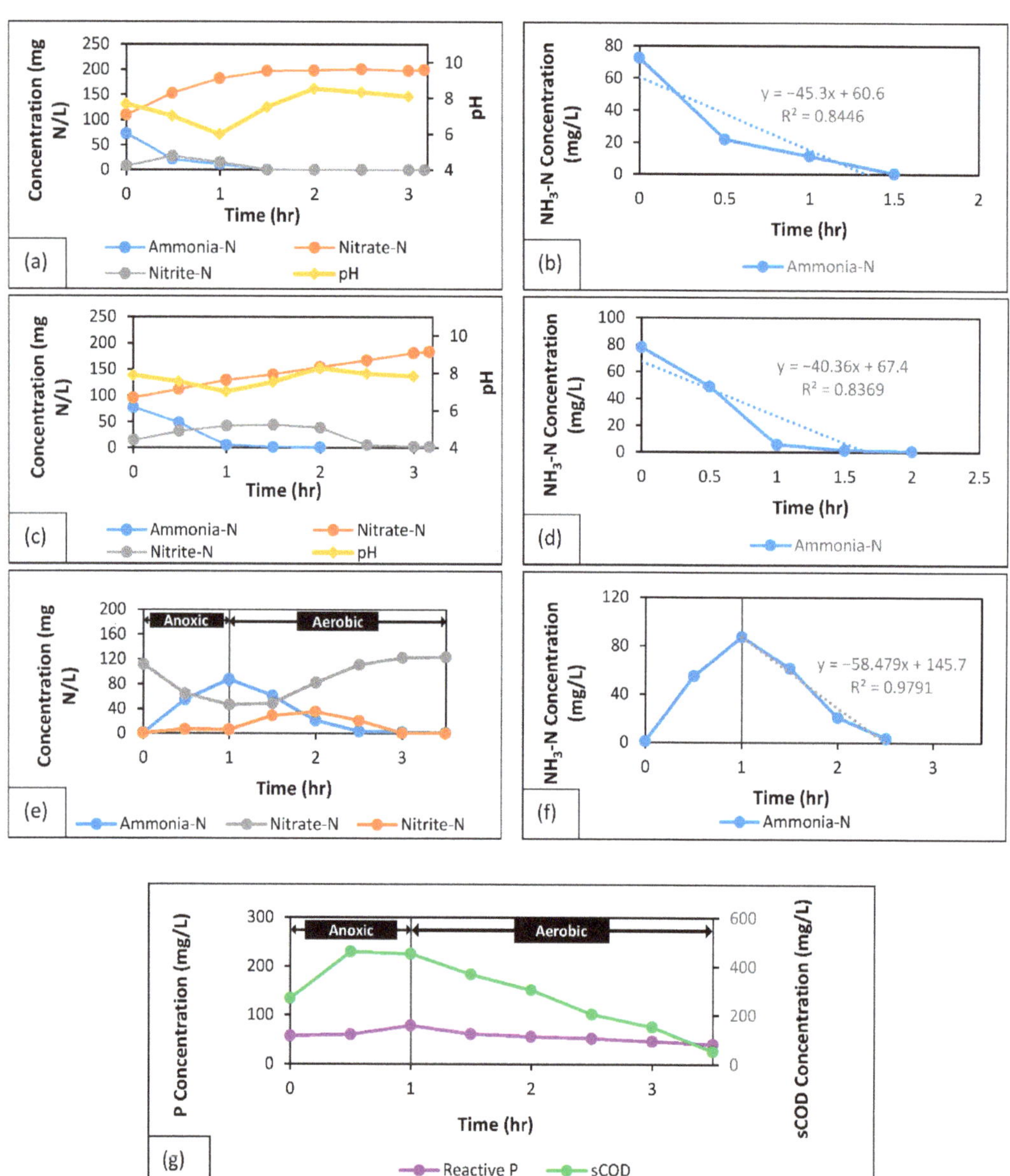

Figure 5. Nitrification profile and ammonia-nitrogen removal rate for (**a**,**b**) AGS-SBR-1, (**c**,**d**) AGS-SBR-2, and (**e**–**g**) AGS-SBR; (**f**) COD profile, and (**g**) Reactive Phosphorus profile for AGS-SBR-3.

3.4.2. Cycle Test: AGS-SBR-2 (Day 72)

In Figure 5c, the depletion of NH$_3$-N is monitored for one complete cycle. In addition to NH$_3$-N, Figure 5c shows how NO$_2$-N is produced and convert to NO$_3$-N. The concentration of NH$_3$-N after 1.5 h is 1.6 mg/L, followed by 0.8 mg/L at the 2-h mark. NH$_3$-N is completed depleted after the 2 h of operation. Interestingly, the concentration of NH_3-N at the 0.5-h mark was 49 mg/L, which is high compared to AGS-SBR-1 where a lower concentration of 22 mg/L was achieved within the same time span. Between 0.5 to 1 h, majority of the NH_3-N in AGS-SBR-1 was oxidized. The production and conversion of NO_2-N to NO_3-N occurred for the entire duration of the cycle, but majority of NO_2-N was converted after the 2.5 h. The concentration of NO_2-N at the 2.5-h mark was 5.37 mg/L. The slow-paced conversion of NO_2-N resulted in a gradual increase in NO_3-N. The NO_3-N curve in Figure 5c increased to a concentration of 184 mg/L. Reactor pH was dependent on the occurrence of nitrification and the addition of $NaHCO_3$ to supplement alkalinity. AGS-SBR-1 and AGS-SBR-2 were fed with $NaHCO_3$ from a common pump which operated on a timer. $NaHCO_3$ would be dosed at the 1-h and 1.5-h mark during a cycle. It can be seen in Figure 5c that pH increased after the alkalinity dosing points for AGS-SBR-2. The pH in AGS-SBR-2 decreased to 7.01 before the first dose of $NaHCO_3$. In AGS-SBR-1, Figure 5a shows that pH decreased to 5.99 before the first dose of $NaHCO_3$. It is evident through the consumption of alkalinity–ability of the system to buffer pH decreases—that AGS-SBR-2 initially experienced a slower nitrification rate compared to AGS-SBR-1, but the rate increased after the 0.5 and 1-h mark. Computational determination of NR and SNR were performed by analyzing the NH_3-N curve in Figure 5d. Irrespective if the slow start to NH_3-N removal in AGS-SBR-2, linear regression of the curve resulted in a NR of 40.36 mg NH_3-N/L·h. The average concentration of MLVSS between the start and end of the cycle was 1.475 $\pm$ 0.26 g/L, therefore SNR was 27.36 mg NH_3-N/g VSS·h. Calculated values for NR and SNR using Equations (4)–(7) were 63.31 mg NH_3-N/L·h and 38.14 mg NH_3-N/g VSS·h, respectively.

3.4.3. Cycle Test: AGS-SBR-3 (Day 128)

The cycle test completed on day 128 as shown in Figure 5e, accurately displays the impact of a low COD/N ratio on NH_3-N removal via the nitrification-denitrification reaction pathway. During the anoxic feeding period, NO_3-N is biologically used as an oxidizer to remove organics–representing denitrification. In Figure 5e, NO_3-N concentration is reduced by 58% during the anoxic period from 111.50 to 46.70 mg/L. NH_3-N concentration continues to increase in the system as the reactor is fed during the anoxic period. NO_2-N in the feed is typically <0.2 mg/L, yet it's generation during the anoxic period is linked to residual oxygen within the system. An extended aeration period from the previous cycle introduces a substantial amount of oxygen in the system that may promote minimal nitrification during the anoxic period. At the 1-h mark, air bubbles are introduced to develop an aerobic environment within the system. NO_2-N is produced and converted to NO_3-N—which continues to increase for the remainder of the cycle. NO_2-N concentration peaked at after 1 h while NH_3-N is decreased to a concentration of 3.59 mg/L within 1.5 h of aeration. The final concentration for NH_3-N recorded at the end of the cycle was 1.23 mg/L. Accumulation of NO_3-N discussed previously was evident in the cycle test conducted on day 128. The COD/N ratio was determined to be 4.77 on the 128th day of operation. Linear regression of the curve resulted in a NR of 58.48 mg NH_3-N/L·h as shown in Figure 5f. The average concentration of MLVSS between the start and end of the cycle was 8.52 $\pm$ 0.26 g/L, therefore SNR was 6.86 mg NH_3-N/g VSS·h. Calculated values for NR and SNR using Equations (4)–(7) were 79.86 mg NH_3-N/L·h and 9.18 mg NH_3-N/g VSS·h, respectively. Nitrification and/or specific nitrification rates for various studies are presented in Table 6.

The concentration profiles of COD and phosphorus for day 128 are shown in Figure 5g. COD increases during the anaerobic feed phase before it is assimilated under aerobic conditions. A fraction of the COD is assimilated under anaerobic conditions for denitrification and phosphorus release [47,48]. Low COD concentrations on day 128 resulted in subpar phosphorus removal as shown in Figure 5g.

Table 6. Summary of nitrification and specific nitrification rates reported in the literature.

Reactor Type	Parameter	References
Aerobic Granular Sludge Sequencing Batch Reactor (AGS-SBR)	Maximum specific ammonia oxidizing rate = 29.8 mg N/g VSS/h	[23]
Aerobic Granular Sludge Sequencing Batch Reactor (AGS-SBR)	Maximum specific ammonia oxidizing rate NIT granules = 31.25 mg N/g VSS/h NDN-PAO granules = 5.83 mg N/g VSS/h	[39]
Activated Sludge Sequencing batch reactor (AS-SBR)	Specific Nitrification rate = 30–45 g NH_4-N/kg MLVSS/h	[49,50]
CAS	Specific nitrification rates in the membrane and conventional mixed liquor: between 3 and 5 mg N/g VSS/h	[51]
CAS	Specific nitrification rates: between 4 and 7 mg N/g VSS/h were measured in both membrane and conventional systems.	[51,52]
Activated Sludge (AS)	Maximum Nitrification rate = 1.0–4.5 mg N/g MLVSS/h	[53]
SBR	Ammonia utilization rate (AUR) = 2.95 ± 0.26 mg $NH4^+$-N/g VSS/h Ammonia utilization rate (AUR) was faster = 6.16 ± 0.34 mg NH_4^+-N/g VSS/h	[54]
AGS-SBRs	Nitrifying AGS-SBR with diluted centrate: Nitrification rate = 45.30 mg N/L·h Specific nitrification rate = 19.19 mg N/g VSS·h Nitrifying AGS-SBR with synthetic centrate: Nitrification rate = 40.36 mg N/L·h Specific nitrification rate = 27.36 mg N/g VSS·h Nitrifying/Denitrifying AGS-SBR: Nitrification rate = 54.48 NH3-N/L·h Specific nitrification rate = 6.86 mg NH3-N/g VSS·h. SND efficiency during periods with negligible NO_x-N accumulation was 87 ± 11%.	(This Study)

4. Conclusions

This study shows that aerobic granular sludge systems are a viable treatment technology for liquid sidestreams generated in a WWTP. Cultivation of nitrifying-denitrifying granules <1000 μm (<1 mm) was observed after 32 days, whereas nitrifying sludge consisted of small aggregates with diameter >100 μm for majority of the study period. Good settleability of nitrifying sludge was observed with SVI_{30} values of 72 ± 12 mL/g (AGS-SBR with real centrate) and 59 ± 11 mL/g (AGS-SBR-2 with synthetic centrate). Nitrifying-denitrifying displayed an average SVI_{30} of 52 ± 23 mL/g for the entire duration of the study. Excellent NH_3-N removal, >90%, was achieved in the nitrifying AGS-SBRs. The nitrifying-denitrifying AGS-SBR displayed good COD (>90%) and P (>80%) removal without NO_x-N accumulation COD/N ratio >11 mg sCOD/mg NH_3-N. Simultaneous nitrification-denitrification (SND) and phosphorus removal was dominant at higher COD/N ratios. NO_x-N accumulation was observed, and P removal was affected when decreasing the COD/N (4–5 mg sCOD/mg NH_3-N). The nitrifying-denitrifying AGS-SBR displayed produced the highest NR 58.48 mg NH_3-N/L·h, while the overall SND during periods with negligible NO_x-N accumulation was 87 ± 11%. With the formation of small aggregates (i.e., microgranules) in the nitrifying systems fed with real and synthetic centrate, the NR was 45.3 mg NH_3-N/L·h and 40.36 mg NH_3-N/L·h, respectively. The AGS-SBRs in this study produced higher nitrification rates compared to that reported in the literature. Future research should focus on enhancing the mechanism of formation for nitrifying granules without an organic carbon source within a short period and implementing SND using nitrifying-denitrifying at pilot scale.

Supplementary Materials: The following supporting information can be downloaded at: https://www.mdpi.com/article/10.3390/pr10091687/s1, Table S1: Performance summary of AGS-SBR-1; Table S2: Performance summary of AGS-SBR-2; Table S3: Performance summary of AGS-SBR-3.

Author Contributions: Conceptualization, R.A.H.; methodology, R.A.H. and G.M.; validation, R.A.H.; formal analysis, R.A.H. and G.M.; investigation, R.A.H. and G.M.; resources, R.A.H.; data curation, G.M.; writing—original draft preparation, G.M.; writing—review and editing, R.A.H.; visualization, R.A.H.; supervision, R.A.H.; project administration, R.A.H.; funding acquisition, R.A.H. All authors have read and agreed to the published version of the manuscript.

Funding: This research was funded by Natural Sciences and Engineering Council of Canada (NSERC), Discovery Grants and Faculty of Engineering and Architectural Science (FEAS) at Ryerson University.

Institutional Review Board Statement: Not Applicable.

Informed Consent Statement: Not Applicable.

Data Availability Statement: Not Applicable.

Acknowledgments: The authors would like to acknowledge the funding support provided by the Natural Sciences and Engineering Council of Canada (NSERC) and the Faculty of Engineering and Architectural Science (FEAS) at Ryerson University (renaming to Toronto Metropolitan University in progress).

Conflicts of Interest: The authors declare no conflict of interest.

References

1. Stamatelatou, K. Novel biological processes for nutrient removal and energy recovery from wastewater. In *Wastewater and Biosolids Management*; IWA Publishing: London, UK, 2020; pp. 27–42.
2. Mihelcic, J.R.; Ren, Z.J.; Cornejo, P.K.; Fisher, A.; Simon, A.J.; Snyder, S.W.; Zhang, Q.; Rosso, D.; Huggins, T.M.; Cooper, W.; et al. Accelerating Innovation that Enhances Resource Recovery in the Wastewater Sector: Advancing a National Testbed Network. *Environ. Sci. Technol.* **2017**, *51*, 7749–7758. [CrossRef] [PubMed]
3. Dev, S.; Saha, S.; Kurade, M.B.; Salama, E.S.; El-Dalatony, M.M.; Ha, G.-S.; Chang, S.W.; Jeon, B.-H. Perspective on anaerobic digestion for biomethanation in cold environments. *Renew. Sustain. Energy Rev.* **2019**, *103*, 85–95. [CrossRef]
4. De Vrieze, J.; Smet, D.; Klok, J.; Colsen, J.; Angenent, L.T.; Vlaeminck, S.E. Thermophilic sludge digestion improves energy balance and nutrient recovery potential in full-scale municipal wastewater treatment plants. *Bioresour. Technol.* **2016**, *218*, 1237–1245. [CrossRef] [PubMed]
5. Ochs, P.; Martin, B.D.; Germain, E.; Stephenson, T.; van Loosdrecht, M.; Soares, A. Ammonia removal from thermal hydrolysis dewatering liquors via three different deammonification technologies. *Sci. Total Environ.* **2020**, *755*, 142684. [CrossRef]
6. Stenström, F.; La Cour Jansen, J. Promotion of nitrifiers through side-stream bioaugmentation: A full-scale study. *Water Sci. Technol.* **2016**, *74*, 1736–1743. [CrossRef]
7. Eskicioglu, C.; Galvagno, G.; Cimon, C. Approaches and processes for ammonia removal from side-streams of municipal effluent treatment plants. *Bioresour. Technol.* **2018**, *268*, 797–810. [CrossRef] [PubMed]
8. Lacroix, A.; Mentzer, C.; Pagilla, K.R. Full-scale N removal from centrate using a sidestream process with a mainstream carbon source. *Water Environ. Res.* **2020**, *92*, 1922–1934. [CrossRef] [PubMed]
9. Bowden, G.; Tsuchihashi, R.; Stensel, H.D. *Technologies for Sidestream Nitrogen Removal*; IWA Publishing: London, UK, 2016; Volume 15.
10. Menkveld, H.W.H.; Broeders, E. Recovery of ammonia from digestate as fertilizer. *Water Pract. Technol.* **2018**, *13*, 382–387. [CrossRef]
11. Lei, C.S.; Ma, J.F.; Li, D.L.; Zhang, W.Y.; Guang, X.H.; Ma, J. Composite denitrification reagent for high concentration ammonia removal by air stripping. *Chin. Sci. Bull.* **2010**, *55*, 2657–2661. [CrossRef]
12. Aljerf, L. Advanced highly polluted rainwater treatment process. *J. Urban Environ. Eng.* **2018**, *12*, 50–58. [CrossRef]
13. Thornton, A.; Pearce, P.; Parsons, S.A. Ammonium removal from digested sludge liquors using ion exchange. *Water Res.* **2007**, *41*, 433–439. [CrossRef] [PubMed]
14. Zeng, Y.; Sun, Z.Y.; Chen, Y.T.; Xia, Z.Y.; Gou, M.; Tang, Y.Q. Performance and bacterial communities in conventional activated sludge and membrane bioreactor systems with Low C/N ratio wastewater for nitrogen removal. *Environ. Eng. Sci.* **2019**, *36*, 1112–1126. [CrossRef]
15. Dold, P.; Du, W.; Burger, G.; Jimenez, J. Is Nitrite-Shunt Happening in the System? Are Nob Repressed? *Proc. Water Environ. Fed.* **2015**, *2015*, 1360–1374. [CrossRef]
16. Kartal, B.; Kuenen, J.G.; Van Loodsdrecht, M.C.M. Sewage Treatment with Anammox. *Am. Assoc. Adv. Sci.* **2010**, *328*, 702–703. Available online: https://www.jstor.org/stable/40655872 (accessed on 14 March 2022). [CrossRef]
17. Pedros, P.B.; Onnis-Hayden, A.; Tyler, C. Investigation of Nitrification and Nitrogen Removal from Centrate in a Submerged Attached-Growth Bioreactor. *Water Environ. Res.* **2008**, *80*, 222–228. [CrossRef] [PubMed]
18. Scaglione, D.; Ficara, E.; Corbellini, V.; Tornotti, G.; Teli, A.; Canziani, R.; Malpei, F. Autotrophic nitrogen removal by a two-step SBR process applied to mixed agro-digestate. *Bioresour. Technol.* **2015**, *176*, 98–105. [CrossRef]
19. López-Palau, S.; Dosta, J.; Pericas, A.; Mata-Álvarez, J. Partial nitrification of sludge reject water using suspended and granular biomass. *J. Chem. Technol. Biotechnol.* **2011**, *86*, 1480–1487. [CrossRef]
20. Zhang, S.; Peng, Y.; Wang, S.; Zheng, S.; Guo, J. Organic matter and concentrated nitrogen removal by shortcut nitrification and denitrification from mature municipal landfill leachate. *J. Environ. Sci.* **2007**, *19*, 647–651. [CrossRef]
21. Świątczak, P.; Cydzik-Kwiatkowska, A. Treatment of Ammonium-Rich Digestate from Methane Fermentation Using Aerobic Granular Sludge. *Water Air. Soil Pollut.* **2018**, *229*, 1–12. [CrossRef]

22. Liu, Y.Q.; Wu, W.W.; Tay, J.H.; Wang, J.L. Formation and long-term stability of nitrifying granules in a sequencing batch reactor. *Bioresour. Technol.* **2008**, *99*, 3919–3922. [CrossRef]
23. Figdore, B.A.; Winkler, M.-K.H.; Stensel, H.D. Bioaugmentation with Nitrifying Granules in Low-SRT Flocculent Activated Sludge at Low Temperature. *Water Environ. Res.* **2017**, *90*, 343–354. [CrossRef] [PubMed]
24. Nancharaiah, Y.V.; Sarvajith, M. Aerobic granular sludge process: A fast growing biological treatment for sustainable wastewater treatment. *Curr. Opin. Environ. Sci. Health* **2019**, *12*, 57–65. [CrossRef]
25. Wei, D.; Shi, L.; Yan, T.; Zhang, G.; Wang, Y.; Du, B. Aerobic granules formation and simultaneous nitrogen and phosphorus removal treating high strength ammonia wastewater in sequencing batch reactor. *Bioresour. Technol.* **2014**, *171*, 211–216. [CrossRef] [PubMed]
26. Kocaturk, I.; Erguder, T.H. Influent COD/TAN ratio affects the carbon and nitrogen removal efficiency and stability of aerobic granules. *Ecol. Eng.* **2016**, *90*, 12–24. [CrossRef]
27. Figdore, B.A.; David Stensel, H.; Winkler, M.K.H. Bioaugmentation of sidestream nitrifying-denitrifying phosphorus-accumulating granules in a low-SRT activated sludge system at low temperature. *Water Res.* **2018**, *135*, 241–250. [CrossRef]
28. Wilén, B.M.; Liébana, R.; Persson, F.; Modin, O.; Hermansson, M. The mechanisms of granulation of activated sludge in wastewater treatment, its optimization, and impact on effluent quality. *Appl. Microbiol. Biotechnol.* **2018**, *102*, 5005–5020. [CrossRef]
29. APHA. *Standard Methods for Examination of Water and Wastewater*, 22nd ed.; American Public Health Association: Washington, DC, USA, 2012.
30. Inc. Metcalf & Eddy; Tchobanoglous, G.; Stensel, H.D.; Tsuchihashi, R.; Burton, F.; Abu-Orf, M.; Bowden, G.; Pfrang, W. *Wastewater Engineering: Treatment and Resource Recovery*; McGraw-Hill Education: New York, NY, USA, 2014.
31. Ren, Y.; Ferraz, F.M.; Yuan, Q. Landfill Leachate Treatment Using Aerobic Granular Sludge. *J. Environ. Eng.* **2017**, *143*, 04017060. [CrossRef]
32. Czarnota, J.; Tomaszek, J.A.; Masłoń, A.; Piech, A.; Łagód, G. Powdered Ceramsite and Powdered Limestone Use in Aerobic Granular Sludge Technology. *Materials* **2020**, *13*, 3894. [CrossRef]
33. Fan, J.; Ji, F.; Xu, X.; Wang, Y.; Yan, D.; Xu, X.; Chen, Q.; Xiong, J.; He, Q. Prediction of the effect of fine grit on the MLVSS/MLSS ratio of activated sludge. *Bioresour. Technol.* **2015**, *190*, 51–56. [CrossRef]
34. Hamza, R.; Rabii, A.; zahra Ezzahraoui, F.; Morgan, G.; Iorhemen, O.T. A review of the state of development of aerobic granular sludge technology over the last 20 years: Full-scale applications and resource recovery. *Case Stud. Chem. Environ. Eng.* **2021**, *5*, 100173. [CrossRef]
35. Law, Y.; Lant, P.; Yuan, Z. The effect of pH on N2O production under aerobic conditions in a partial nitritation system. *Water Res.* **2011**, *45*, 5934–5944. [CrossRef] [PubMed]
36. Zhang, B.; Long, B.; Cheng, Y.; Wu, J.; Zhang, L.; Zeng, Y.; Zeng, M.; Huang, S. Rapid domestication of autotrophic nitrifying granular sludge and its stability during long-term operation. *Environ. Technol.* **2021**, *42*, 2587–2598. [CrossRef]
37. Shi, X.Y.; Sheng, G.P.; Li, X.Y.; Yu, H.Q. Operation of a sequencing batch reactor for cultivating autotrophic nitrifying granules. *Bioresour. Technol.* **2010**, *101*, 2960–2964. [CrossRef]
38. Tsuneda, S.; Nagano, T.; Hoshino, T.; Ejiri, Y.; Noda, N.; Hirata, A. Characterization of nitrifying granules produced in an aerobic upflow fluidized bed reactor. *Water Res.* **2003**, *37*, 4965–4973. [CrossRef] [PubMed]
39. Figdore, B.A.; Stensel, H.D.; Winkler, M.K.H. Comparison of different aerobic granular sludge types for activated sludge nitrification bioaugmentation potential. *Bioresour. Technol.* **2018**, *251*, 189–196. [CrossRef] [PubMed]
40. He, Q.; Chen, L.; Zhang, S.; Wang, L.; Liang, J.; Xia, W.; Wang, H.; Zhou, J. Simultaneous nitrification, denitrification and phosphorus removal in aerobic granular sequencing batch reactors with high aeration intensity: Impact of aeration time. *Bioresour. Technol.* **2018**, *263*, 214–222. [CrossRef]
41. Klein, K.; Tenno, T. Estimating the impact of inhibitory substances on activated sludge denitrification process. *Water Pract. Technol.* **2019**, *14*, 863–871. [CrossRef]
42. Luiz de Sousa Rollemberg, S.; Queiroz de Oliveira, L.; Nascimento de Barros, A.; Igor Milen Firmino, P.; Bezerra dos Santos, A. Pilot-scale aerobic granular sludge in the treatment of municipal wastewater: Optimizations in the start-up, methodology of sludge discharge, and evaluation of resource recovery. *Bioresour. Technol.* **2020**, *311*, 123467. [CrossRef] [PubMed]
43. Zeng, R.J.; Lemaire, R.; Yuan, Z.; Keller, J. Simultaneous nitrification, denitrification, and phosphorus removal in a lab-scale sequencing batch reactor. *Biotechnol. Bioeng.* **2003**, *84*, 170–178. [CrossRef]
44. Kuba, T.; Smolders, G.; van Loosdrecht, M.C.M.; Heijnen, J.J. Biological phosphorus removal from wastewater by anaerobic-anoxic sequencing batch reactor. *Water Sci. Technol.* **1993**, *27*, 241–252. [CrossRef]
45. Pronk, M.; de Kreuk, M.K.; de Bruin, B.; Kamminga, P.; Kleerebezem, R.; van Loosdrecht, M.C.M. Full scale performance of the aerobic granular sludge process for sewage treatment. *Water Res.* **2015**, *84*, 207–217. [CrossRef] [PubMed]
46. Gerardi, M.H. Alkalinity and pH. In *The Microbiology of Anaerobic Digesters*; John Wiley and Sons, Inc.: Hoboken, NJ, USA, 2003; pp. 99–103. [CrossRef]
47. Sarvajith, M.; Nancharaiah, Y.V. Enhancing biological nitrogen and phosphorus removal performance in aerobic granular sludge sequencing batch reactors by activated carbon particles. *J. Environ. Manag.* **2022**, *303*, 114134. [CrossRef] [PubMed]
48. Callado, N.H.; Foresti, E. Removal of organic carbon, nitrogen and phosphorus in sequential batch reactors integrating the aerobic/anaerobic processes. *Water Sci. Technol.* **2001**, *44*, 263–270. [CrossRef] [PubMed]

49. Leu, S.-Y.; Stenstrom, M.K. Bioaugmentation to Improve Nitrification in Activated Sludge Treatment. *Water Environ. Res.* **2010**, *82*, 524–535. [CrossRef] [PubMed]
50. Mossakowska, A.; Reinius, L.-G.; Hultman, B. Nitrification reactions in treatment of supernatant from dewatering of digested sludge. *Water Environ. Res.* **1997**, *69*, 1128–1133. [CrossRef]
51. Monti, A.; Hall, E.R. Comparison of Nitrification Rates in Conventional and Membrane-Assisted Biological Nutrient Removal Processes. *Water Environ. Res.* **2008**, *80*, 497–506. [CrossRef] [PubMed]
52. Soriano, G.A.; Erb, M.; Garel, C.; Audic, J.M. A Comparative Pilot-Scale Study of the Performance of Conventional Activated Sludge and Membrane Bioreactors under Limiting Operating Conditions. *Water Environ. Res.* **2003**, *75*, 225–231. [CrossRef]
53. Choubert, J.M.; Racault, Y.; Grasmick, A.; Beck, C.; Heduit, A. Maximum nitrification rate in activated sludge processes at low temperature: Key parameters, optimal value. *Eur. Water Manag. Online* **2005**, 1–13.
54. Dytczak, M.A.; Londry, K.L.; Oleszkiewicz, J.A. Nitrifying Genera in Activated Sludge May Influence Nitrification Rates. *Water Environ. Res.* **2008**, *80*, 388–396. [CrossRef]

Article

Phosphorus Removal from Aerobic Granular Sludge: Proliferation of Polyphosphate-Accumulating Organisms (PAOs) under Different Feeding Strategies

Oliver Terna Iorhemen [1,2], Sandra Ukaigwe [2], Hongyu Dang [2] and Yang Liu [2,*]

[1] School of Engineering, University of Northern British Columbia, Prince George, BC V2N 4Z9, Canada; oliver.iorhemen@unbc.ca

[2] Department of Civil and Environmental Engineering, University of Alberta, Edmonton, AB T6G 2R3, Canada; ukaigwe@ualberta.ca (S.U.); hdang2@ualberta.ca (H.D.)

* Correspondence: yang.liu@ualberta.ca; Tel.: +1-780-492-5115

Citation: Iorhemen, O.T.; Ukaigwe, S.; Dang, H.; Liu, Y. Phosphorus Removal from Aerobic Granular Sludge: Proliferation of Polyphosphate-Accumulating Organisms (PAOs) under Different Feeding Strategies. *Processes* **2022**, *10*, 1399. https://doi.org/10.3390/pr10071399

Academic Editors: Joo Hwa Tay and Yongqiang Liu

Received: 14 June 2022
Accepted: 14 July 2022
Published: 18 July 2022

Publisher's Note: MDPI stays neutral with regard to jurisdictional claims in published maps and institutional affiliations.

Abstract: Aerobic granular sludge (AGS) is known for high phosphorus removal from wastewaters, and phosphorus can be recovered from high phosphorus-containing waste sludge granules. This study aimed at determining the feeding strategy that provides the best performance in terms of the proliferation of polyphosphate-accumulating organisms (PAOs) and phosphorus removal. Using three AGS bioreactors, this study compared phosphorus removal and the proliferation dynamics of PAOs under three different feeding strategies: anaerobic slow feeding (R1), pulse feeding + anaerobic mixing (R2), and pulse feeding (R3). Results indicate that R1 and R2 achieved significantly higher phosphorus removal (97.6 ± 3% for R1 and 98.3 ± 1% for R2) than R3 (55 ± 11%). The anaerobic slow feeding procedure (R1) achieved the highest specific phosphorus release rate (SPRR) and specific phosphorus uptake rate (SPUR) as compared to the other two feeding conditions. 16S ribosomal ribonucleic acid (rRNA) gene sequencing assay of the microbial community for the three feeding strategies indicated that although the feeding strategy impacted reactor performance, it did not significantly alter the microbial community. The bacteria community composition maintained a similar degree of diversity. *Proteobacteria*, *Bacteroidetes*, and *Verrucomicrobia* were the dominant bacterial phyla in the system. Dominant PAOs were from the class *Betaproteobacteria* and the genera *Paracoccus* and *Thauera*. Glycogen-accumulating organisms were significantly inhibited while other less-known bacteria such as *Wandonia* and *Hyphomonas* were observed in all three reactors.

Keywords: aerobic granular sludge; biological wastewater treatment; phosphorus removal

1. Introduction

The growing demand for phosphorus in the fertilizer industry, as well as the textile, rubber, leather, ceramics, and detergent industries [1,2], makes the recovery of phosphorus profitable. The primary source of phosphorus, phosphate rock, is under threat of depletion due to the world's increasing population [3,4]. Phosphorus recovery from wastewater treatment facilities can reduce our reliance on phosphate rock to meet phosphorus demands.

Aerobic granular sludge (AGS) has been demonstrated to achieve outstanding phosphorus removal from wastewaters [5–7]. Each sludge granule presents a layered structure with an aerobic outer layer, a transient anoxic zone, and an anaerobic inner core [6], which is made possible by oxygen diffusion limitation of large sized granules. The aerobic and anaerobic conditions provided by the granule structure alongside the sequencing batch operation mode allow for high phosphorus removal. The mechanisms for phosphorus removal in AGS-based treatment systems include (i) enhanced biological phosphorus removal (EBPR) and (ii) phosphorus precipitation inside the granular sludge matrix [8–14]. It has been shown that of the total phosphorus removed in AGS reactors, 55–73.7% is through EBPR [10,15].

Enhanced biological phosphorus removal (EBPR) is widely regarded as the most economical and sustainable method for the removal of phosphorus from wastewater. In EBPR, polyphosphate accumulating organisms (PAOs) within the sludge release phosphorus to the bulk liquid during anaerobic conditions, subsequently uptake phosphorus from the bulk liquid during the aerobic phase, and store phosphorus as intracellular polyphosphates inside their cells and use for energy generation [14]. This implies the success of EBPR depends on the enrichment of PAOs in the AGS bioreactor. Numerous factors affect the EBPR performance of AGS systems, including carbon source, pH, carbon-to-phosphorus ratio, feeding strategy, temperature, and free ammonia concentration.

The feeding strategy is an operational parameter that impacts both aerobic granule formation and stability. Five different feeding strategies are used in AGS bioreactors: aerated feeding, pulse (fast static) feeding, anaerobic slow (slow static) feeding, pulse feeding followed by anaerobic mixing, and split anaerobic–aerobic feeding [16]. While the aerated feeding strategy (feeding and aeration/mixing taking place simultaneously) leads to fast granulation, the resulting granules are not capable of removing nutrients since the dominant microorganisms (COD degraders) outcompete slow-growing nitrifiers [17]. Granules formed through the aerated feeding strategy are also not stable due to the weak anaerobic core [18]. The pulse feeding strategy poses no problem with granule formation, but the granules formed are not stable [19]. The anaerobic slow feeding strategy results in compact and stable granules that are capable of removing both organic matter and nutrients [20,21]. Pulse feeding followed by anaerobic mixing allows for the formation of stable granules due to the stable growth rate, which allows for a suitable balance between the growth of heterotrophic and autotrophic microorganisms [22]. Nevertheless, while one study found the split anaerobic–aerobic feeding strategy suitable for aerobic granule formation and stable granules at low organic loading rate [23], another study reported the contrary [24].

Currently, there is no information in the scientific literature on the comparative study of different feeding strategies in terms of the proliferation of PAOs and phosphorus removal in AGS systems. Thus, the aim of this study was to fill this important research gap. Using synthetic wastewater, the present study used three different feeding strategies to compare phosphorus removal and the proliferation dynamics of PAOs in AGS bioreactors: anaerobic slow feeding, pulse feeding + anaerobic mixing, and pulse feeding. The specific phosphorus uptake rate (SPUR), the specific phosphorus release rate (SPRR), and phosphorus distribution were determined for the seed sludge and for the three reactors at steady state conditions.

2. Materials and Methods

2.1. Experimental Set-Up

Three reactors (R1, R2, R3), fabricated using acrylic (Plexiglas), were used for the experiment. Each reactor had a working volume of 4.5 L. The reactors were of diameter 9 cm for R1 and R2 and 10 cm for R3, and effective height-to-diameter (H/D) ratios of 7.56 for R1 and R2 and 5.75 for R3. Figure 1 shows the three reactors in operation.

A sequencing batch reactor (SBR) mode of operation was adopted as it is common with AGS systems. The details of the SBR operation are shown in Table 1. Influent chemical oxygen demand (COD) was 2092 ± 198 mg/L (day 1–27) and 971 ± 114 mg/L (day 28–84). The reactors were operated at a COD:N:P ratio of 100:5:1. Hence, the influent ammonia–nitrogen concentrations were 97 ± 7 mg/L (day 1–27) and 45 ± 3 mg/L (day 28–84) and the influent phosphorus concentrations were 20 ± 4 mg/L (day 1–27) and 11 ± 2 mg/L (day 28–84). Pentair ceramic diffusers (AS4) were used to provide fine air bubbles at the bottom of the reactors at superficial upflow velocities of 1.48 cm/s in R1 and R2, and at 1.2 cm/s in R3. Effluent was withdrawn via a port located around the middle of the reactor height, resulting in the volumetric exchange ratio (VER) shown in Table 1. In R2, anaerobic mixing was done by recirculating the wastewater in the bioreactor from the top to the bottom of the bioreactor using a peristaltic pump.

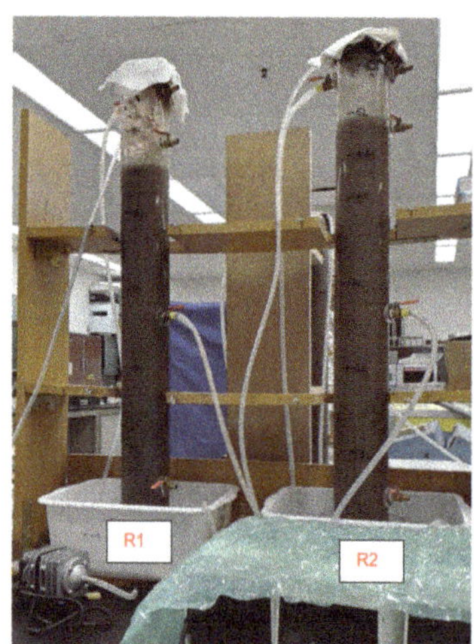

Figure 1. AGS bioreactors (R1, R2, R3) in operation.

Table 1. SBR operational conditions.

Reactors	Feeding Time (min)	Anaerobic Mixing Time (min)	Aeration Time (min)	Settling Time (min)	Decanting Time (min)	Idle Time (min)	VER (%)	Cycle Time (h)
R 1	60	-	143–163	10–30	6	1	40	4
R 2	6	54	143–163	10–30	6	1	40	4
R 3	6	-	197–217	10–30	6	1	30	4

2.2. Seed Sludge and Media

Return activated sludge was obtained from a full-scale biological nutrient removal wastewater treatment plant in Alberta. The activated sludge was acclimated to the substrate for 3 d in a batch mode before being used to start the bioreactors. Experiments were conducted at room temperature (20 ± 2 °C).

Synthetic wastewater was prepared using the following compounds: sodium acetate, sodium propionate, NH_4Cl, KH_2PO_4, K_2HPO_4, $CaCl_2 \cdot 2H_2O$, $MgSO_4 \cdot 7H_2O$, and $FeSO_4 \cdot 7H_2O$. Micronutrients were prepared from: H_3BO_3 0.05 g/L, $ZnCl_2$ 0.05 g/L, $CuCl_2$ 0.03 g/L, $MnSO_4 \cdot H_2O$ 0.05 g/L, $(NH_4)_6 \cdot Mo_7O_{24} \cdot 4H_2O$ 0.05 g/L, $AlCl_3$ 0.05 g/L, $CoCl_2 \cdot 6H_2O$ 0.05 g/L, and $NiCl_2$ 0.05 g/L as detailed elsewhere (Tay et al., 2002).

2.3. Analytical Methods

Biomass characteristics in reactors 1–3—mixed liquor suspended solids (MLSS), mixed liquor volatile suspended solids (MLVSS), COD, orthophosphate, SVI_{30}, and SVI_5—were determined according to standard methods [25]. SVI refers to the amount of sludge (milliliters) that has settled, and the SVI subscript refers to the settling time (minutes). Ammonia nitrogen (NH_3 N) was analysed using the Hach Ammonia Nessler Method 8038. Nitrite-nitrogen was determined using Hach NitriVer®3 Nitrite Reagent. Nitrate-nitrogen was analysed using Hach TNT kits (TNT835). Total phosphorus was determined using Hach kits (TNT844).

2.4. Specific Phosphorus Uptake and Release Rates

The specific phosphorus release rate (SPRR) (phosphorus release solely due to PAOs activities) in sludge samples from each reactor was measured near the end of the settling phase when the reactors attained steady state (day 67 of the experiment) when the influent COD concentration was 971 ± 114 mg/L. The SPRR was determined as follows: 100 mL of the sludge was transferred into a 160 mL test bottle and the bottle was sparged with nitrogen gas for 15 min to create anaerobic conditions. Sodium acetate stock solution (0.625 mL) (COD = 10 g/L) was added to the test bottle, and the bottle was inverted three times and quickly sealed. Samples were then taken from the test bottle at 5, 10, 15, 20, 30, 45, and 60 min and filtered at 0.45 μm. The orthophosphate concentration in solution (as PO_4-P) was measured in each sample, and the dissolved phosphorus (orthophosphate) concentration versus time was plotted. The MLVSS concentration of the withdrawn sample was also determined. The dissolved phosphorus concentration at 60 min was divided by the MLVSS (in g/L) to obtain the SPRR in mg P/g VSS.h.

To measure the specific phosphorus uptake rate (SPUR) of PAOs from wastewater, a sample from each SPRR test bottle at 60min was transferred to an open beaker and stone diffusers were employed to aerate the mixed liquor. Samples were taken at 0, 5, 10, 15, 20, 30, 45, and 60 min, filtered at 0.45 μm, and the dissolved phosphorus concentration vs. time was plotted for each sample. The dissolved phosphorus concentration at 60 min was divided by the mixed liquor volatile suspended solids MLVSS (in g/L) to obtain the SPUR in mg P/g VSS.h.

2.5. Phosphorus Distribution within the Sludge

Sludge samples were withdrawn from each reactor toward the end of the settling phase and their MLSS and MLVSS concentrations were determined. The total phosphorus in the sludge samples was measured using the digestion method (Hach TNT844): 40 mL of sludge was transferred to a 50 mL test bottle; the bottle was sparged with nitrogen gas for 15 min, then a sample (S_0) was taken. Sodium acetate stock solution (COD 10 g/L) (25 mL) was added to the test bottle, and the bottle was inverted three times for thorough mixing and quickly sealed. Samples from the bottle were then taken at 5, 10, 15, 20, 30, 45, and 60 min and filtered at 0.45 μm. The dissolved phosphorus concentration in the 60 min filtrate was divided by the MLVSS (in g/L) to obtain the phosphorus contribution from PAOs in mg P/g VSS.

To measure precipitated phosphorus, the test bottle from each reactor was opened and the pH was adjusted to 6.0–6.5 with hydrochloric acid (HCl). Liquid samples were withdrawn from the bottles after 10 min to measure the dissolved phosphorus concentration. The test bottles were then sonicated for 20 min at 100 W and 42 kHz. Liquid samples were withdrawn to measure dissolved phosphorus from the microbial cell surfaces. To measure phosphorus inside microbial cells, 1 mL of the sludge sample was transferred to a 2 mL microcentrifuge tube and treated with a cell disruptor (Branson Ultrasonics™ Sonifier™ SFX150 Cell Disruptor, Brookfield, WI, USA) for 10 min. Liquid samples were withdrawn from the bottles, filtered at 0.45 μm, and the dissolved phosphorus was estimated using the formula for the composition of a bacteria cell ($C_{60}H_{87}O_{23}N_{12}P$)-22.53 mg/g VSS.

2.6. Deoxyribonucleic Acid (DNA) Extraction

Deoxyribonucleic acid (DNA) was extracted from the sludge during granule formation and maturation using a DNeasy PowerSoil Kit following the manufacturer's protocol. Extraction was performed on the seed sludge, and in all reactors on days 11, 27, 39, 56, and 58. The purity and concentration of the DNA extracted from each sample were measured with NanoDrop One (ThermoFisher, Waltham, MA). Microbial communities in the samples were analyzed for the 16S ribosomal ribonucleic acid (rRNA) gene sequence. The sequence was amplified by the polymerase chain reaction (PCR) using primer sets with the sequencing adaptors 515F (GTGCCAGCMGCCGCGG) and 806R (GGACTACHVGGGTWTCTAAT) (Apprill et al., 2015, Parada et al., 2016). DNA amplicon samples were stored at −20 °C and sent to the Génome Québec Innovation Centre (Montréal, QC, Canada) for barcoding and sequencing on the Illumina Miseq PE250 platform using primer pair 515F/806R. Forward and reverse reads of the raw sequence were paired and screened, and chimera were removed with the "DADA2" algorithm in the QIIME2 pipeline (Callahan et al., 2016).

Taxonomy was determined with 99% similarity in the GreenGenes database, version 13_8 (McDonald et al., 2012, Werner et al., 2012).

3. Results and Discussion

3.1. Granule Development and Biomass Characteristics

Biomass characteristics for R1, R2, and R3 are presented in Figure 2a–c. It took about 40 d for flocculent sludge to mature to dense granular sludge in all three reactors, as can be seen by the SVI_{30} values in Figure 2a–c. From this point, the SVI_{30} values were about 80 mL/g or lower, meeting the requirement for granular sludge. During the AGS maturation phase (day 41–84), the SVI_{30} was 69 ± 14 mL/g for R1, 38 ± 8 mL/g for R2, and 41 ± 6 mL/g for R3. The SVI_5 during this period was 113 ± 18 mL/g for R1, 44 ± 10 mL/g for R2, and 47 ± 8 mL/g for R3. The SVI_5 is a great indicator of the quality of granulation—a full granular sludge system is attained when SVI_5 is comparable to SVI_{30} or when SVI_{30}/SVI_5 ratio approaches 0.9 (Liu et al., 2010, Nancharaiah and Reddy, 2018). The SVI_{30}/SVI_5 ratio ranged from 0.5 to 0.8 for R1, 0.8 to 0.9 for R2, and 0.8 to 1.0 for R3 during the AGS maturation phase. While the SVI_{30}/SVI_5 ratio is relatively smaller in R1 compared to R2 and R3, the SVI_{30} in R1 is 69 ± 14 mL/g is still below the 80 mL/g threshold.

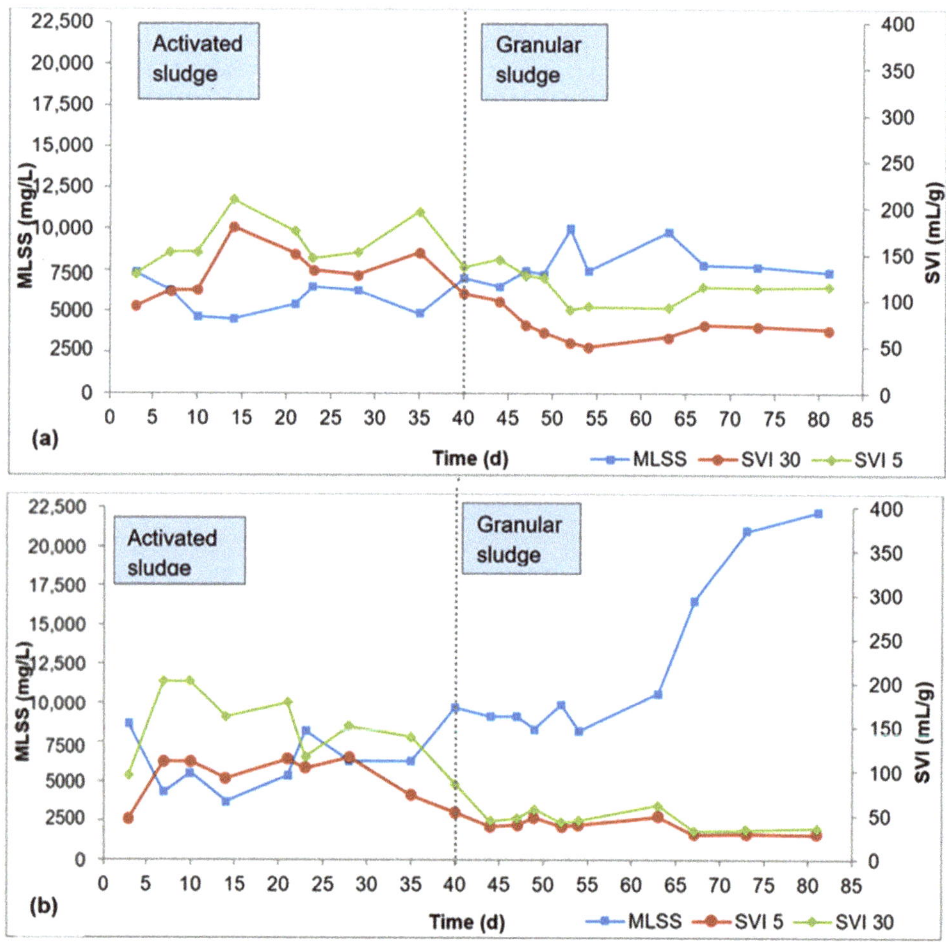

Figure 2. *Cont.*

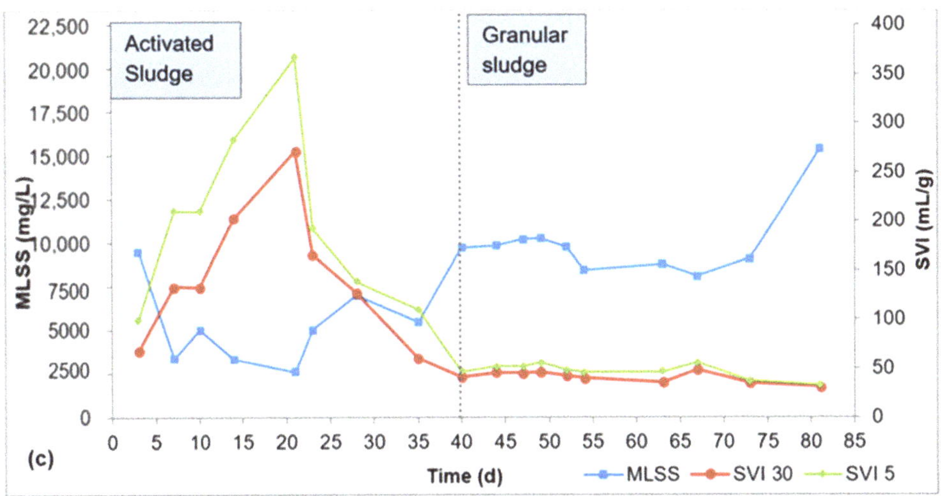

Figure 2. Profiles of biomass concentration and sludge settleability: (**a**) R1, (**b**) R2, (**c**) R3.

In the AGS maturation phase (day 41–84), the average biomass concentration was 8000 ± 1200 mg/L, 13,000 ± 5600 mg/L, and 10,000 ± 2000 mg/L mg/L in R1, R2, and R3, respectively. AGS bioreactors exhibit very high biomass concentration, in the range 8000–15,000 mg/L, due to the high biomass density and outstanding settleability of the granules [26]. The biomass concentrations obtained in this study all fall within this range. While there are spikes in biomass concentration in R2 (day 67–84) and R3 (day 73–84), these elevated biomass concentrations showed no effect on phosphorus removal.

3.2. Reactor Performance

3.2.1. Profiles of Organic Matter Degradation

All three reactors exhibited high COD degradation at the steady state, as shown in Figure 3. R1, R2, and R3 had mean removal efficiencies of 96 ± 4%, 95 ± 2%, and 97 ± 2%, respectively. R3 had a longer aeration period compared to the R1 and R2 (54 min longer than the aeration period in R1 and R2), hence the slightly higher COD removal in R3. This implies that the feeding strategy had no impact on COD removal, and the COD concentration did not impact the treatment performance in the reactors. The high COD degradation attained in this study is consistent with other AGS systems used for both municipal and industrial wastewaters [5,27,28]. The high biomass concentration, the dense nature of the AGS, and the good settling properties of the biomass allowed for a high biomass retention, which enhanced the performance of the bioreactors.

3.2.2. Profiles of Nitrogen Removal

The nitrogen profile in each reactor is presented in Figure 4a–c. At steady state (in the granule maturation phase), ammonia-N removal was 98 ± 3%, 98 ± 3%, and 99 ± 0.6% for R1, R2, and R3, respectively. The high ammonia-N removal indicates the occurrence of nitrification in all three reactors. Nitrite-N exhibited low values across all reactors, demonstrating complete nitrification. This implies that the rate-limiting step for nitrification, that is, the transformation of ammonia to nitrite, was completed, and that the second nitrification step—the conversion of nitrite to nitrate—was also successful. The high retention of nitrifying biomass in AGS [29] facilitates nitrification.

Within an aerobic granule, there is a layered structure with an outer aerobic layer (for nitrification) and an inner anoxic zone (for denitrification) [6,30]. Denitrification is normally the limiting step for total nitrogen removal, but an anoxic interior is required [31].

Denitrification seems to have occurred due to this stratified structure. However, there appeared to be nitrate accumulation in all three reactors.

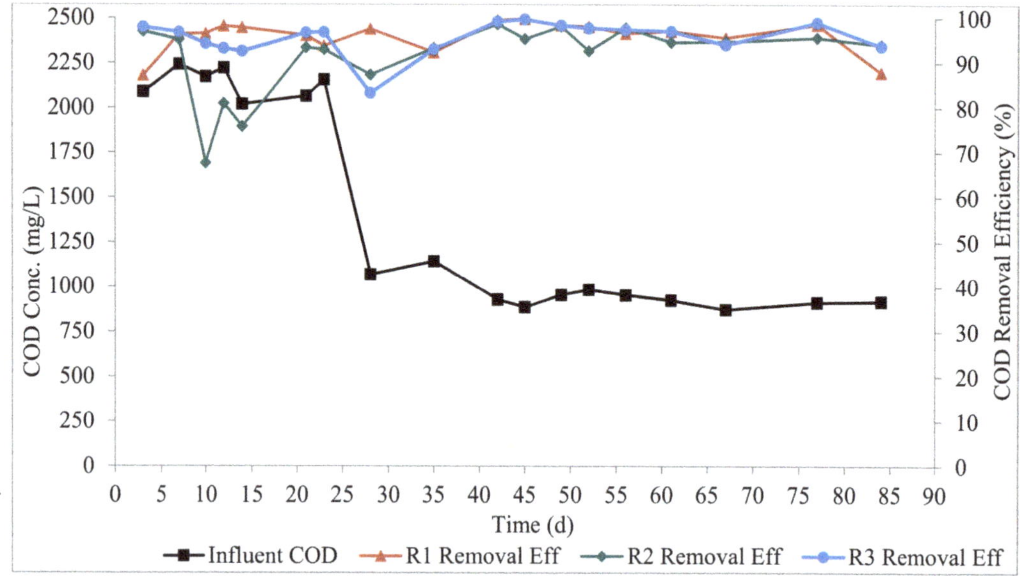

Figure 3. Profiles of organic matter degradation.

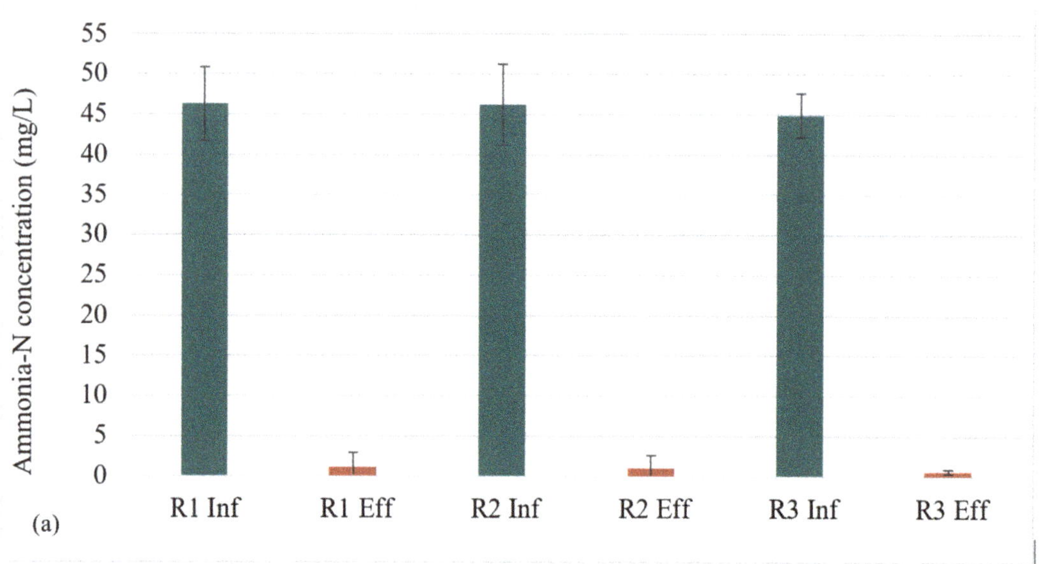

(a)

Figure 4. *Cont.*

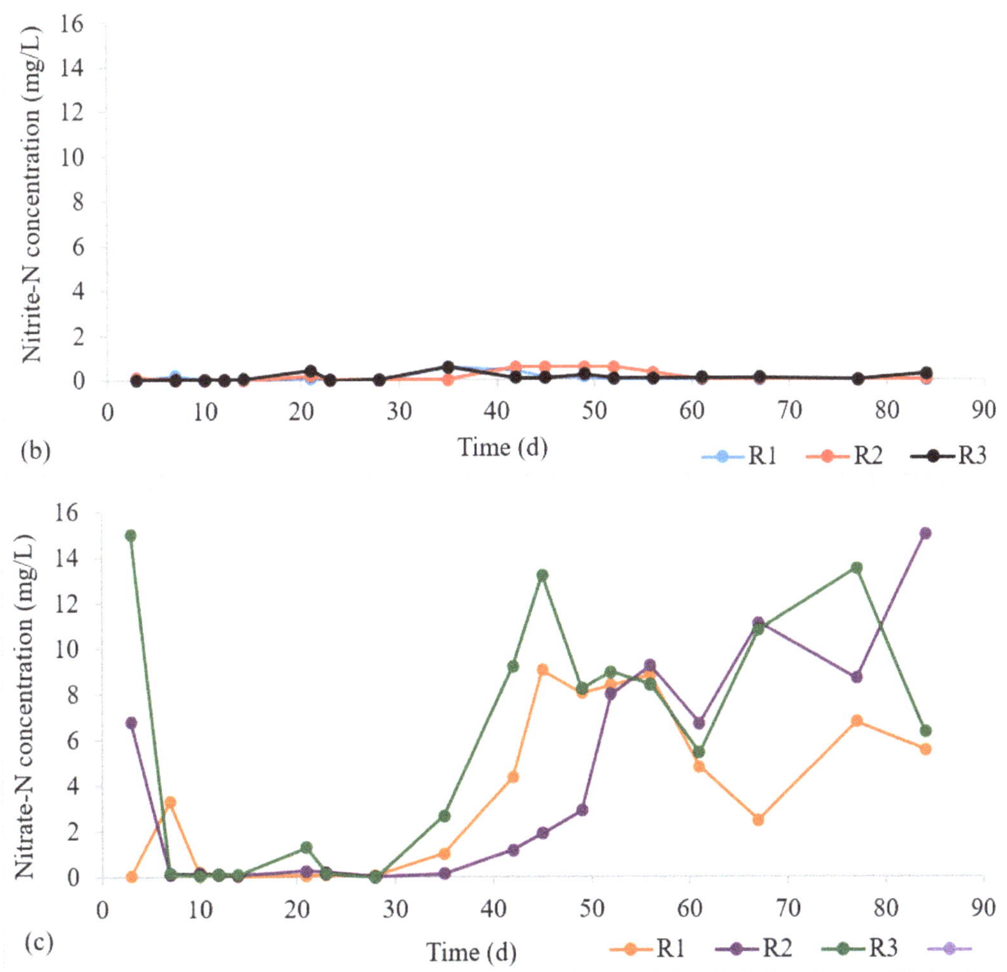

Figure 4. Profiles of nitrogen removal: (**a**) NH$_3$-N influent and effluent concentrations at steady state, (**b**) Effluent NO$_2$-N profile, (**c**) Effluent NO$_3$-N profile.

3.2.3. Profiles of Phosphorus Removal

Phosphorus removal profiles are presented in Figure 5. At the steady state, phosphorus removals in R1, R2, and R3 were 97.6 $\pm$ 3, 98.3 $\pm$1, and 55 $\pm$ 11%, respectively. This shows that the slow anaerobic feeding in R1 and the pulse feeding followed by anaerobic mixing in R2 facilitated the proliferation of PAOs, presenting an opportunity for high phosphorus recovery. High phosphorus removal efficiencies have been reported for AGS systems in the scientific literature [6,14,30]. This is due to the layered structure of the aerobic granule, which provides both aerobic and anaerobic conditions within the granule structure, as a result of oxygen diffusion limitation, which enable EBPR by PAOs. In R1 and R2, there is an anaerobic phase of about 1 h (filling for R1, and filling/mixing for R2) followed by aeration, which provides the desired alternating anaerobic–aerobic conditions in the SBR cycle in addition to the layered structure of the granule. This explains the outstanding phosphorus removal achieved in R1 and R2. High phosphorus removal has been previously reported in AGS systems with an anaerobic phase in the SBR cycle [8,12,29,32]. In R3, there is no anaerobic phase within the SBR cycle, and EBPR was only due to the stratified structure of the granule. The lower phosphorus removal in R3 implies that the stratified structure of aerobic granules (with an aerobic outer layer and an anaerobic core) alone is inadequate to allow anaerobic phosphorus release and potent phosphorus uptake. Low phosphorus removal was previously reported in AGS systems without the anaerobic phase [33]. It can be inferred that both an anaerobic phase and the layered structure of the granule are essential for high phosphorus removal in AGS systems.

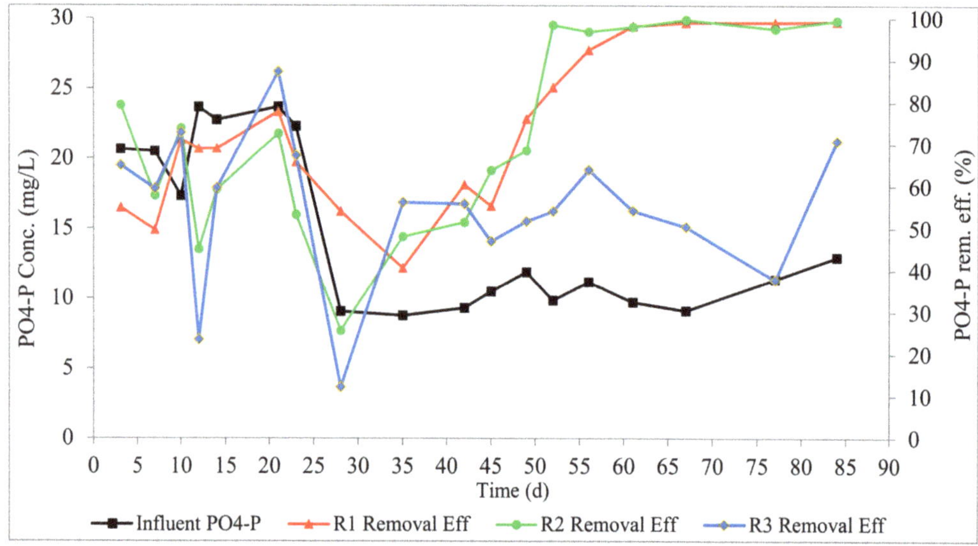

Figure 5. Profiles of phosphorus removal.

3.3. Specific Phosphorus Release and Uptake Rates

The specific phosphorus release rate (SPRR) for the seed sludge and in R1, R2, and R3 are presented in Table 2 below. The seed sludge used in the reactors had a high SPRR (23.33 mg P/g Vss.h) because it was the return activated sludge from a biological nutrient removal wastewater treatment plant. At steady state, the SPRR changed to 11.51 mg P/g VSS.h, 7.42 mg P/g VSS.h, and 2.60 mg P/g VSS.h in R1, R2, and R3, respectively. These values imply that in the course of granulation, several microbial species increased in relative abundance within the sludge. We conclude that the anaerobic feeding, the pulse feeding, and the anaerobic mixing applied in R1 and R2 were responsible for the high SPRR and the high phosphorus removal.

Table 2. SPRR and SPUR values for seed sludge and granular sludge in R1, R2, and R3 at steady state.

	Seed Sludge	R1	R2	R3
SPRR (mg P/g VSS.h)	23.33	11.51	7.42	2.60
SPUR (mg P/g VSS.h)	−12.83	−9.79	−7.05	−0.55

The specific phosphorus uptake rate (SPUR) in the seed sludge was −12.83 mg P/g Vss.h. At steady state, the SPUR changed to −9.79 mg P/g VSS.h, −7.05 mg P/g VSS.h, and −0.55 mg P/g VSS.h in R1, R2, and R3, respectively. Similar to the SPRR values, the SPUR values imply that some microbial species increased in relative abundance in the course of granulation. R3, with the lowest SPUR, had no anaerobic phase to allow for PAOs. Therefore, R3 had a low phosphorus uptake. R1 and R2 exhibited high SPUR values because the feeding phase was anaerobic.

3.4. Phosphorus Distribution in the Sludge

Figure 6a–d shows the phosphorus distribution in the biomass of the wastewater treatment plant seed sludge and in the R1 sludge, the R2 sludge, and the R3 sludge at the steady state condition. Similar to the rates of phosphorus release and phosphorus uptake reported in Section 3.3, the percentage of phosphorus released by PAOs decreased in the course of sludge granulation. The percentage of P released by PAOs decreased from 42.8% in the seed sludge to 24.8%, 8.8%, and 7.2% in R1, R2, and R3, respectively. This indicates that there was a change in the microbial composition in the sludge in the course of granulation from flocculent sludge to granular sludge. Similarly, the percentage of precipitated P increased from 0.73% in the seed sludge to 11.1%, 5%, and 10% in R1, R2, and R3, respectively. This indicates that the granular sludge matrix encapsulated precipitated phosphorus, whereas the seed sludge, with no granule structure, could not encapsulate precipitated phosphorus.

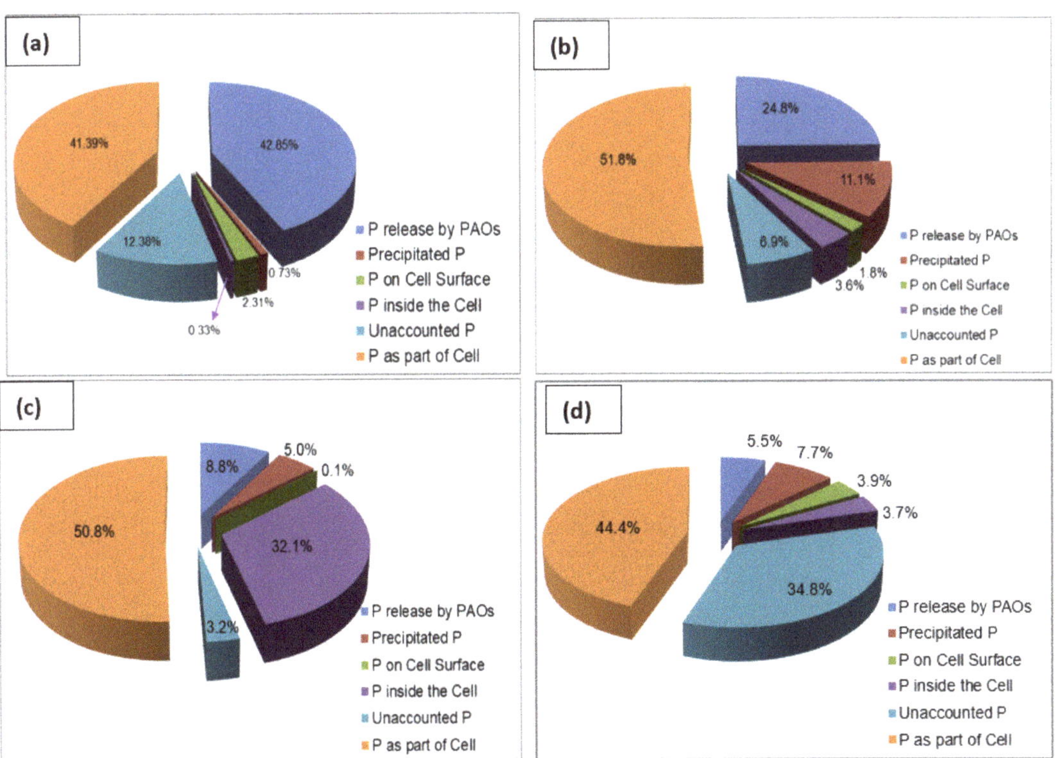

Figure 6. Phosphorus distribution in the sludge: (**a**) seed sludge, (**b**) R1 sludge at steady state, (**c**) R2 sludge at steady state, (**d**) R3 sludge at steady state.

3.5. Microbial Analysis

3.5.1. Shift of Microbial Communities

Nonmetric multidimensional scaling (NMDS) plots, alpha diversity measurements, and phylum-level classifications under the different operating conditions are presented in Figure 7. NMDS analysis suggested that microbial composition differed more with respect to sampling time than with respect to feeding strategy. The seed sludge had the highest microbial diversity. The microbial diversity decreased in the first 27 days of reactor operation, then increased. This transition was depicted at the phylum level. Proteobacteria increased significantly across all feeding conditions. Firmicutes and Verrucomicrobia increased at least four times in R1 and R2. At the last sampling point, the phyla distribution was similar in all reactors, consistent with the results obtained for biomass concentration. Figure 8 shows the dominant amplicon sequence variants (ASVs) in the microbial community, indicating an apparent microbial shift. In the seed sludge, the dominant ASVs belonged to *Methylotenera*, *Simplicispira*, and *Rhodoferax* genera, which decreased over time in all three reactors. In contrast, several ASVs of *Paracoccus*, *Prosthecobacter*, and *Pedobacter* genera, and unclassified genera from the class Betaproteobacteria, increased.

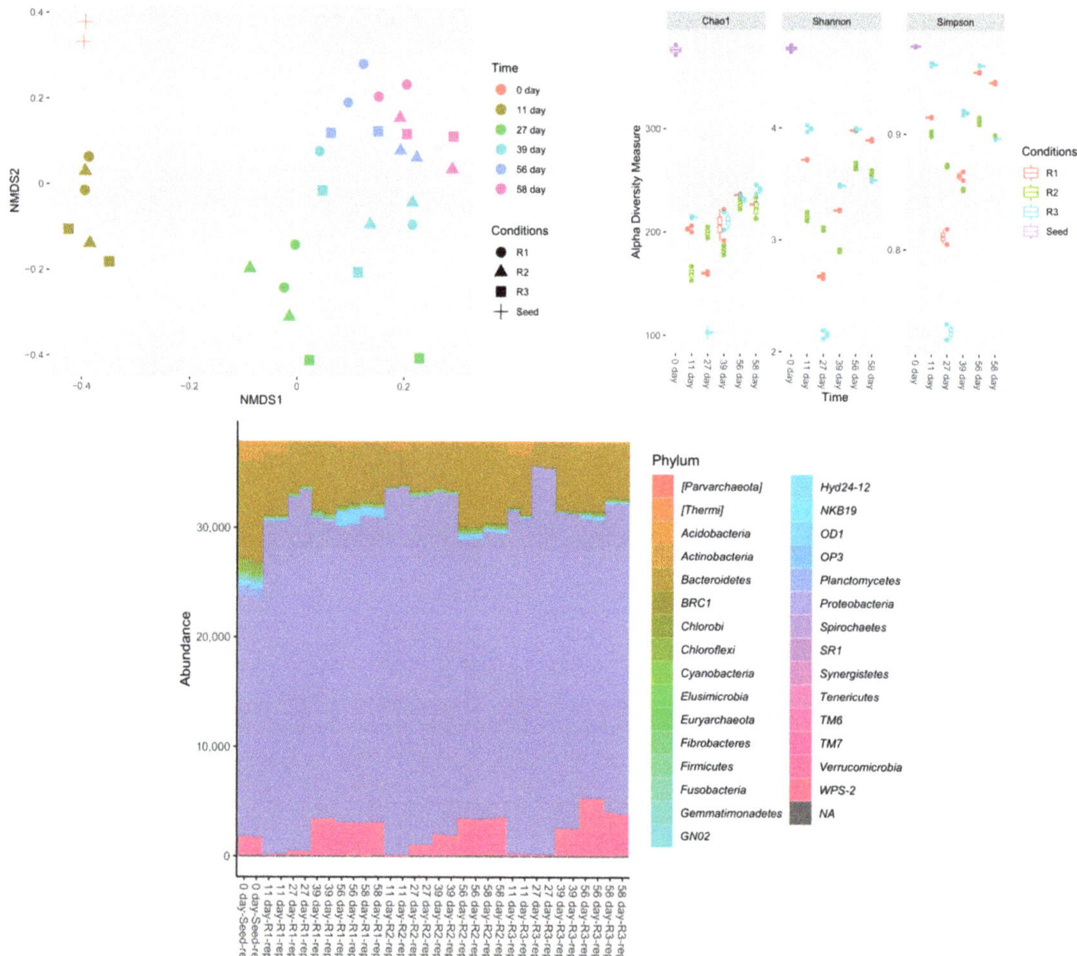

Figure 7. Nonmetric multidimensional scaling (NMDS) plots, alpha diversity measurements, and phylum-level classification with a rarefied-even-depth function of 95% of the minimum sum of amplicon sequence variants (ASVs) for samples under different conditions.

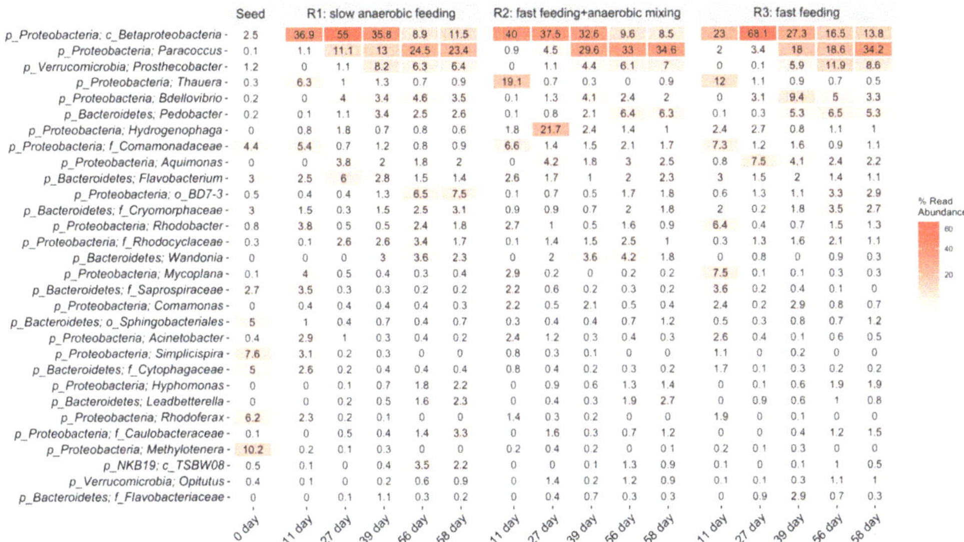

	Seed	R1: slow anaerobic feeding					R2: fast feeding+anaerobic mixing					R3: fast feeding				
	0 day	11 day	27 day	39 day	56 day	58 day	11 day	27 day	39 day	56 day	58 day	11 day	27 day	39 day	56 day	58 day
p_Proteobacteria; c_Betaproteobacteria	2.5	36.9	55	35.8	8.9	11.5	40	37.5	32.6	9.6	8.5	23	68.1	27.3	16.5	13.8
p_Proteobacteria; Paracoccus	0.1	1.1	11.1	13	24.5	23.4	0.9	4.5	29.6	33	34.6	2	3.4	18	18.6	34.2
p_Verrucomicrobia; Prosthecobacter	1.2	0	1.1	8.2	6.3	6.4	0	1.1	4.4	6.1	7	0	0.1	5.9	11.9	8.6
p_Proteobacteria; Thauera	0.3	6.3	1	1.3	0.7	0.9	19.1	0.7	0.3	0	0.9	12	1.1	0.9	0.7	0.5
p_Proteobacteria; Bdellovibrio	0.2	0	4	3.4	4.6	3.5	0.1	1.3	4.1	2.4	2	0	3.1	9.4	5	3.3
p_Bacteroidetes; Pedobacter	0.2	0.1	1.1	3.4	2.5	2.6	0.1	0.8	2.1	6.4	6.3	0.1	0.3	5.3	6.5	5.3
p_Proteobacteria; Hydrogenophaga	0	0.8	1.8	0.7	0.8	0.6	1.8	21.7	2.4	1.4	1	2.4	2.7	0.8	1.1	1
p_Proteobacteria; f_Comamonadaceae	4.4	5.4	0.7	1.2	0.8	0.9	6.6	1.4	1.5	2.1	1.7	7.3	1.2	1.6	0.9	1.1
p_Proteobacteria; Aquimonas	0	0	3.8	2	1.8	2	0	4.2	1.8	3	2.5	0.8	7.5	4.1	2.4	2.2
p_Bacteroidetes; Flavobacterium	3	2.5	6	2.8	1.5	1.4	2.6	1.7	1	2	2.3	3	1.5	2	1.4	1.1
p_Proteobacteria; o_BD7-3	0.5	0.4	0.4	1.3	6.5	7.5	0.1	0.7	0.5	1.7	1.8	0.6	1.3	1.1	3.3	2.9
p_Bacteroidetes; f_Cryomorphaceae	3	1.5	0.3	1.5	2.5	3.1	0.9	0.9	0.7	2	1.8	2	0.2	1.8	3.5	2.7
p_Proteobacteria; Rhodobacter	0.8	3.8	0.5	0.5	2.4	1.8	2.7	1	0.5	1.6	0.9	6.4	0.4	0.7	1.5	1.3
p_Proteobacteria; f_Rhodocyclaceae	0.3	0.1	2.6	2.6	3.4	1.7	0.1	1.4	1.5	2.5	1	0.3	1.3	1.6	2.1	1.1
p_Bacteroidetes; Wandonia	0	0	0	3	3.6	2.3	0	2	3.6	4.2	1.8	0	0.8	0	0.9	0.3
p_Proteobacteria; Mycoplana	0.1	4	0.5	0.4	0.3	0.4	2.9	0.2	0	0.2	0.2	7.5	0.1	0.1	0.3	0.3
p_Bacteroidetes; f_Saprospiraceae	2.7	3.5	0.3	0.3	0.2	0.2	2.2	0.6	0.2	0.3	0.2	3.6	0.2	0.4	0.1	0
p_Proteobacteria; Comamonas	0	0.4	0.4	0.4	0.4	0.3	2.2	0.5	2.1	0.5	0.4	2.4	0.2	2.9	0.8	0.7
p_Bacteroidetes; o_Sphingobacteriales	5	1	0.4	0.7	0.4	0.7	0.3	0.4	0.4	0.7	1.2	0.5	0.3	0.8	0.7	1.2
p_Proteobacteria; Acinetobacter	0.4	2.9	1	0.3	0.4	0.2	2.4	1.2	0.3	0.4	0.3	2.6	0.4	0.1	0.6	0.5
p_Proteobacteria; Simplicispira	7.6	3.1	0.2	0.3	0	0	0.8	0.3	0.1	0	0	1.1	0	0.2	0	0
p_Bacteroidetes; f_Cytophagaceae	5	2.6	0.2	0.4	0.4	0.4	0.8	0.4	0.2	0.3	0.2	1.7	0.1	0.3	0.2	0.2
p_Proteobacteria; Hyphomonas	0	0	0.1	0.7	1.8	2.2	0	0.9	0.6	1.3	1.4	0	0.1	0.6	1.9	1.9
p_Bacteroidetes; Leadbetterella	0	0	0.2	0.5	1.6	2.3	0	0.4	0.3	1.9	2.7	0	0.9	0.6	1	0.8
p_Proteobacteria; Rhodoferax	6.2	2.3	0.2	0.1	0	0	1.4	0.3	0.2	0	0	1.9	0	0.1	0	0
p_Proteobacteria; f_Caulobacteraceae	0.1	0	0.5	0.4	1.4	3.3	0	1.6	0.3	0.7	1.2	0	0	0.4	1.2	1.5
p_Proteobacteria; f_Methylotenera	10.2	0.2	0.1	0.3	0	0	0.2	0.4	0.2	0	0.1	0.2	0.1	0.3	0	0
p_NKB19; c_TSBW08	0.5	0.1	0	0.4	3.5	2.2	0	0	0.1	1.3	0.9	0.1	0	0.1	1	0.5
p_Verrucomicrobia; Opitutus	0.4	0.1	0	0.2	0.6	0.9	0	1.4	0.2	1.2	0.9	0.1	0.1	0.3	1.1	1
p_Bacteroidetes; f_Flavobacteriaceae	0	0	0.1	1.1	0.3	0.2	0	0.4	0.7	0.3	0.3	0	0.9	2.9	0.7	0.3

Figure 8. The 30 most abundant ASVs (by means of the replicates) classified to the lowest taxonomic rank (majority are genus, shown with phylum) across all samples. p_, c_, and f_ represent phylum, class, and family, respectively.

3.5.2. Distribution of the Microbial Profile

The dominant ASVs in all three reactors, each treated with a different feeding strategy, comprised the class of Betaproteobacteria, (29–35.84%) in R1, (25.7–32.6%) in R2, and (29.8–23%) in R3. Betaproteobacteria are in a class that belongs to the phylum *Proteobacteria*, the members of which have extreme metabolic diversity and are the most important, and often the most dominant, of the PAOs in lab-scale reactors and in full-scale wastewater treatment facilities [2,34–36]. The consistent increase in the relative abundance of *Proteobacteria* phylum during the stabilization phases in R1, R2, and R3, suggests quick adaptation of the microbes to the operating condition and also indicates that the operational conditions had insignificant effect on diversity and abundance of the microbial communities. This finding agrees with the result of Wang et al. [37], who reported an increase in the relative abundance of phylum *Proteobacteria* in AGS reactors under both continuous and alternating aeration conditions. Xu et al. [38] also reported an increase in the relative abundance of *Proteobacteria* during granulation in a continuous-flow integrated oxidation ditch AGS system with two-zone clarifiers. The second most abundant bacteria in the three reactors were *Paracoccus*: 1.1–24.5% in R1, 0.9–34.6% in R2, and 2–34.2% in R3. *Paracoccus* are denitrifiers with the ability to use both oxygen and nitrogenous oxides. This enables them to survive in aerobic or anaerobic conditions, which explains their ongoing abundance in all three reactors. *Paracoccus* are well known for their denitrification and phosphorus removal capabilities [39,40]. *Thauera* were identified at 6.3% in R1, 19% in R2, and 12% in R3 on day 11, and decreased after that. *Thauera* are anaerobic, facultative denitrifying bacteria with versatile metabolic potentials, including the ability to remove nitrogen, nitrates, nitrites, and phosphorus from the reactors [41,42]. Hamza et al. [41] reported a decline in the relative abundance of Thauera upon reactor stabilization. *Proteobacteria* belonging to the Bacteroidetes and Verrucomicrobia phyla that have been reported in AGS bioreactors were present in small amounts in our reactors; their slow growth provided a low competitive advantage. Proteobacteria thrive better than Bacteroidetes and Verrucomicrobia in an anaerobic environment [43]. Several investigators have reported *Proteobacteria* and *Bacteroidetes* as the predominant phyla in AGS systems, and these phyla include a number of important functional bacteria comprising of *Rhodobacter*, *Bdellovibrio*, and *Rhodocyclaceae*, which play important roles in AGS granulation. For instance, *Rhodocyclaceae*, an EPS-secreting microbe, has denitrification potential as well. EPS secretion promotes self-aggregation of flocculent sludge needed in granulation and also contributes to maintaining the operational stability of the AGS system [44]. *Rhodobacter* also acts as a stabilizer, but possesses denitrification abilities. *Proteobacteria* and *Bacteroidetes* bacteria have the ability to degrade carbon, including recalcitrant organic carbon, nitrogen, and phosphorus [37]. Some less-known bacteria such as *Wandonia* and *Hyphomonas* were observed to consistently increase in the three reactors, thus, their identification and characterization should be the focus of future work.

Predominant bacterial families present in R1, R2, and R3 included: Comamonadaceae—0.7–5.4% in R1, 1.4–6.6% in R2, 0.9–7.3% in R3 and Saprospiraceae—0.2–3.5% in R1, 0.2–2.2% in R2, and 0–3.6% in R3. Comamonadaceae and other bacterial families within the Proteobacteria phylum are commonly involved in nitrification and denitrification (Adav et al., 2010), while members of the Saprospiraceae family are acetotrophic denitrifiers [43]. *Rhodobacter*—0.5–3.8% in R1, 0.9–2.7% in R2, and 0.4–6.4% in R3—are known to accumulate phosphorus during denitrification [45]. *Acinetobacter*—0.2–2.4% in R1, 0.3–2.4% in R2, and 0.0–2.6% in R3—persistent but usually low in abundance in AGS reactors [46], were also found. Environmental conditions control the abundance of each bacterial species [47].

3.5.3. Distribution of Putative PAOs

The heat map showing the evolution of putative PAOs in R1, R2, and R3 at phyla and genus levels is presented in Figure 9.

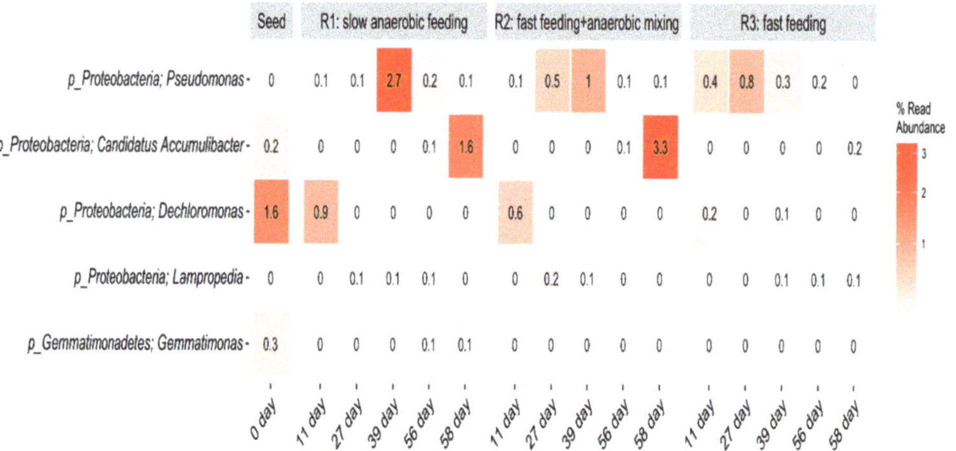

	Seed	R1: slow anaerobic feeding						R2: fast feeding+anaerobic mixing					R3: fast feeding				
p_Proteobacteria; Pseudomonas	0	0.1	0.1	2.7	0.2	0.1	0.1	0.5	1	0.1	0.1	0.4	0.8	0.3	0.2	0
p_Proteobacteria; Candidatus Accumulibacter	0.2	0	0	0	0.1	1.6	0	0	0	0.1	3.3	0	0	0	0	0.2
p_Proteobacteria; Dechloromonas	1.6	0.9	0	0	0	0	0.6	0	0	0	0	0.2	0	0.1	0	0
p_Proteobacteria; Lampropedia	0	0	0.1	0.1	0.1	0	0	0.2	0.1	0	0	0	0	0.1	0.1	0.1
p_Gemmatimonadetes; Gemmatimonas	0.3	0	0	0	0.1	0.1	0	0	0	0	0	0	0	0	0	0
	0 day	11 day	27 day	39 day	56 day	58 day	11 day	27 day	39 day	56 day	58 day	11 day	27 day	39 day	56 day	58 day

Figure 9. Heat map showing the evolution of putative PAOs in R1, R2, and R3 at phyla and genus levels.

Low percentage abundance of well-known and putative PAOs including, *Pseudomonas*, *Accumulibacter*, *Dechloromonas*, and *Lampropedia* [36,48,49] were dictated at genus level within this AGS system. *Pseudomonas* was the most abundant with average read abundance of 0.64% in R1, 0.36% in R2, and 0.34% in R3. All three reactors recorded small but steady increments and eventual decline of *Pseudomonas* population over time. Within the first 39 days of reactor operation, *Pseudomonas* within R1 (slow anaerobic feeding) increased from 0–2.7%, before declining to 0.1%. In R2 (pulse feeding + anaerobic mixing), they increased from 0–1.0% before declining to 0.1% too, while in R3 (pulse feeding), they increased to 0.8% on day 27 before steady decline and eventual complete decay by day 58. This could imply that this bacterial community responded to factors other than the feeding strategies within this system. Günther et al. [50] identified members of the genus *Pseudomonas* as the predominant PAOs in an SBR-EBPR system. The second most abundant putative PAO genus in the present study was Ca. *Accumulibacter* with an average read abundance of 0.14% in R1, 0.68% in R2, and 0.04% in R3. This group of microbes also gradually but steadily increased over time. Ca. *Accumulibacter* is the most studied of the PAO genera and is often assumed to be the most important PAO [48]. *Dechloromonas* genus was also found in small quantities in the seed sludge and decayed over time in all three reactors. The dramatic decrease of the read abundance of PAOs within this system is difficult to explain, as there were no apparent substrate limitations within the system. Consistent decline over time shows lack of resilience. However, the core species may have been retained within the reactor or induced within the granules, which ensured the accomplishment of the biological removal processes and process stability. This aligns with the work of Adler and Holliger [48], who reported that with fermentative bacteria within a phosphorus removing system, various sugars and amino acids can be anaerobically fermented to release fermentation products, which allow classical PAOs to remain in the system. Wang et al. [37] reported a similar situation including a decrease in microbial communities at the class level, in an AGS system using an alternating aeration strategy. Minute quantities of the genus *Lampropedia* developed in the three reactors over time, while Gemmatimonas, though present in the seed sludge, did not strive in the reactors. *Lampropedia* are unlikely numerically important in phosphorus removal processes, with the only evidence suggesting their role as PAOs being the ability to store visible polyphosphate granules [36]. Further analysis of this result shows that R1, R2, and R3 have similar microbial community compositions despite the different feeding strategies. Although the reasons for this are not clear, it is known that certain operational parameters can affect both the diversity and the stability of bacterial populations [51]. The abundance and diversity of PAOs are critical for an assessment of the P-removal efficiency in the AGS. Thus, based on the small percentages of the PAOs found within this system, it can be inferred that some unidentified bacterial species may have also contributed to the performances achieved by R1, R2, and R3.

4. Conclusions

Phosphorus removal and PAO proliferation were studied in AGS bioreactors under three different feeding strategies: anaerobic slow feeding (R1), pulse feeding + anaerobic mixing (R2), and pulse feeding (R3). The results show that feeding strategy significantly impacts AGS bioreactor performance in terms of phosphorus removal but has no impact on organic matter degradation and ammonia removal. High phosphorus recovery from wasted aerobic sludge granules is possible when the reactor is operated using slow anaerobic feeding or fast feeding + anaerobic mixing strategies. With these feeding strategies, active PAOs can be accumulated and GAOs can be significantly inhibited within the bioreactor, leading to high SPRR, SPUR, and significantly higher phosphorus removal rates.

Author Contributions: Conceptualization, O.T.I.; methodology, O.T.I.; formal analysis, O.T.I.; investigation, O.T.I. and S.U.; resources, Y.L.; data curation, H.D.; writing—original draft preparation, O.T.I.; writing—review and editing, O.T.I., S.U., H.D. and Y.L.; supervision, Y.L.; funding acquisition, O.T.I. and Y.L. All authors have read and agreed to the published version of the manuscript.

Funding: This research received no external funding.

Institutional Review Board Statement: Not applicable.

Acknowledgments: The authors acknowledge financial support from the Natural Sciences and Engineering Research Council of Canada (NSERC) Industrial Research Chair (IRC) Program in Sustainable Urban Water Development through the support by EPCOR Water Services, EPCOR Drainage Operation and Alberta Innovates, an NSERC Discovery project, the Canada Research Chair (CRC) in Future Community Water Services (L.Y.), and an NSERC Postdoctoral Fellowship (I.O.T.).

Conflicts of Interest: The authors declare no conflict of interest.

References

1. Heckenmüller, M.; Narita, D.; Klepper, G. *Global Availability of Phosphorus and Its Implications for Global Food Supply: An Economic Overview*; Kiel Working Paper, No. 1897; Kiel Institute for the World Economy (IfW): Kiel, Germany, 2014.
2. Lu, Y.-Z.; Wang, H.-F.; Kotsopoulos, T.A.; Zeng, R.J. Advanced phosphorus recovery using a novel SBR system with granular sludge in simultaneous nitrification, denitrification and phosphorus removal process. *Appl. Microbiol. Biotechnol.* **2016**, *100*, 4367–4374. [CrossRef] [PubMed]
3. Cordell, D.; Rosemarin, A.; Schröder, J.; Smit, A. Towards global phosphorus security: A systems framework for phosphorus recovery and reuse options. *Chemosphere* **2011**, *84*, 747–758. [CrossRef] [PubMed]
4. Chrispim, M.C.; Scholz, M.; Nolasco, M.A. Phosphorus recovery from municipal wastewater treatment: Critical review of challenges and opportunities for developing countries. *J. Environ. Manag.* **2019**, *248*, 109268. [CrossRef] [PubMed]
5. de Kreuk, M.; Heijnen, J.; van Loosdrecht, M. Simultaneous COD, nitrogen, and phosphate removal by aerobic granular sludge. *Biotechnol. Bioeng.* **2005**, *90*, 761–769. [CrossRef]
6. Yilmaz, G.; Lemaire, R.; Keller, J.; Yuan, Z. Simultaneous nitrification, denitrification, and phosphorus removal from nutrient-rich industrial wastewater using granular sludge. *Biotechnol. Bioeng.* **2007**, *100*, 529–541. [CrossRef]
7. Khan, M.Z.; Mondal, P.K.; Sabir, S. Aerobic granulation for wastewater bioremediation: A review. *Can. J. Chem. Eng.* **2013**, *91*, 1045–1058. [CrossRef]
8. Cassidy, D.; Belia, E. Nitrogen and phosphorus removal from an abattoir wastewater in a SBR with aerobic granular sludge. *Water Res.* **2005**, *39*, 4817–4823. [CrossRef]
9. Gao, D.; Liu, L.; Liang, H.; Wu, W. Aerobic granular sludge: Characterization, mechanism of granulation and application to wastewater treatment. *Crit. Rev. Biotechnol.* **2010**, *31*, 137–152. [CrossRef]
10. Mañas, A.; Biscans, B.; Spérandio, M. Biologically induced phosphorus precipitation in aerobic granular sludge process. *Water Res.* **2011**, *45*, 3776–3786. [CrossRef]
11. Filali, A.; Mañas, A.; Mercade, M.; Bessière, Y.; Biscans, B.; Spérandio, M. Stability and performance of two GSBR operated in alternating anoxic/aerobic or anaerobic/aerobic conditions for nutrient removal. *Biochem. Eng. J.* **2012**, *67*, 10–19. [CrossRef]
12. Henriet, O.; Meunier, C.; Henry, P.; Mahillon, J. Improving phosphorus removal in aerobic granular sludge processes through selective microbial management. *Bioresour. Technol.* **2016**, *211*, 298–306. [CrossRef]
13. Sarma, S.; Tay, J.-H. Carbon, nitrogen and phosphorus removal mechanisms of aerobic granules. *Crit. Rev. Biotechnol.* **2018**, *38*, 1077–1088. [CrossRef] [PubMed]
14. Purba, L.D.A.; Ibiyeye, H.T.; Yuzir, A.; Mohamad, S.E.; Iwamoto, K.; Zamyadi, A.; Abdullah, N. Various applications of aerobic granular sludge: A review. *Environ. Technol. Innov.* **2020**, *20*, 101045. [CrossRef]
15. Huang, W.; Huang, W.; Li, H.; Lei, Z.; Zhang, Z.; Tay, J.H.; Lee, D.-J. Species and distribution of inorganic and organic phosphorus in enhanced phosphorus removal aerobic granular sludge. *Bioresour. Technol.* **2015**, *193*, 549–552. [CrossRef] [PubMed]
16. Iorhemen, O.T.; Liu, Y. Effect of feeding strategy and organic loading rate on the formation and stability of aerobic granular sludge. *J. Water Process Eng.* **2020**, *39*, 101709. [CrossRef]
17. Winkler, M.-K.H.; Meunier, C.; Henriet, O.; Mahillon, J.; Suárez-Ojeda, M.E.; Moro, G.D.; Sanctis, M.D.; Iaconi, C.D.; Weissbrodt, D.G. An integrative review of granular sludge for the biological removal of nutrients and recalcitrant organic matter from wastewater. *Chem. Eng. J.* **2018**, *336*, 489–502. [CrossRef]
18. Yuan, Q.; Gong, H.; Xi, H.; Xu, H.; Jin, Z.; Ali, N.; Wang, K. Strategies to improve aerobic granular sludge stability and nitrogen removal based on feeding mode and substrate. *J. Environ. Sci.* **2019**, *84*, 144–154. [CrossRef]
19. Martins, A.M.; Heijnen, J.J.; van Loosdrecht, M.C. Effect of feeding pattern and storage on the sludge settleability under aerobic conditions. *Water Res.* **2003**, *37*, 2555–2570. [CrossRef]
20. Liu, Y.; Yang, S.-F.; Tay, J.-H. Improved stability of aerobic granules by selecting slow-growing nitrifying bacteria. *J. Biotechnol.* **2004**, *108*, 161–169. [CrossRef]
21. Lochmatter, S.; Holliger, C. Optimization of operation conditions for the startup of aerobic granular sludge reactors biologically removing carbon, nitrogen, and phosphorous. *Water Res.* **2014**, *59*, 58–70. [CrossRef]
22. Rocktäschel, T.; Klarmann, C.; Helmreich, B.; Ochoa, J.; Boisson, P.; Sørensen, K.; Horn, H. Comparison of two different anaerobic feeding strategies to establish a stable aerobic granulated sludge bed. *Water Res.* **2013**, *47*, 6423–6431. [CrossRef] [PubMed]
23. Thwaites, B.J.; Reeve, P.; Dinesh, N.; Short, M.D.; Akker, B.V.D. Comparison of an anaerobic feed and split anaerobic–aerobic feed on granular sludge development, performance and ecology. *Chemosphere* **2017**, *172*, 408–417. [CrossRef]
24. McSwain, B.; Irvine, R.; Wilderer, P. The effect of intermittent feeding on aerobic granule structure. *Water Sci. Technol.* **2004**, *49*, 19–25. [CrossRef] [PubMed]
25. APHA/AWWA/WEF. *Standard Methods for the Examination of Water and Wastewater*; American Public Health Association/American Water Works Association/Water Environment Federation: Washington, DC, USA, 2012.
26. De Kreuk, M.; Van Loosdrecht, M. Selection of slow growing organisms as a means for improving aerobic granular sludge stability. *Water Sci. Technol.* **2004**, *49*, 9–17. [CrossRef]
27. Show, K.-Y.; Lee, D.-J.; Tay, J.-H. Aerobic Granulation: Advances and Challenges. *Appl. Biochem. Biotechnol.* **2012**, *167*, 1622–1640. [CrossRef] [PubMed]
28. Franca, R.D.; Pinheiro, H.M.; van Loosdrecht, M.C.; Lourenço, N.D. Stability of aerobic granules during long-term bioreactor operation. *Biotechnol. Adv.* **2018**, *36*, 228–246. [CrossRef] [PubMed]

29. Iorhemen, O.T.; Zaghloul, M.S.; Hamza, R.A.; Tay, J.H. Long-term aerobic granular sludge stability through anaerobic slow feeding, fixed feast-famine period ratio, and fixed SRT. *J. Environ. Chem. Eng.* **2020**, *8*, 103681. [CrossRef]

30. Nancharaiah, Y.; Reddy, G.K.K. Aerobic granular sludge technology: Mechanisms of granulation and biotechnological applications. *Bioresour. Technol.* **2018**, *247*, 1128–1143. [CrossRef] [PubMed]

31. Chen, F.-Y.; Liu, Y.-Q.; Tay, J.-H.; Ning, P. Alternating anoxic/oxic condition combined with step-feeding mode for nitrogen removal in granular sequencing batch reactors (GSBRs). *Sep. Purif. Technol.* **2013**, *105*, 63–68. [CrossRef]

32. He, Q.; Wang, H.; Chen, L.; Gao, S.; Zhang, W.; Song, J.; Yu, J. Elevated salinity deteriorated enhanced biological phosphorus removal in an aerobic granular sludge sequencing batch reactor performing simultaneous nitrification, denitrification and phosphorus removal. *J. Hazard. Mater.* **2020**, *390*, 121782. [CrossRef]

33. Iorhemen, O.T.; Hamza, R.A.; Sheng, Z.; Tay, J.H. Submerged aerobic granular sludge membrane bioreactor (AGMBR): Organics and nutrients (nitrogen and phosphorus) removal. *Bioresour. Technol. Rep.* **2019**, *6*, 260–267. [CrossRef]

34. Crocetti, G.R.; Hugenholtz, P.; Bond, P.L.; Schuler, A.; Keller, J.; Jenkins, D.; Blackall, L.L. Identification of Polyphosphate-Accumulating Organisms and Design of 16S rRNA-Directed Probes for Their Detection and Quantitation. *Appl. Environ. Microbiol.* **2000**, *66*, 1175–1182. [CrossRef] [PubMed]

35. Wong, M.-T.; Mino, T.; Seviour, R.J.; Onuki, M.; Liu, W.-T. In situ identification and characterization of the microbial community structure of full-scale enhanced biological phosphorous removal plants in Japan. *Water Res.* **2005**, *39*, 2901–2914. [CrossRef] [PubMed]

36. Stokholm-Bjerregaard, M.; McIlroy, S.J.; Nierychlo, M.; Karst, S.M.; Albertsen, M.; Nielsen, P.H. A Critical Assessment of the Microorganisms Proposed to be Important to Enhanced Biological Phosphorus Removal in Full-Scale Wastewater Treatment Systems. *Front. Microbiol.* **2017**, *8*, 718. [CrossRef]

37. Wang, B.; Liu, S.-G.; Yao, X.; Liu, Y.; Qi, G. Alternating aeration strategy to reduce aeration energy demand for aerobic granular sludge and analysis of microbial community dynamics. *Environ. Sci. Water Res. Technol.* **2022**, *8*, 1111–1125. [CrossRef]

38. Xu, D.; Li, J.; Ma, T. Rapid aerobic sludge granulation in an integrated oxidation ditch with two-zone clarifiers. *Water Res.* **2020**, *175*, 115704. [CrossRef] [PubMed]

39. Lee, H.W.; Park, Y.K. Characterizations of denitrifying polyphosphate-accumulating bacterium Paracoccus sp. Strain YKP-9. *J. Microbiol. Biotechnol.* **2008**, *18*, 1958–1965.

40. Medhi, K.; Mishra, A.; Thakur, I.S. Genome Sequence of a Heterotrophic Nitrifier and Aerobic Denitrifier, Paracoccus denitrificans Strain ISTOD1, Isolated from Wastewater. *Genome Announc.* **2018**, *6*, e00210-18. [CrossRef]

41. Hamza, R.A.; Sheng, Z.; Iorhemen, O.T.; Zaghloul, M.S.; Tay, J.H. Impact of food-to-microorganisms ratio on the stability of aerobic granular sludge treating high-strength organic wastewater. *Water Res.* **2018**, *147*, 287–298. [CrossRef] [PubMed]

42. Wu, X.; Wu, X.; Li, J.; Wu, Q.; Ma, Y.; Sui, W.; Zhao, L.; Zhang, X. Cross-Feeding between Members of *Thauera* spp. and *Rhodococcus* spp. Drives Quinoline-Denitrifying Degradation in a Hypoxic Bioreactor. *mSphere* **2020**, *5*, e00246-20. [CrossRef]

43. Wiątczak, P.; Cydzik-Kwiatkowska, A. Performance and microbial characteristics of biomass in a full-scale aerobic granular sludge wastewater treatment plant. *Environ. Sci. Pollut. Res.* **2018**, *25*, 1655–1669. [CrossRef] [PubMed]

44. Zou, J.; Pan, J.; Wu, S.; Qian, M.; He, Z.; Wang, B.; Li, J. Rapid control of activated sludge bulking and simultaneous acceleration of aerobic granulation by adding intact aerobic granular sludge. *Sci. Total Environ.* **2019**, *674*, 105–113. [CrossRef] [PubMed]

45. Zheng, D.; Sun, Y.; Li, H.; Lu, S.; Shan, M.; Xu, S. Multistage A-O Activated Sludge Process for Paraformaldehyde Wastewater Treatment and Microbial Community Structure Analysis. *J. Chem.* **2016**, *2016*, 2746715. [CrossRef]

46. Xia, J.; Ye, L.; Ren, H.; Zhang, X.-X. Microbial community structure and function in aerobic granular sludge. *Appl. Microbiol. Biotechnol.* **2018**, *102*, 3967–3979. [CrossRef]

47. Zhang, Y.; Islam, S.; McPhedran, K.N.; Dong, S.; Rashed, E.M.; El-Shafei, M.M.; Noureldin, A.M.; El-Din, M.G. A comparative study of microbial dynamics and phosphorus removal for a two side-stream wastewater treatment processes. *RSC Adv.* **2017**, *7*, 45938–45948. [CrossRef]

48. Adler, A.; Holliger, C. Multistability and Reversibility of Aerobic Granular Sludge Microbial Communities Upon Changes From Simple to Complex Synthetic Wastewater and Back. *Front. Microbiol.* **2020**, *11*, 574361. [CrossRef] [PubMed]

49. Nierychlo, M.; Andersen, K.S.; Xu, Y.; Green, N.; Jiang, C.; Albertsen, M.; Dueholm, M.S.; Nielsen, P.H. MiDAS 3: An ecosystem-specific reference database, taxonomy and knowledge platform for activated sludge and anaerobic digesters reveals species-level microbiome composition of activated sludge. *Water Res.* **2020**, *182*, 115955. [CrossRef] [PubMed]

50. Günther, S.; Trutnau, M.; Kleinsteuber, S.; Hause, G.; Bley, T.; Röske, I.; Harms, H.; Müller, S. Dynamics of Polyphosphate-Accumulating Bacteria in Wastewater Treatment Plant Microbial Communities Detected via DAPI (4',6'-Diamidino-2-Phenylindole) and Tetracycline Labeling. *Appl. Environ. Microbiol.* **2009**, *75*, 2111–2121. [CrossRef] [PubMed]

51. Mielczarek, A.T.; Nguyen, H.T.T.; Nielsen, J.L.; Nielsen, P.H. Population dynamics of bacteria involved in enhanced biological phosphorus removal in Danish wastewater treatment plants. *Water Res.* **2013**, *47*, 1529–1544. [CrossRef]

Article

Reproducibility of Aerobic Granules in Treating Low-Strength and Low-C/N-Ratio Wastewater and Associated Microbial Community Structure

Hongxing Zhang [1], Yong-Qiang Liu [2], Shichao Mao [1], Christain E. W. Steinberg [3], Wenyan Duan [1] and Fangyuan Chen [1,*]

[1] Yunnan Key Lab of Soil Carbon Sequestration and Pollution Control, Faculty of Environmental Science and Engineering, Kunming University of Science and Technology, Kunming 650500, China; hong_xing2207080@126.com (H.Z.); mscsuperman123@163.com (S.M.); duanwenyan0405@gmail.com (W.D.)

[2] Faculty of Engineering and Physical Sciences, University of Southampton, Southampton SO17 1BJ, UK; y.liu@soton.ac.uk

[3] Institute of Biology, Faculty of Life Sciences, Humboldt Universität zu Berlin, 10117 Berlin, Germany; christian_ew_steinberg@web.de

* Correspondence: chenfy@kust.edu.cn

Abstract: Long-term stability of the aerobic granular sludge system is essentially based on the microbial community structure of the biomass. In this study, the physicochemical and microbial characteristics of sludge and wastewater treatment performance were investigated regarding formation, maturation, and long-term maintenance of granules in two parallel sequencing batch reactors (SBR), R1 and R2, under identical conditions. The aim was to explore the linkage between microbial community structure of the aerobic granules, their long-term stability, as well as the reproducibility of granulation and long-term stability. The two reactors were operated with a COD concentration of 400 mg/L and a chemical oxygen demand to nitrogen (COD/N) ratio of 4:1 under anoxic–oxic conditions. It was found that although SVI_{30}, sludge size, and distributions in R1 and R2 were different, aerobic granules were formed, and they maintained long-term stability in both reactors for 320 days, implying that a certain level of randomness of granulation does not affect the long-term stability and performance for COD and N removal. In addition, a significant reduction in the richness and diversity of microbial production was observed after the sludge was converted from inoculum or flocs to granules, but this did not negatively affect the performance of wastewater treatment. Among the predominant microbial species in aerobic granules, Zoogloea was identified as the most important bacteria present during the whole operation with the highest abundance, while Thauera was the important genus in the formation and maturation of the aerobic granules, but it cannot be maintained long-term due to the low food-to-microorganisms ratio (F/M) in the system. In addition, some species from Ohtaekwangia, Chryseobacterium, Taibaiella, and Tahibacter were found to proliferate strongly during long-term maintenance of aerobic granules. They may play an important role in the long-term stability of aerobic granules. These results demonstrate the reproducibility of granulation, the small influence of granulation on long-term stability, and the robustness of aerobic granulation for the removal of COD and N. Overall, our study contributes significantly to the understanding of microbial community structure for the long-term stability of aerobic granular sludge in the treatment of low-COD and low-COD/N-ratio wastewater in practice.

Keywords: aerobic granular sludge; microbial community structure; reproducibility; low-strength COD; low COD/N ratio

Citation: Zhang, H.; Liu, Y.-Q.; Mao, S.; Steinberg, C.E.W.; Duan, W.; Chen, F. Reproducibility of Aerobic Granules in Treating Low-Strength and Low-C/N-Ratio Wastewater and Associated Microbial Community Structure. *Processes* **2022**, *10*, 444. https://doi.org/10.3390/pr10030444

Academic Editor: Francesca Raganati

Received: 18 January 2022
Accepted: 18 February 2022
Published: 23 February 2022

Publisher's Note: MDPI stays neutral with regard to jurisdictional claims in published maps and institutional affiliations.

1. Introduction

Aerobic granules are dense and compact aggregates immobilized by microorganisms themselves without a carrier medium. Due to the high compactness and large size, the aerobic granules have high settling capacity, biomass retention, and tolerance to toxic and shock loads during treatment of various wastewaters. However, aerobic granular sludge is still far from widespread in practice due to a concern of low stability in long-term operation, during which the main reasons for the instability of aerobic granules could be core hydrolysis [1,2], overgrowth of filament bacteria [3,4], and reduction in the excretion of functional extracellular polymer substrates [5]. In recent years, many efforts have been made to address this problem, but it remains a challenge because the factors involved are interrelated, manipulation capabilities are limited, and trials for long-term studies are very time-consuming. The stability of the aerobic granules in practice still needs to be investigated thoroughly.

The stability of aerobic granular sludge is essentially determined by the structure of microbial populations residing in granules, such as predominant bacteria groups, the richness and diversity of microbial community, and microbial dynamics in response to operational and environmental changes in wastewater treatment. The structure of microbial populations in aerobic granules is greatly dependent on influent wastewater compositions, operation parameters, and environmental conditions. For example, Cydzik-Kwiatkowska et al. (2015) found that the bacterial structure of aerobic granules is significantly affected by the aeration mode and nitrogen loading in a nitrogen removal system [6]. Specifically, Swiatczak et al. (2018) proved that the variables, i.e., temperature, time of aerobic granular sludge adjustment at start-up, and reactor conditions, contributed 92.3% to the variability of the microbial community of a full-scale granular sludge system [7]. However, all the above adjustable factors cannot be used to directly and precisely regulate the microbial structure, because some uncontrollable factors due to theoretical and technical limitations, such as size distribution, biomass growth, and microenvironment in aerobic granules, also affect the microbial structure. The growth of microbial species inside the aerobic granules is controlled by the microenvironment, which is affected by the granule size, compactness, and other physical properties of the granules, resulting in different performances and stabilities of the aerobic granules during the long-term operation period, even under the same operating conditions. Currently, there is limited information on the effects of these uncontrollable conditions on the microbial community in aerobic granular sludge, which hinders a deep understanding and effective control of the aerobic granular sludge system for long-term stable operation. Cultivation of aerobic granules using feeds with high ammonia concentration and low carbon-to-nitrogen (C/N) ratio under anoxic–oxic conditions has been proven to be effective in increasing the stability of aerobic granules [8]. The high ammonia concentration favors the enrichment and accumulation of slow-growing nitrifying bacteria in the granules, while the anoxic–oxic conditions inhibit the overgrowth of ordinary heterotrophic organisms (OHO) to promote floc formation [9]. Liu et al. (2004) reported that the nitrifying granules at a C/N ratio of 100/30 had the smallest average size, i.e., about 0.3 mm, compared to the other granules cultured at higher C/N ratios [10]. Cha et al. (2021) reported that the mean size of aerobic granules in the system with a C/N of 4 was less than 0.5 mm during operation for over 300 days [8]. It has also been reported that the diffusivity of the substrate directly affects microbial competition for the substrate and thus granulation [11,12]. When the size of the aerobic granules is less than 0.7 mm, the substrate and oxygen distribute more evenly in the granules than in the large granules because the mass transfer resistance is lower due to the size [13]. Therefore, it could be assumed that the microbial population structure in smaller aerobic granules is mainly affected by the population of microorganisms and their mutual interactions, rather than by the physical properties of the granules such as size and compactness. This would greatly simplify the mechanism of microbial community formation and facilitate the understanding of microbial development in practical aerobic granule applications.

In this study, aerobic granular sludge treating synthetic wastewater with COD of 400 mg/L and ammonia nitrogen of 100 mg N/L under anoxic–oxic conditions was cultured to investigate the microbial structure of the aerobic granular sludge during the long-term operation period. During more than 300 days of operation, the wastewater treatment performance and the physical and microbial properties of the aerobic granules were monitored at selected stages. The specific objective of this study was to characterize the relationships between the performance, microbial structure, and stability of the aerobic granular sludge during start-up, maturation, and continued long-term maintenance of the aerobic granular sludge operation. The results of the study provide insight into the key factors affecting the long-term stability of an aerobic granular system and clues for better control of the system in practical applications.

2. Materials and Methods

2.1. Experimental Setup and Operation

The experiment was carried out in two parallel bubble columns (R1 and R2) with an internal diameter of 5 cm and a working volume of 2 L. The two reactors were operated sequentially with a cycle time of 4 h, including 5 min of anaerobic feeding, 55 min of static condition without mixing, 145 to 170 min of aeration, 30 to 5 min of settling, and 5 min of effluent discharging. The feed was pumped into the reactors from the top port, while the effluents were discharged from the middle port of the reactor with a volumetric exchange ratio of 50%. During the aeration phases, fine air bubbles were pumped into the reactors through an air sparger at the bottom of the reactors with a flowrate of 2 L/min. The inoculated activated sludge was collected from a local domestic wastewater treatment plant with a sludge volume index (SVI_{30}) of 106.2 mL/g. The experiment took place at a temperature ranging from 18 to 25 °C.

2.2. Medium

The influent was synthetic wastewater with compositions of $NaCH_3COO$, $(NH_4)_2SO_4$, KH_2PO_4, $NaHCO_3$, and micronutrients. $NaCH_3COO$ and $(NH_4)_2SO_4$ provided carbon and nitrogen sources, respectively, while $NaHCO_3$ provided inorganic carbon and pH buffer for nitrification. The experiment was conducted in two stages: the first stage was from day 1 to 32 with influent concentrations of COD and NH_4^+-N of 400 mg/L and 50 mg N/L, respectively; while the second stage lasted from day 32 until the end of the operation, i.e., day 320, with influent concentrations of COD and NH_4^+-N of 400 mg/L and 100 mg N/L, respectively. With these compositions of synthetic wastewater, OLR was constant at 1.21 kg $COD/m^{-3}d^{-1}$ during the two stages, while the ammonia loading rate increased from 0.15 to 0.3 kg $N/m^{-3}d^{-1}$ from the first to the second stage. pH was controlled in the range of 7.5–8.5 during the whole operation period. Micronutrients were supplied: $CaCl_2 \cdot 2H_2O$ 25 mg/L, $MgSO_4 \cdot 7H_2O$ 20 mg/L, $FeSO_4 \cdot 7H_2O$ 10 mg/L, EDTA-2Na 10 mg/L, $MnCl_2 \cdot 4H_2O$ 0.12 mg/L, $ZnSO_4 \cdot 7H_2O$ 0.12 mg/L, $CuSO_4 \cdot 5H_2O$ 0.03 mg/L, $(NH_4)_6Mo_7O_{24} \cdot 4H_2O$ 0.05 mg/L, $NiCl_2 \cdot 6H_2O$ 0.1 mg/L, $CoCl_2 \cdot 6H_2O$ 0.1 mg/L, $AlCl_3 \cdot 6H_2O$ 0.05 mg/L, and H_3BO_3 0.05 mg/L.

2.3. Analytical Method

COD, NH_4^+-N, NO_3^--N, NO_2^--N, SVI, mixed liquor suspended solids (MLSS) and mixed liquor volatile suspended solids (MLVSS) were analyzed according to standard methods (APHA) [14]. Average particle size was determined by a laser particle size analysis system with a measuring range from 0 to 2000 μm (Malvern MasterSizer Series 2600, Malvern Instruments Ltd., Malvern, UK). Morphometry of the aerobic granules was observed by an optical microscope equipped with a digital camera (Leica Microsystems Wetzlar GmbH.DM100.DEU, Wetzlar, Germany). Surface profile was evaluated by means of a scanning electron microscope (Hitachi SU-8010, Tokyo, Japan).

2.4. Microbiological Analysis of Granular Sludge

For biomolecular analyses of the microorganisms, sludge from the two reactors was collected at different days of operation and stored at −80 °C. Sample DNA was extracted using a PowerSoil® DNA Isolation kit. The purity and concentration of the isolated

DNA was measured by a Qubit2.0 DNA kit (Life Technologies, Carlsbad, CA, USA). The 515F/805R primer set (GTGCCAGCMGCCGCGGTAA/GACTACHVGGGTATCTAATCC) was used to amplify the V4 regions of the bacterial 16SrDNA gene. Amplification began with an initial denaturation at 95 °C for 3 min, followed by 5 cycles of denaturation at 94 °C for 20 s, annealing at 55 °C for 20 s and extension at 72 °C for 30 s. It ended with a final extension step at 72 °C for 5 min. The PCR products were then subjected to 1.8% agarose gel electrophoresis for detection. Samples with a bright main band of approximately 450 bp were chosen and mixed in equidensity ratios. The mixture of PCR products was purified using a VAHTSTM DNA Clean Beads (Vazyme Biotech Co., Ltd., Nanjing, China). After quantification, the amplicons were sequenced using the MiSeq platform (Illumina, LA, USA). The sequences were validated by software cutadapt (version 1.2.1), pear (version 0.9.6) and Prinseq (version 0.20.4), among which low-quality sequences were treated by Prinseq. Chimeras were detected and removed by software Usearch (version 5.2.236) and identified by Uchime (version 4.2.40). An identity of 97% was adopted to cluster the normalized sequences into operational taxonomic units (OTUs), which were annotated based on the Silva taxonomic database. The diversity was assessed within a community (α-diversity) and was calculated and displayed with Mothur (Version 1.30.1). The Chao-1, Shannon, and Simpson indices were analyzed according to Hill et al. [15], and the abundance-based coverage estimator (ACE) was analyzed according to Westphal et al. [16].

3. Results and Discussion

3.1. Characteristics of Aerobic Granular Sludge during Long-Term Operation

Figure 1 shows the morphology of sludge in R1 and R2, respectively, during the operation time. It can be seen that tiny aggregates appeared on day 30 in the two reactors and mixed with flocs. On day 79, the aerobic granules formed in R1 with clear-cut outlines and fewer flocs, while in R2 there were only small granules with a large number of flocs. On day 106, aerobic granules of similar size were present in both reactors and gradually developed into large and dense granules with small flocs by day 300. Figure 2 shows the surface morphometry of the aerobic granules in R1 and R2 examined by SEM. It can be seen that the surface of the aerobic granules with cocci, bacilli, and filaments are quite similar.

The size of aerobic granules throughout the whole operation is shown in Figure 3, which reveals a different development of granules in R1 and R2. The mean size of the aerobic granules in R1 increased with a large fluctuation, while that in R2 increased stably, until the end of the operation. In R1, the mean size of aerobic granules increased to 380 μm on day 136, while it decreased to 276 μm on day 240, which was combined with variations in the settling ability of aerobic granules. On day 294, the size appeared to recover and increased to 462 μm. In R2, the mean size of aerobic granules gradually increased to 662 μm on day 294. Meanwhile, the size distribution of aerobic granules was found to be highly variable, with a narrower size distribution in R1 than in R2. These discrepancies in the size development and morphometry of the sludge in R1 and R2 over time indicate that the formation and growth of aerobic granules are different even under seemingly identical operating conditions, suggesting some degree of randomness unrelated to the apparent operating conditions related to granule formation and development. In addition, it should be noted that the mean size of the aerobic granules in the two reactors was less than 700 μm, implying that the overall mass transfer resistance in the granules may be negligible.

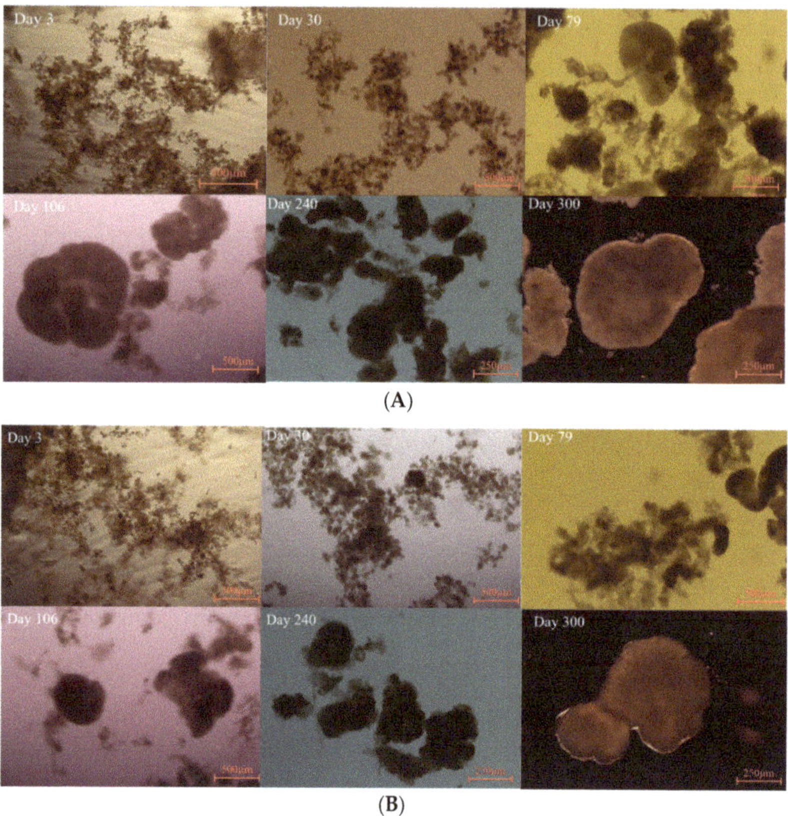

Figure 1. Morphology of the sludge on different days of the operation. (**A**) in R1; (**B**) in R2.

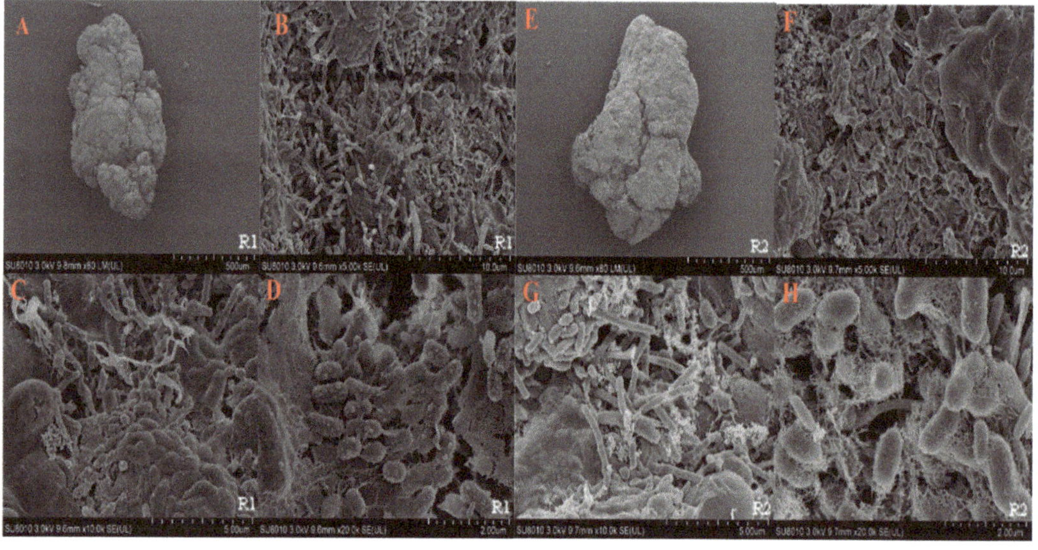

Figure 2. SEM of the aerobic granules in R1 and R2. (**A–D**), in R1; (**E–H**), in R2.

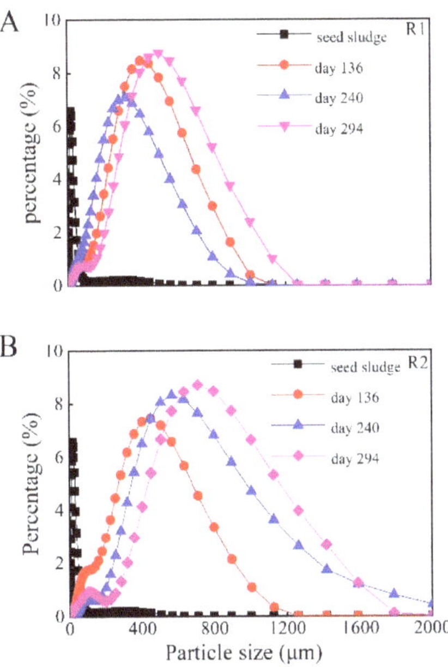

Figure 3. Size distribution and development of aerobic granules in R1 and R2 over the operating time. (**A**) R1; (**B**) R2.

SVI is normally used as a direct indicator of both sludge settling ability and aerobic granule stability. An SVI_{30} below 80 mL/g is usually considered indicative of increased sludge stability with good settling ability. At the same time, the ratio of SVI_{30} to SVI_5 close to 1 is often used as an indicator of aerobic granule dominance. Figure 4A,B shows the development of SVI in R1 and R2 during the operating period. It can be seen that the sludge in both reactors was subject to large variations in settling ability and stability during the formation period, i.e., from the inoculation of the activated sludge to the formation of the aerobic granules. In R1, SVI_{30} decreased sharply in the first 20 days and then increased abruptly to 148.6 mL/g within the next 20 days due to sludge bulking, indicating low settling ability and stability of the sludge. Thereafter, SVI_{30} began to decrease and reached 50 mL/g on day 65, which was considered to be the time when a dominant aerobic granule formed in R1, as SVI_{30}/SVI_5 was close to 1 at this time. Similarly, SVI_{30} in R2 first decreased and then increased due to sludge bulking until day 90, showing a twofold increase and decrease. Thereafter, aerobic granules began to form in R2 with SVI_{30}/SVI_5 close to 1.

After the formation of the aerobic granules in each reactor, the SVI_{30} remained in the range of 60–80 mL/g until day 140, which can be considered as the maturity stage of the aerobic granules in the two reactors. From day 140 to the end of the operation period on day 320, the SVI_{30} of the aerobic granules in both R1 and R2 remained around 40 mL/g, and no floc competition was observed, as reported by Liu et al. (2012) on the real wastewater treatment during the long-term operation period [17], demonstrating the high stability of the aerobic granules.

Again, the difference in the change of SVI_{30} in two identical reactors indicates some randomness of granule formation especially during the granule formation phase. In contrast to the size development of aerobic granules, it is noted that the settling ability and stability of aerobic granules in R1 and R2 were quite similar during maturation and long-term maintenance, regardless of the differences in the mean size and distribution of aerobic granules in the two reactors. This indicates that aerobic granules with different granule formation with respect to SVI_{30} can still achieve similar settling ability and stability after maturation.

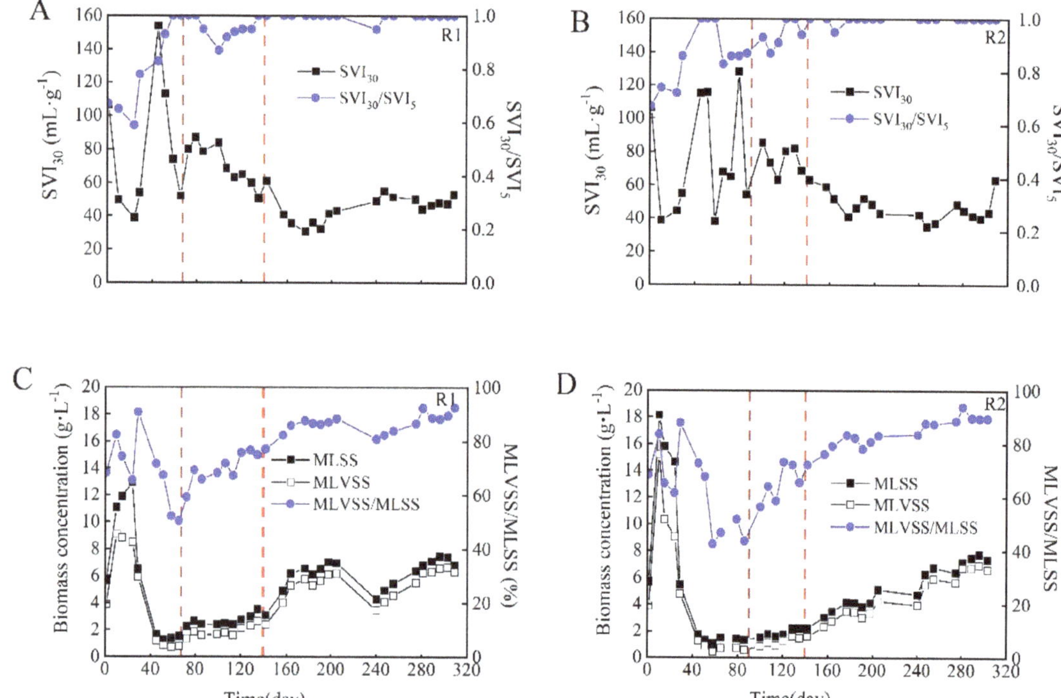

Figure 4. Profiles of SVI and MLSS during the operation period. (**A,C**), in R1; (**B,D**), in R2.

The biomass concentration, indicated by MLVSS/MLSS over the whole operation time, is shown in Figure 4C,D, which depicts the growth and accumulation of aerobic granules as a whole. MLSS in the two reactors initially decreased, stagnated at the lowest concentration after the granules had formed for a period of time, and increased again by the end of the operation. In R1, the MLSS gradually decreased from 5.6 g/L of inoculum to 0.7 g/L on day 58, which was caused by the washout of the flocs due to the increased SVI and the shortened settling time. After this low MLSS was maintained for about 7 days, aerobic granules began to form and MLSS began to increase due to retention and growth of high-settling biomass. In R2, the only difference in MLSS development from R1 was that the low biomass period (0.69 g/L) lasted longer, approximately 40 days. This was due to the longer sludge bulking and SVI_{30} fluctuating twice in R2, resulting in less biomass retention than in R1. After aerobic granules formed in R2 on day 90, MLSS began to increase. After the aerobic granules were formed in both reactors, the MLSS gradually increased to 6.8 g/L in R1 and 5.1 g/L in R2 at the end of the operation.

At the same time, it is observed that MLSS shows an inverse trend with SVI, i.e., lowest MLSS in the formation phase, slightly increased MLSS in the maturation phase, and further increased MLSS in the long-term maintenance of aerobic granules. The MLVSS can be used to detect the amount of microbial organisms in the biomass, which represents the organic composition of the MLSS. During the development of the aerobic granules, the MLVSS/MLSS changed according to the MLSS: it decreased from 80% to 60% and 50% in R1 and R2, respectively, after the formation of the aerobic granules, and then gradually increased to 80% in both reactors during the maturation of the aerobic granules. Subsequently, the MLVSS/MLSS remained in the range of 80–90% until the end of the operation, indicating high microbial richness and activity in the long-term maintenance of the aerobic granules in R1 and R2.

3.2. Performance of the Aerobic Granular Sludge during the Long-Term Operation Period

Figure 5 shows the performance of the aerobic granules in R1 and R2, respectively. It can be seen that the COD removal efficiency in both reactors was above 90% throughout the operating period, although sludge bulking and extremely low biomass concentrations occurred in the first operating phase. This indicates that the heterotrophic microorganisms responsible for organic matter removal were very active in the aerobic granules to handle the added COD. In comparison, the efficiency of ammonium removal fluctuated when the aerobic granules were formed, which can be attributed to the low biomass concentration at the beginning due to the leaching of the nitrifying bacteria.

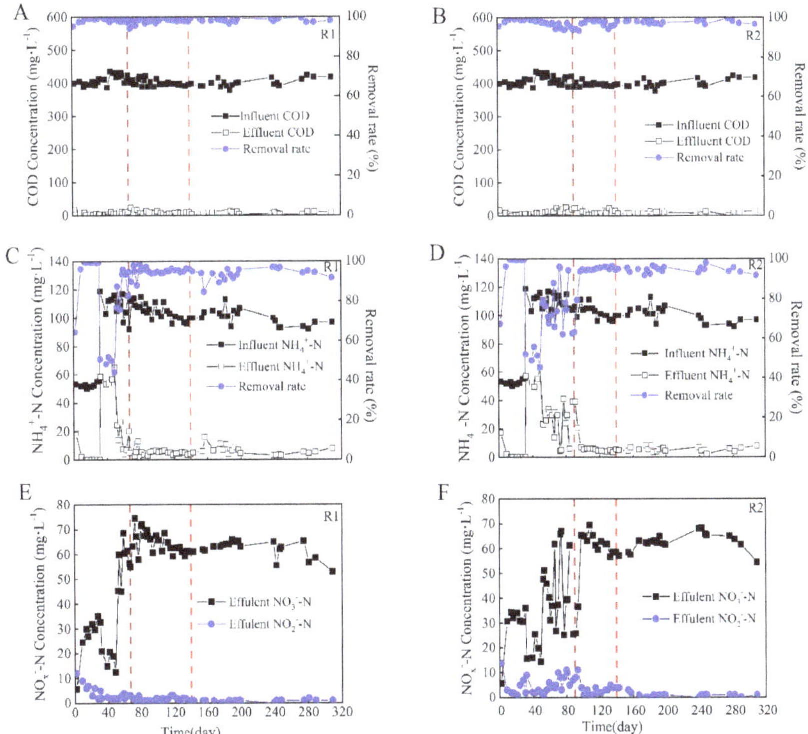

Figure 5. Performance of the aerobic granules over the operation period in R1 and R2. (**A,B**), COD removal in R1 and R2, respectively; (**C,D**), ammonium removal in R1 and R2, respectively; (**E,F**), NO$_X$ formation in R1 and R2, respectively.

After the formation of aerobic granules in the two reactors, the ammonia removal efficiency increased to over 90% as the nitrifying bacteria in the granules re-accumulated with increasing biomass. Throughout the operating period, ammonium was mostly converted to nitrate in R1 and R2, except during the period of aerobic granule formation with fluctuations in ammonium removal. The similar performances of the aerobic granules in R1 and R2 indicate that the different average size and size distribution of the aerobic granules does not have a large effect on the ability to biodegrade when the granule size is smaller and there is no obvious mass transfer resistance.

3.3. The Richness and Diversity of Microbial Community in the Aerobic Granular Sludge

The richness and diversity of the microbial population throughout the operating period are presented in Figure 6. The number of operational taxonomic units (OTUs), the abundance-based coverage estimator (ACE) index, and the Chao1 index are used to indicate the richness of the microbial population.

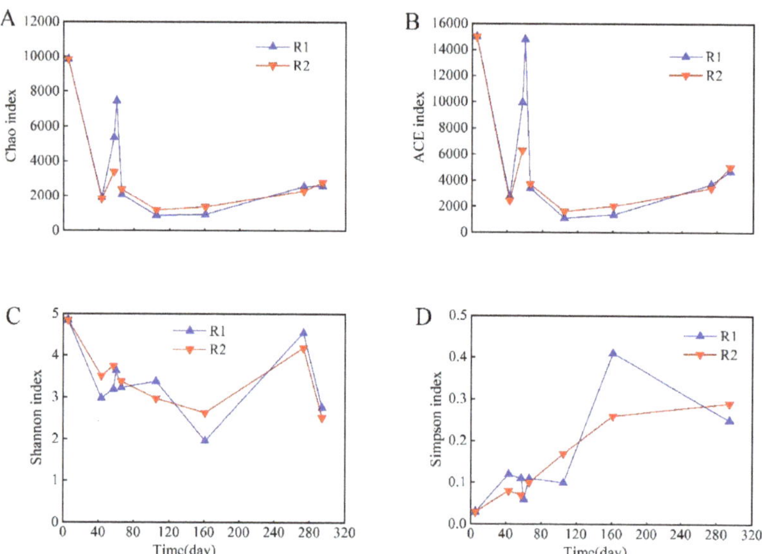

Figure 6. Richness and diversity indices of microbial populations in sewage sludge over the entire operating period in R1 and R2. (**A**) Chao1; (**B**) ACE; (**C**) Shannon; (**D**) Simpson.

It can be seen that there was no obvious difference in terms of the richness of the microbial population in both reactors during the different phases of operation. In addition, it was found that the richness of the microbial population decreased sharply after the formation of the aerobic granules, although there was a small fluctuation before the aerobic granules became dominant. The richness of the microbial population remained at its lowest level during the maturation period of the aerobic granules. After the maturation period, microbial richness increased slightly, but was still far below that of the original microbial population in the inoculum.

The Shannon and Simpson indices, which are positively and negatively related to diversity, respectively, are standard indicators of microbial diversity [15]. The Shannon index decreased during aerobic granule formation and maturation and recovered somewhat after maturation, while the Simpson index showed the opposite trend. All the indices shown in Figure 4 indicate that the richness and diversity of the microbial population decreased sharply during the formation of aerobic granules due to the selection of granule-favoring microbial populations and the leaching of flocs-favoring microbial populations, while it increased slightly during long-term maintenance due to the embedding and growth of more species in the granules. The coverage rate of the detected genera was 0.99 for both reactors, indicating the high credibility of the statistical data obtained from the microbial analysis of the aerobic granule samples.

3.4. Microbial Population Dynamics and the Predominant Functional Groups

Tracking the microbial composition of aerobic granules at various stages of growth and development can provide important information for identifying key microbial organisms and their associated specialized functions in aerobic granules, which could be used as meaningful indicators for monitoring and controlling the operating process. Tables 1 and 2 show the main composition of the microbial populations in R1 and R2, respectively, during the entire operating period. A significant shift in microbial structure from the inoculum to the stable granule was observed, which was consistent with the report of aerobic granule microbial population dynamics during the start-up and steady state [18]. In addition, it is clear that the dominant genera in the two reactors are quite similar on the different days of operation, although the abundance of each genus is different. This indicates that the microbial structure in aerobic granules with small average size is determined by the operating conditions, which can be duplicated by setting the same operating parameters. Of all the genera of aerobic granules identified in Tables 1 and 2, most microbial organisms have at least one of the functions of EPS excretion, nitrification, denitrification, and organic matter degradation. EPS excretion is one of the most important microbial properties in granules, which can help microbial organisms aggregate and maintain a stable structure during long-term operation. Denitrifying bacteria or slow-growing autotrophic nitrifying bacteria promote a dense structure of aerobic granules by minimizing overgrowth of more general heterotrophic organisms that hinder aerobic granule formation and performance by promoting loosely structured flocs rather than compact granules [11]. A comparison with the predominant genera in the inoculum also shows that most of the bacteria in the aerobic granules originated from the genera in the inoculum during the formation and growth periods, and only 5 of the total 30 genera in the aerobic granules were new. This could show that most of the microbial genera in the aerobic granules are not exclusive and can be selected and enriched from the inoculated activated sludge.

Table 1. Predominant microbial genera in R1.

Genus	Operation Day						
	1	43	74	105	161	273	294
Alkanindiges	6.88%						
Dechloromonas	4.57%	26.43%	1.95%	0.45%	0.15%	0.05%	0.11%
Acidovorax	1.52%			0.31%	0.66%	0.86%	1.53%
Stenotrophomonas	2.72%	0.03%					
Propionivibrio	1.94%						0.04%
Nitrosomonas	1.63%		4.25%	3.09%	0.15%		0.23%
Flavobacterium	1.61%	0.42%	0.09%	0.1%	0.06%	0.22%	0.05%
Thauera	0.13%	21.6%	5.69%	16.06%	5.20%	0.04%	0.12%
Zoogloea	0.38%	8.15%	18.32%	26.00%	64.26%	11.9%	50.13%
Sideroxydans		3.99%		0.02%		0.14%	
Bdellovibrio	0.11%	1.69%	0.96%	0.26%	0.08%	0.10%	
Ferruginibacter	0.26%	1.07%	0.66%	0.39%	0.11%	1.58%	0.95%
Simplicispira	0.05%	0.64%	0.05%			0.01%	0.03%
Phreatobacter		0.12%	0.14%	0.06%	0.01%		
Prosthecobacter			3.43%	2.57%	1.16%	2.76%	1.1%
Aquimonas		0.83%	1.24%	0.37%	0.05%		
Sphingopyxis			1.8%	1.02%		0.04%	0.01%
Chryseolinea	0.05%	0.33%	2.64%	2.26%	0.18%	0.83%	0.42%
Terrimonas	0.45%	0.57%	0.75%	0.3%	0.32%	1.03%	0.48%
Lacibacter	0.05%			0.06%	0.04%	0.01%	
Azoarcus		0.06%	0.74%	2.08%	0.03%	0.72%	0.09%
Phaeodactylibacter	0.04%			4.83%		0.12%	
Ohtaekwangia	0.02%		0.11%	0.55%	1.09%	8.67%	2.61%
Aggregicoccus	0.11%	0.01%	0.02%	0.04%	0.13%	0.04%	
Chryseobacterium	0.04%		0.26%	0.01%	0.36%	2.65%	5.63%
Nitrospira	0.47%		0.01%	0.01%	0.41%	4.06%	0.76%
Taibaiella	0.98%				0.36%	2.43%	10.68%
Tahibacter	0.01%	0.01%	0.13%	0.14%	0.42%	4.21%	2.33%
Sediminibacterium	0.01%	0.13%	0.49%	0.27%	0.01%	1.32%	0.49%
Luteimonas	0.06%	0.04%				0.1%	0.38%

Table 2. Predominant microbial genera in R2.

Genus	Operation Day						
	1	43	74	105	161	273	294
Alkanindiges	6.88%						
Dechloromonas	4.57%	26.01%	2.2%	1.11%	0.43%	0.02%	0.10%
Acidovorax	1.52%			0.64%	1.13%	0.76%	1.22%
Stenotrophomonas	2.72%	0.02%					
Propionivibrio	1.94%						0.03%
Nitrosomonas	1.63%		5.45%	5.84%	0.66%		0.24%
Flavobacterium	1.61%	0.22%	0.24%	0.45%	0.29%	0.76%	0.05%
Thauera	0.13%	3.9%	4.96%	6.59%	2.19%	0.01%	0.07%
Zoogloea	0.38%	8.53%	26.11%	38.93%	60.09%	21.13%	54.16%
Sideroxydans		3.81%		0.01%		0.06%	
Bdellovibrio	0.11%	2.09%	0.26%	1.10%	0.06%	0.13%	0.01%
Ferruginibacter	0.26%	1.19%	0.38%	0.08%	0.04%	1.26%	0.86%
Simplicispira	0.05%	2.11%	0.08%			0.01%	0.01%
Phreatobacter		1.53%	0.16%	0.16%	0.01%	0.02%	
Prosthecobacter					1.89%	2.4%	1.01%
Aquimonas		0.07%	1.5%	0.81%	0.51%		
Sphingopyxis			0.44%	0.10%	0.07%	0.07%	
Chryseolinea	0.05%	1.17%	3.97%	2.30%	1.04%	0.14%	0.03%
Terrimonas	0.45%	0.81%	1.43%	0.97%	0.1%	2.34%	0.43%
Lacibacter	0.05%			3.82%	3.64%	0.05%	
Azoarcus		0.44%	0.31%	1.92%	0.06%	0.29%	0.1%
Phaeodactylibacter	0.04%	0.01%		0.14%		0.06%	
Ohtaekwangia	0.02%			0.02%	0.28%	2.96%	2.71%
Aggregicoccus	0.11%	0.17%	0.01%		1.04%		
Chryseobacterium	0.04%		0.19%	0.29%	0.02%	5.26%	6.02%
Nitrospira	0.47%	0.07%	0.01%		0.01%	2.08%	0.55%
Taibaiella	0.98%				0.09%	2.45%	10.58%
Tahibacter	0.01%	0.03%	0.36%	0.03%	0.1%	0.01%	1.79%
Sediminibacterium	0.01%	0.07%	0.56%	0.35%	0.02%	0.07%	0.34%
Luteimonas	0.06%	0.06%				2.54%	0.38%

The microbial genera from the inoculum that persisted and thrived in the aerobic granules can be characterized by their good aggregation ability combined with a short settling time. Among these species, *Zoogloea*, *Thauera*, and *Dechloromonas* were predominant and had a high percentage in the different stages of the sludge in both reactors (R1, R2). *Zoogloea* is much more abundant in the aerobic granules than *Thauera* and *Dechloromonas*, which increased significantly from 0.38% in the inoculum to over 50% in the granules at the end of the operation in both reactors, although there were fluctuations in the long-term maintenance of the aerobic granules. *Zoogloea* is a filamentous denitrifier that improves sludge adhesion and hydrophobicity by secreting EPS, thereby promoting microbial aggregation and granule stability [19]. Consequently, it becomes obvious that *Zoogloea* in aerobic granules play an important role in the formation and long-term stability of aerobic granules. Thus, *Zoogloea* can be considered as one of the major indicators of the stability of long-term operation of aerobic granule system for wastewater treatment with low strength and C/N ratio. *Thauera* is also a typical denitrifier with the ability to excrete EPS in an aerobic granule system [20]. The abundance of *Thauera* gradually increased from 0.13% in the inoculum to maxima of 21.6% in R1 and 6.59% in R2, respectively, during the formation and maturation phases of the aerobic granule, but declined sharply to a level similar to that of the inoculum during long-term maintenance of the aerobic granules. Therefore, it can be hypothesized that *Thauera* is a very important functional genus for the formation and maturation of aerobic granules, but an unimportant functional genus for maintaining the long-term operation of the aerobic granular system. From the comparison of the abundance of *Thauera* in the two reactors, it can be seen that the abundance of *Thauera* in the granules in R2 with a much looser structure and a longer formation time was much lower than in R1. This confirms that *Thauera* plays a key role in the formation and maturation of the aerobic granules, the abundance of which determines the compactness of the aerobic granules and the rate of formation of the aerobic granules. In addition, *Dechloromonas* was found to be transiently abundant at 26.43% in R1 and 26.01% in R2 on day 43, when sludge bulking occurred in both reactors. After the aerobic granules were formed, the abundance decreased sharply and remained at a much lower level than in the inoculum (4.57%) during the remaining days of operation. *Dechloromonas* is a type of denitrifier with the ability to reduce nitrate to nitrite. The huge accumulation of *Dechloromonas* in the sludge bulking and the decrease in the established aerobic granule system indicate that *Dechloromonas* is closely related to the sludge formation and plays an unimportant role in the formation and maintenance of aerobic granules.

It should be noted that some genera, including *Ohtaekwangia*, *Chryseobacterium*, *Taibaiella*, and *Tahibacter*, had very low abundance during most of the operating time, but had relatively higher abundance during longer operation of the mature aerobic granules. The abundance of these bacteria ranged from 2 to 10% in both reactors, compared to the low abundance of 0.01 to 0.5% during most of the operating time. *Ohtaekwangia* has been described as a type of structure-forming bacteria in aerobic granules that excrete EPS and whose abundance increases with the growth of aerobic granules [7]. As reported, *Chryseobacterium* can live in a nutrient-poor environment [21]. Although a low organic concentration and a high ammonium concentration were used in this study, the low F/M in the long-term maintenance phase of the aerobic granules actually resulted in a nutrient-poor environment. This could explain the existence of *Chryseobacterium* in the aerobic granules. *Tahibacter* and *Taibaiella* are both aerobic, and it has been reported that they can reduce nitrate to nitrite [22] and nitrate to gaseous nitrogen [23], respectively. Their increased abundance may indicate that they play an important role in denitrification in the long-term maintenance phase of aerobic granules. Thus, from the above, it is clear that these genera perform different functions in aerobic granules, and their structure supports their proliferation and accumulation in the granules during the long-term maintenance phase.

In addition, there were also some microbial genera such as *Acidovorax*, *Bdellovibrio*, *Ferruginibacter*, *Terrimonas*, and *Chryseolinea* which showed very low abundance throughout the entire operating period. Their abundances were less than 2%, indicating that they may play an unimportant role in the aerobic granules. Upon closer inspection, it turns out that these bacteria are all heterotrophic, but have several adaptive characteristics that allow them to live in aerobic granular sludge. Specifically, *Acidovorax* is a type of facultative denitrifier that occurs in low-COD conditions [24,25] and is able to self-aggregate and produce EPS. *Bdellovibrio* is aerobically heterotrophic and occurs in well-treated wastewater where it is able to prey on Gram-negative bacteria and utilize COD such as polymers, and ammonium [26]. *Ferruginibacter* and *Terrimonas* can degrade organic matter and excrete EPS for better sludge aggregation [27,28]. *Chryseolinea* can degrade small organic molecules as well as large polymers such as proteins and polysaccharides [29]. Of all the genera identified in Tables 1 and 2, *Nitrosomonas* and *Nitrospira* are the ammonia-oxidizing bacteria in R1 and R2 that originated from the inoculum and remained in the aerobic granules throughout the operating period. The abundance of the two types of bacteria showed opposite development trends during operation. *Nitrosomonas* was more abundant in the newly formed aerobic granules, with a maximum of 4.25% in R1 and 5.84% in R2, respectively, while its abundance was very low in the long-term maintenance phase. Conversely, the abundance of *Nitrospira* was higher in the long-term maintenance phase, with a maximum of 4.06% in R1 and 2.08% in R2, respectively, but a very low abundance in the granule formation phase. This is due to the different ammonium affinity of the two bacteria, which determines their proliferation and accumulation in the stages with different available ammonium concentrations. *Nitrosomonas* tends to grow at high ammonium concentrations but is inhibited by low ammonium concentrations, while *Nitrospira* is adaptive only to conditions of low ammonium concentration. In both R1 and R2, although the ammonium concentrations in the influent were constant throughout the operating period, the F/M of ammonium changed greatly due to the development of biomass, with MLSS ranging from less than 1 g/L to about 6 g/L. During the formation phase of the aerobic granules, the F/M of ammonium was comparatively high with extremely low MLSS, while after the maturation phase, the F/M of ammonium was comparatively low due to the highly accumulated biomass in the reactors. This explains well why *Nitrosomonas* predominated during the period of aerobic granule formation when the ammonium concentration was relatively high for a long time, while *Nitrospira* predominated after the maturation of aerobic granules when the ammonium concentration in the microenvironment was lower.

In addition to the above genera that originated from the inoculum, there were some bacteria that were not detectable in the inoculum but that formed during granule formation. These were *Sideroxydans*, *Phreatobacter*, *Prosthecobacter*, *Aquimonas*, *Sphingopyxis*, and *Azoarcus*. The abundance of these bacteria was less than 3.5% in R1 and R2, but they were present in different stages of aerobic granule formation, showing the possible role of each bacterium in the corresponding stages. In particular, *Phreatobacter* accounted for 3.43% and 2.76% in aerobic granules at days 76 and 273 in R1, respectively, while it accounted for 2.4% at day 273 but did not appear in R2 until after 161. *Phreatobacter* is a strictly aerobic, oligotrophic denitrifier capable of excreting extracellular protein for sludge aggregation [30,31]. Its presence in the maturation and long-term maintenance phases of aerobic granules in R1 and R2 may indicate that it is particularly adapted to the long-term maintenance environment. Moreover, *Azoarcus* occurred in all stages of aerobic granules since its first appearance on day 43 in both R1 and R2. *Azoarcus* is a denitrifier similar to *Thauera* [32,33], but may play only a minor role in aerobic granules at such low abundance.

3.5. Food to Microbial Biomass Ratio (F/M) and its Relationship with the Predominant Genus in the Aerobic Granules

Food-to-microbial biomass ratio (F/M) is the biomass loading rate, which refers to the substrate loading per unit biomass in a unit time for wastewater treatment. In the aerobic granular sludge system, F/M affects the average available substrate in the microenvironment, which could influence the microbial growth of the aerobic granules. From Figure 7, F/M is high during aerobic granule formation, while it decreases sharply during maturation and remains at a very low level in the long-term maintenance phase. Specifically, F/M increased to 1.83 g COD/g VSS-d at the maximum in formation and remained in a narrow range of 0.2–0.4 g COD/g VSS-d in the long-term maintenance of aerobic granules in R1; whereas F/M increased to 2.90 g COD/g VSS-d at the maximum in formation and remained in the same range as in R1 in the long-term maintenance of aerobic granules in R2. The changing trends of F/M are consistent with the previous assumption that high F/M stimulates the formation while low F/M maintains the long-term stability of aerobic granules [8]. In addition, it is noted that the much higher F/M in R2 causes a much longer formation time than in R1, which may indicate that there may be a suitable range of F/M for a better balance between formation stimulation and formation rate of aerobic granules.

F/M was the main variable in the operation as other adjustable operating parameters were set unchanged in this study. The change of F/M in the operation period may lead to changes in the microenvironment of the bioprocess and thus to the response of the predominant bacteria. In order to clarify the role of the predominate bacteria in the bioprocess, the relationship between F/M and the predominant bacteria during the whole operation period was analyzed as shown in Figure 8. From Figure 8A,B, it can be seen that denitrifier *Zoogloea* is one of the most predominant genera throughout the whole F/M range while it is an absolutely dominant genus when the F/M was below 0.6 in both R1 and R2. Denitrifier *Thauera* and *Dechloromonas* are adaptive to the high F/M range but cannot survive in the low F/M range, especially for *Dechloromonas*, which presented as a transit. Based on the characteristics of these genera, it can be assumed that *Zoogloea* has a fundamental role in the aerobic granules, whereas *Thauera* plays important roles at high F/M values due to their great ability for EPS excretion. *Dechloromonas* may play unimportant roles in aerobic granules due to inability for EPS excretion, though they are active at high F/M values. Conversely, *Ohtaekwangia, Chryseobacterium*, and *Taibaiella* are only adaptive in the low-F/M ranges, which might be connected with their ability for EPS excretion, adaption to nutrient-deficient environment, and aerobic denitrification, respectively. Besides, ammonia-oxidation and nitrification are also very important metabolism *processes* in the aerobic granules, with *Nitrosomonas* in dominance in the high F/M ranges while *Nitrospira* in dominance in the low F/M ranges. In addition, from Figure 8C,D, it can be seen that some genera with very low abundance could contribute to the diversity of the microbial community in the aerobic granules, but their microbial functions are unclear.

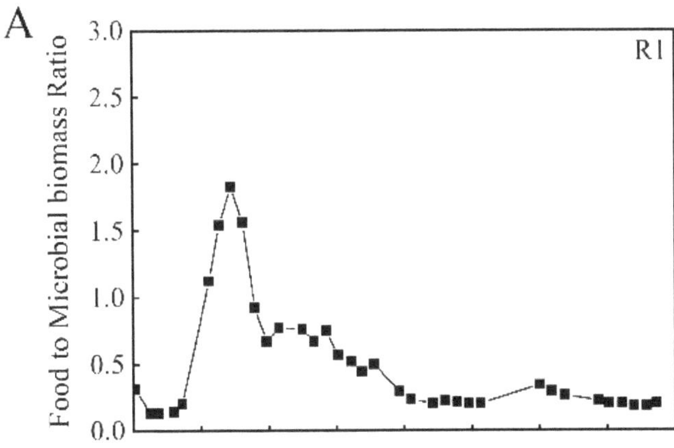

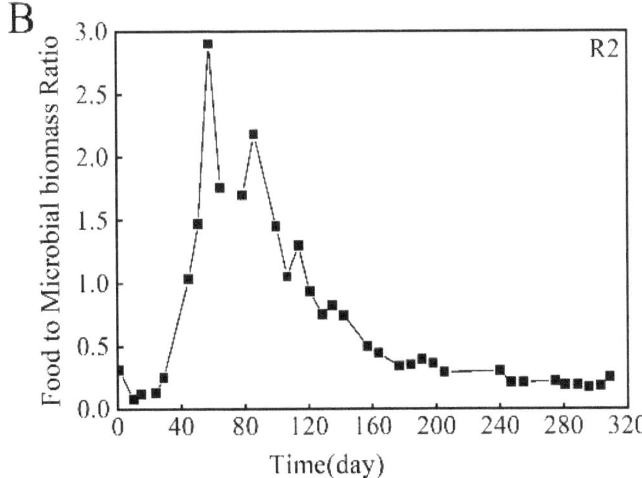

Figure 7. Food-to-Microbial Biomass Ratio in the whole operation period. (**A**) R1; (**B**) R2.

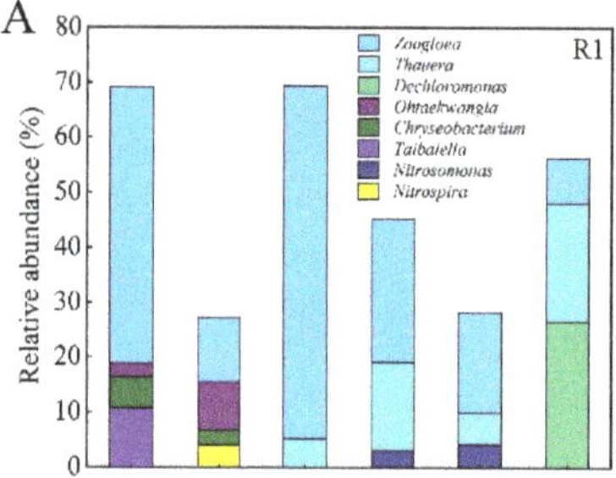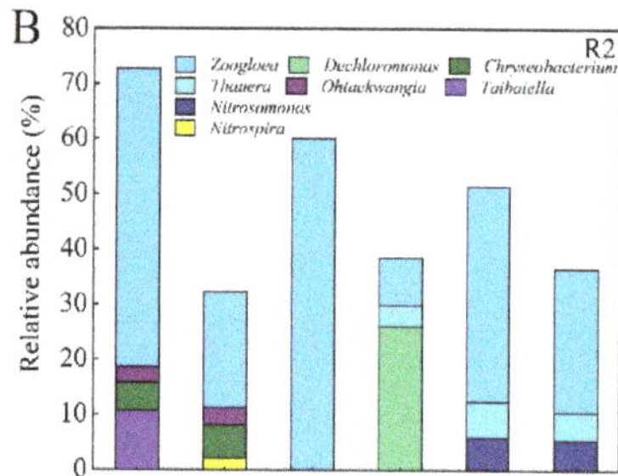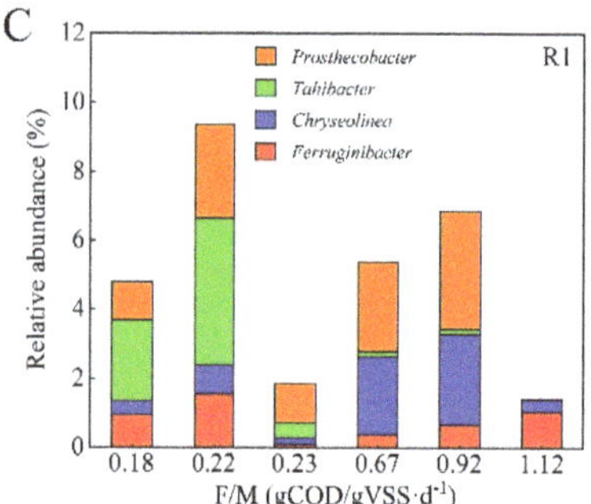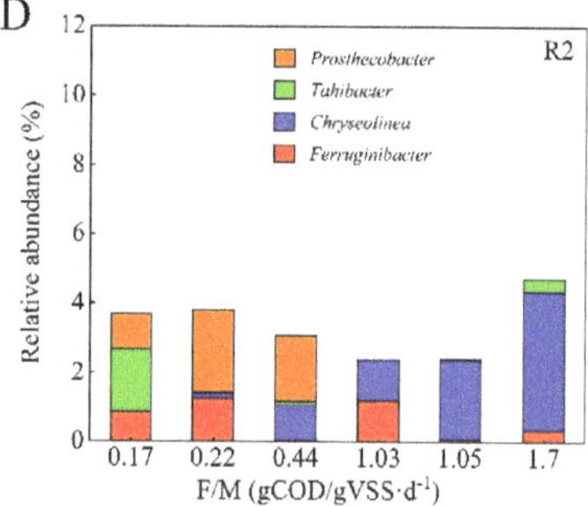

Figure 8. The relationship between F/M and the relative abundance of the predominant genus in the aerobic granules in R1 and R2. (**A**,**B**), genera with maximum relative abundance higher than 5% in R1 and R2, respectively; (**C**,**D**), genera with maximum relative abundance lower than 5% in R1 and R2, respectively.

4. Conclusions

In this study, two parallel aerobic granular sludge SBR reactors, R1 and R2, were operated for the treatment of synthetic wastewater with a COD concentration of 400 mg/L and a chemical-oxygen-demand-to-nitrogen (COD/N) ratio of 4:1 under anoxic–oxic conditions. All other operating conditions were also identical. The main conclusions are as follows.

- The SVI_{30} before and during granulation was different in the two reactors, although it reached similar values after granule maturation. In addition, the mean size and size distribution of the sludge were quite different in the two reactors, although both reactors were operated under exactly the same conditions. These differences indicate some degree of randomness in granule formation and size development even under identical conditions. However, the similarity of the physicochemical and microbial properties of the granules, as well as the performance of the wastewater treatment after granule maturation, indicate that the operating conditions can produce consistent results, implying predictability of stable operation in practice.
- A high F/M value promotes the formation of aerobic granules, while a low F/M value in the range of 0.2–0.4 g COD/g VSS·d facilitates the long-term stability of aerobic granules.
- The richness of the microbial population of the granules was much lower than that of the inoculum and the flocs with bulking, although the richness may increase slightly during the long-term operation period. The diversity of the microbial structure decreased over time. However, the reduction in the richness and diversity of the microbial population due to the conversion of flocs to granules did not affect the wastewater

treatment performance and long-term stability of the sludge, indicating the robustness of different microbial structures to achieve the same function in wastewater treatment.

- Among the dominant genera of the sludge, *Zoogloea* played a key role in maintaining the stable structure of aerobic granules throughout the operation period, while *Thauera* is an important genus for the formation and maturation of aerobic granules, but not for long-term maintenance. In addition, *Ohtaekwangia*, *Chryseobacterium*, *Taibaiella*, and *Tahibacter* can play an important role in the long-term stability of aerobic granules.

These results demonstrate the reproducibility of granulation, the small influence of granulation on long-term stability, and the robustness of aerobic granulation for the removal of COD and N. Overall, our study contributes significantly to the understanding of microbial community structure for the long-term stability of aerobic granular sludge in the treatment of low-COD and low-COD/N-ratio wastewater in practice.

Author Contributions: Conceptualization, F.C. and Y.-Q.L.; methodology, H.Z. and S.M.; validation, F.C. and Y.-Q.L. and Sternberg, C.E.W.S.; formal analysis, H.Z. and S.M.; investigation, H.Z., S.M. and W.D.; data curation, H.Z.; writing—original draft preparation, H.Z. and F.C.; writing—review and editing, F.C., Y.-Q.L. and C.E.W.S.; supervision, F.C.; project administration, S.M. and W.D.; funding acquisition, F.C. All authors have read and agreed to the published version of the manuscript.

Funding: This research was funded by the National Nature Science Foundation of China (grant number 41763016).

Informed Consent Statement: Information consent was obtained from all subjects involved in the study.

Conflicts of Interest: The authors declare no conflict of interest.

References

1. Wang, L.; Liu, X.; Lee, D.-J.; Tay, J.-H.; Zhang, Y.; Wan, C.-L.; Chen, X.-F. Recent advances on biosorption by aerobic granular sludge. *J. Hazard. Mater.* **2018**, *357*, 253–270. [CrossRef] [PubMed]
2. Lee, D.-J.; Chen, Y.-Y.; Show, K.-Y.; Whiteley, C.G.; Tay, J.-H. Advances in aerobic granule formation and granule stability in the course of storage and reactor operation. *Biotechnol. Adv.* **2010**, *28*, 919–934. [CrossRef] [PubMed]
3. Liu, Y.; Liu, Q.S. Causes and control of filamentous growth in aerobic granular sludge sequencing batch reactors. *Biotechnol. Adv.* **2006**, *24*, 115–127. [CrossRef] [PubMed]
4. Bengtsson, S.; de Blois, M.; Wilen, B.-M.; Gustavsson, D. Treatment of municipal wastewater with aerobic granular sludge. *Crit. Rev. Environ. Sci. Technol.* **2018**, *48*, 119–166. [CrossRef]
5. Adav, S.S.; Lee, D.-J.; Show, K.-Y.; Tay, J.-H. Aerobic granular sludge: Recent advances. *Biotechnol. Adv.* **2008**, *26*, 411–423. [CrossRef]
6. Cydzik-Kwiatkowska, A.; Wojnowska-Barya, I. Nitrogen-converting communities in aerobic granules at different hydraulic retention times (HRTs) and operational modes. *World J. Microbiol. Biotechnol.* **2015**, *31*, 75–83. [CrossRef]
7. Swiatczak, P.; Cydzik-Kwiatkowska, A. Performance and microbial characteristics of biomass in a full-scale aerobic granular sludge wastewater treatment plant. *Environ. Sci. Pollut. Res.* **2018**, *25*, 1655–1669. [CrossRef]
8. Cha, L.; Liu, Y.-Q.; Duan, W.; Sternberg, C.E.W.; Yuan, Q.; Chen, F. Fluctuation and Re-Establishment of Aerobic Granules Properties during the Long-Term Operation Period with Low-Strength and Low C/N Ratio Wastewater. *Process* **2021**, *9*, 1290. [CrossRef]
9. Pronk, M.; de Kreuk, M.K.; de Bruin, B.; Kamminga, P.; Kleerebezem, R.; van Loosdrecht, M.C.M. Full scale performance of the aerobic granular sludge process for sewage treatment. *Water Res.* **2015**, *84*, 207–217. [CrossRef]
10. Yang, S.F.; Tay, J.H.; Liu, Y. Effect of substrate nitrogen/chemical oxygen demand ratio on the formation of aerobic granules. *J. Environ. Eng.* **2005**, *131*, 86–92. [CrossRef]
11. de Kreuk, M.K.; Kishida, N.; Tsuneda, S.; van Loosdrecht, M.C.M. Behavior of polymeric substrates in an aerobic granular sludge system. *Water Res.* **2010**, *44*, 5929–5938. [CrossRef] [PubMed]
12. Layer, M.; Adler, A.; Reynaert, E.; Hernandez, A.; Pagni, M.; Morgenroth, E.; Holliger, C.; Derlon, N. Organic substrate diffusibility governs microbial community composition, nutrient removal performance and kinetics of granulation of aerobic granular sludge. *Water Res. X* **2019**, *4*, 100033. [CrossRef] [PubMed]
13. Liu, Y.Q.; Liu, Y.; Tay, J.H. Relationship between size and mass transfer resistance in aerobic granules. *Lett. Appl. Microbiol.* **2005**, *40*, 312–315. [CrossRef] [PubMed]
14. *APHA, Standard Methods for the Examination of Water and Wastewater*, 23rd ed.; American Public Health Association, American Water Works Association, and Water Environmental Federation: Washington, DC, USA, 2017.

15. Hill, T.C.J.; Walsh, K.A.; Harris, J.A.; Moffett, B.F. Using ecological diversity measures with bacterial communities. *FEMS Microbiol. Ecol.* **2003**, *43*, 1–11.
16. Westphal, C.; Bommarco, R.; Carré, G.; Lamborn, E.; Morison, N.; Petanidou, T.; Potts, S.G.; Roberts, S.P.M.; Szentgyörgyi, H.; Tscheulin, T.; et al. Measuring bee diversity in different European habitats and biogeographical regions. *Ecol. Monogr.* **2008**, *78*, 653–671. [CrossRef]
17. Liu, Y.-Q.; Tay, J.-H. The competition between flocculent sludge and aerobic granules during the long-term operation period of granular sludge sequencing batch reactor. *Environ. Technol.* **2012**, *33*, 2619–2626. [CrossRef]
18. Liu, Y.Q.; Kong, Y.H.; Zhang, R.; Zhang, X.; Wong, F.S.; Tay, J.H.; Zhu, J.R.; Jiang, W.J.; Liu, W.T. Microbial population dynamics of granular aerobic sequencing batch reactors during start-up and steady state periods. *Water Sci. Technol.* **2010**, *62*, 1281–1287. [CrossRef]
19. Wang, L.; Zhan, H.; Wang, Q.; Wu, G.; Cui, D. Enhanced aerobic granulation by inoculating dewatered activated sludge under short settling time in a sequencing batch reactor. *Bioresour. Technol.* **2019**, *286*, 121386. [CrossRef]
20. He, Q.; Chen, L.; Zhang, S.; Chen, R.; Wang, H. Hydrodynamic shear force shaped the microbial community and function in the aerobic granular sequencing batch reactors for low carbon to nitrogen (C/N) municipal wastewater treatment. *Bioresour. Technol.* **2019**, *271*, 48–58. [CrossRef]
21. Yoon, J.-H.; Kang, S.-J.; Oh, T.-K. Chryseobacterium daeguense sp nov., isolated from wastewater of a textile dye works. *Int. J. Syst. Evol. Microbiol.* **2007**, *57*, 1355–1359. [CrossRef]
22. Makk, J.; Homonnay, Z.G.; Keki, Z.; Lejtovicz, Z.; Marialigeti, K.; Sproeer, C.; Schumann, P.; Toth, E.M. Tahibacter aquaticus gen. nov., sp nov., a new gammaproteobacterium isolated from the drinking water supply system of Budapest (Hungary). *Syst. Appl. Microbiol.* **2011**, *34*, 110–115. [CrossRef] [PubMed]
23. Chhetri, G.; Kim, I.; Kim, J.; Kang, M.; Seo, T. Taibaiella lutea sp. nov., Isolated from Ubiquitous Weedy Grass. *Curr. Microbiol.* **2021**, *78*, 2799–2805. [CrossRef] [PubMed]
24. Gonzalez-Gil, G.; Holliger, C. Dynamics of microbial community structure of and enhanced biological phosphorus removal by aerobic granules cultivated on propionate or acetate. *Appl. Environ. Microbiol.* **2011**, *77*, 8041–8051. [CrossRef] [PubMed]
25. Ehsani, E.; Jauregui, R.; Geffers, R.; Jarek, M.; Boon, N.; Pieper, D.H.; Vilchez-Vargas, R. First Draft Genome Sequence of the Acidovorax caeni sp. nov. Type Strain R-24608 (DSM 19327). *Genome Announc.* **2015**, *3*, e01378-15. [CrossRef] [PubMed]
26. Feng, S.; Tan, C.H.; Constancias, F.; Kohli, G.S.; Cohen, Y.; Rice, S.A. Predation by Bdellovibrio bacteriovorus significantly reduces viability and alters the microbial community composition of activated sludge flocs and granules. *FEMS Microbiol. Ecol.* **2017**, *93*, fix020. [CrossRef] [PubMed]
27. Han, X.; Zhou, Z.; Mei, X.; Ma, Y.; Xie, Z. Influence of fermentation liquid from waste activated sludge on anoxic/oxic-membrane bioreactor performance: Nitrogen removal, membrane fouling and microbial community. *Bioresour. Technol.* **2018**, *250*, 699–707. [CrossRef]
28. Ummalyma, S.B.; Gnansounou, E.; Sukumaran, R.K.; Sindhu, R.; Pandey, A.; Sahoo, D. Bioflocculation: An alternative strategy for harvesting of microalgae—An overview. *Bioresour. Technol.* **2017**, *242*, 227–235. [CrossRef]
29. Yu, H.; Meng, W.; Song, Y.; Tian, Z. Understanding bacterial communities of partial nitritation and nitratation reactors at ambient and low temperature. *Chem. Eng. J.* **2017**, *337*, 755–763. [CrossRef]
30. Staley, J.T.; Bouzek, H.; Jenkins, C. Eukaryotic signature proteins of Prosthecobacter dejongeii and Gemmata sp Wa-1 as revealed by in silico analysis. *FEMS Microbiol. Lett.* **2005**, *243*, 9–14. [CrossRef]
31. Baek, K.; Choi, A. Complete Genome Sequence of Phreatobacter sp. Strain NMCR1094, a Formate-Utilizing Bacterium Isolated from a Freshwater Stream. *Microbiol. Resour. Announc.* **2019**, *8*, e00860-19. [CrossRef]
32. Raittz, R.T.; de Pierri, C.R.; Maluk, M.; Batista, M.B.; Carmona, M.; Junghare, M.; Faoro, H.; Cruz, L.M.; Battistoni, F.; Souza, E.D.; et al. Article comparative genomics provides insights into the taxonomy of *Azoarcus* and reveals separate origins of *nif* genes in the proposed *Azoarcus* and *Aromatoleum* genera. *Genes* **2021**, *12*, 71. [CrossRef] [PubMed]
33. Xi, L.; Liu, D.; Huang, W. Effect of acetate and propionate on the production and characterization of soluble microbial products (SMP) in aerobic granular sludge system. *Toxicol. Environ. Chem.* **2018**, *100*, 175–190. [CrossRef]

Article

Hydroxyapatite Precipitation and Accumulation in Granules and Its Effects on Activity and Stability of Partial Nitrifying Granules at Moderate and High Temperatures

Yong-Qiang Liu * and Simone Cinquepalmi

Faculty of Engineering and Physical Sciences, University of Southampton, Southampton SO17 1BJ, UK;
S.Cinquepalmi@soton.ac.uk
* Correspondence: Y.Liu@soton.ac.uk; Tel.: +44-02380592843

Abstract: Precipitation and accumulation of calcium phosphate in granular sludge has attracted research attention recently for phosphate removal and recovery from wastewater. This study investigated calcium phosphate accumulation from granulation stage to steady state by forming heterotrophic granules at different COD/N ratios at 21 and 32 °C, respectively, followed by the transformation of heterotrophic granules to partial nitrifying granules. It was found that mature granules accumulated around 60–80% minerals in granules, much higher than young granules with only around 30% ash contents. In addition, high temperature promoted co-precipitation of hydroxyapatite and calcite in granules with more calcite than hydroxyapatite and only 4.1% P content, while mainly hydroxyapatite was accumulated at the moderate temperature with 7.7% P content. The accumulation of minerals in granules at the high temperature with 75–80% ash content also led to the disintegration and instability of granules. Specific ammonium oxidation rates were reduced, as well, from day 58 to day 121 at both temperatures due to increased mineral contents. These results are meaningful to control or manipulate granular sludge for phosphorus removal and recovery by forming and accumulating hydroxyapatite in granules, as well as for the maintenance of microbial activities of granules.

Keywords: nitrifying granules; hydroxyapatite; calcite; stability; microbial activity; phosphorus removal and recovery

Citation: Liu, Y.-Q.; Cinquepalmi, S. Hydroxyapatite Precipitation and Accumulation in Granules and Its Effects on Activity and Stability of Partial Nitrifying Granules at Moderate and High Temperatures. *Processes* **2021**, *9*, 1710. https://doi.org/10.3390/pr9101710

Academic Editor: Maria Jose Martin de Vidales

Received: 19 July 2021
Accepted: 15 September 2021
Published: 24 September 2021

Publisher's Note: MDPI stays neutral with regard to jurisdictional claims in published maps and institutional affiliations.

1. Introduction

Aerobic granular sludge is a promising technology to replace suspended sludge for wastewater treatment. Basically, granular sludge is a kind of self-immobilized biofilm without carriers, which was developed in the 1980s in Upflow Anaerobic Sludge Blanket (UASB) systems for anaerobic wastewater treatment, and in the 2000s in Sequential Batch Reactor (SBR) systems for aerobic wastewater treatment. Compared with suspended sludge, granular sludge possesses a compact structure and large size, which is believed to be favorable to inorganic precipitation and accumulation in granules. High inorganic content in aerobic granules, anaerobic granules, and ANAMMOX granules under some operational conditions have been widely observed and reported [1–3]. Reactor operation mode might affect inorganic mineral accumulation in granules. It was found that, after the operation mode was changed from sequential batch to continuous, the ash content of aerobic granules increased from 20 to 84% with inorganic minerals identified by XRD analysis as $CaCO_3$, although Ca^{2+} concentration in wastewater as low as 19 mg L^{-1} [4]. Since heavy and large granules are more easily retained in reactors, particularly at the bottom of the reactors, the ash content of ANAMMOX granules at the reactor bottom was 60%, while it was only 10% at the reactor top [5]. Juang et al. [6] hypothesized that calcium and/or iron precipitates in the granule interior substantially enhanced the structural stability of aerobic granules under continuous operation mode, while Liu et al. [7] reported a significantly reduced microbial activity of biomass under continuous operation mode with the excessive precipitation of $CaCO_3$, although granules maintained excellent structural stability. From these studies, it looks as if the inorganic precipitation and accumulation in granules are conducive to the structural stability of granular sludge. Furthermore, it was reported that the granulation period was shortened by adding extra metal ions, such as Ca^{2+} [8] Mg^{2+}, Al^{3+} [9], Fe^{3+} [10], and Mn^{2+}. It was speculated that metal ions could form metal precipitates which could become the nucleus for bacteria to attach to form granules with a similar mechanism for biofilm growth. This theory seems to be partially supported by the phenomenon that inorganic precipitates were mostly found in the core of granules [11,12]. However, so far, there are no reports about the comparison between the newly formed granules and mature granules regarding the spatial distribution of inorganic precipitates to fully validate biofilm growth theory.

Biologically induced precipitation in granules is another mechanism proposed to explain mineral precipitates in granules [3,13]. It was believed that a three-dimensional compact structure working together with biological activities inside, such as nitrification, denitrification, and biological phosphorus removal, could change the chemical environment in granules, thus benefiting metal precipitation [13]. Although the mechanism of mineral precipitation in granules is unclear, there is no doubt that divalent or trivalent metal ions could accelerate granulation, increase granule compactness, and decrease SVI. The long-term stability of granules with high mineral precipitates is unclear. Reduced microbial activities of granules with high mineral precipitates have been reported [4,14–16], but it is unclear how high mineral precipitate contents affect microbial activities of nitrifying bacteria in granules with varied pH in SBR cycles.

Temperature is another factor that needs to be investigated for its effect on precipitation in granules because temperature affects biomass growth rate, nitrification rate, precipitation speed, the solubility of mineral precipitates, and the stability of granules. Aerobic granulation, granule stability, COD, N, and P removal could be achieved from a wide range of temperatures, i.e., from 8 °C to 50 °C, although sludge characteristics and treatment performance could be different [17,18]. With the development and implementation of ANAMMOX for nitrogen removal, partial nitrification under high temperatures, such as 30–35 °C, was demanded before ANAMMOX. How mineral precipitates in nitrifying granules at different temperatures affect granule stability and microbial activities of nitrifying bacteria is still unknown.

Furthermore, in practice, it is less likely to dose extra Ca^{2+}, Mg^{2+}, Al^{3+}, or Fe^{3+} as research was done in labs to stimulate granulation or maintain granule stability because it involves the extra cost of chemicals, handling, and storage. However, in different regions, water hardness levels are different with a wide range of Ca^{2+} and Mg^{2+} concentrations. For example, in the east and south of England, Ca^{2+} concentration in water reaching as high as 150 mg L^{-1} and 50–100 mg L^{-1} of Ca^{2+} in water is very common. How a high water hardness level with a high Ca^{2+} concentration affects mineral precipitation in granules and reactor performance is important for the implementation of granular sludge technology in these regions.

Thus, this study aims to investigate mineral precipitates and their effects on the formation, activity, and stability of nitrifying granules at moderate and high temperatures in regions with a high water hardness level. It is expected that this study would provide some practical guidance on the operation of nitrifying granule reactors. Meanwhile, it could shed light on the possible mechanisms involved in mineral precipitation and accumulation in granules, as well as granule stability.

2. Materials and Methods

2.1. Reactors Operation and Inoculum

Four Perspex columns (R1, R2, R3, and R4), each with a working volume of 2.6 L and a column height to diameter ratio of 20 (Ø 60 mm), were operated sequentially with a cycle time of 4 h. The cycle consisted of 10-min feeding, 227–199 min aeration, 2–30 min settling time, and 1-min discharging. The effluent was discharged from the middle port of the reactors, corresponding to a volumetric exchange ratio of 50%. The air was provided from the reactor bottom with a flow rate of 5 L min^{-1}, corresponding to a superficial velocity of 3.0 cm s^{-1}.

The reactors were started up at two different temperatures, i.e., ambient temperature at 21 °C and high temperature at 32 °C, by inoculating 3.2 g L^{-1} activated sludge from Portswood Wastewater Treatment Plant, Southampton, UK, which was only for COD removal without nitrification capability. The reactor operation was divided into four phases: (i) In phase 1, the inoculated activated sludge was converted into heterotrophic granules by providing high OLR and short settling time; (ii) in phase 2, the nitrifying bacteria were enriched within heterotrophic granules by decreasing COD/N ratio until no organic carbon was provided; (iii) in phase 3, NH_4^+ concentration was quickly elevated to speed-up the AOBs enrichment; and, (iv) in phase 4, long-term stability of nitritation systems was tested. The specific operational conditions for four reactors are listed in Table 1.

Table 1. Specific operational conditions of four reactors for granulation and nitrification.

		Units	R1	R2	R3	R4
	Temperature	°C	21 ± 3	32 ± 2	21 ± 3	32 ± 2
Phase 1 (0–21)	COD	mg L^{-1}	1000	1000	1000	1000
	NH_4^+	mg N L^{-1}	50	50	200	200
	COD/N		20	20	5	5
Phase 2 (22–37/49 [a])	COD	mg L^{-1}	1000 to 0	1000 to 0	1000 to 0	1000 to 0
	NH_4^+	mg N L^{-1}	50 to 400	50 to 850	200 to 150	200 to 750
Phase 3 (38/50 [a]–107)	NH_4^+	mg N L^{-1}	Varied influent ammonium concentration [b]			
Phase 4 (108–171)	NH_4^+	mg N L^{-1}	500	65	650	850

[a] Phase 2 lasted until day 37 for R2 and R4 and until day 49 for R1 and R3. [b] The influent NH_4^+ concentrations to 4 reactors were varied to ensure more than 90% ammonium-nitrogen removal until the values in phase 4 (R2 collapsed so influent NH_4^+ concentration was reduced).

By designing experiments for the operation of 4 reactors under conditions listed in Table 1, we can compare granulation and stability of nitrifying granules for (i) effects of COD/N ratios (i.e., 20 and 5) on heterotrophic granulation at ambient temperature, i.e., 21 ± 3 °C, between R1 and R3; (ii) effects of COD/N ratios (i.e., 20 and 5) on heterotrophic granulation at high temperature, i.e., 32 ± 2 °C, between R2 and R4; (iii) effects of temperature (i.e., 21 ± 3 °C and 32 ± 2 °C) on heterotrophic granulation at COD/N of 20 between R1 and R2; and (iv) effects of temperature (i.e., 21 ± 3 °C and 32 ± 2 °C) on heterotrophic granulation at COD/N of 5 between R3 and R4. In addition, once heterotrophic granules were completely converted into nitrifying granules, we can compare (i) activity and the long-term stability of nitrifying granules at two different temperatures, between R3 and R4 or between R1 and R2; and (ii) effects of initial COD/N ratios on the long-term stability of nitrifying granules between R1 and R3 or between R2 and R4.

2.2. Medium

Synthetic wastewater was prepared with tap water, in which sodium acetate, ammonium sulphate, and mono-potassium phosphate were added as carbon, nitrogen, and phosphorous sources, respectively, with COD, NH_4^+-N concentrations, as shown in Table 1, and phosphorous concentration as 15 mg PO_4^{3-}-P L^{-1}. In phase 2, for nitrifying bacteria enrichment, the organic carbon concentration was reduced stepwise, along with a concurrent increase in ammonium and $NaHCO_3$ concentrations with a ratio of 1 mg NH_4^+-N to 14 mg $NaHCO_3$. The bicarbonate was used as the inorganic carbon source for autotrophic nitrifying bacteria, as well as a buffer for pH. In phase 4, the ammonium concentrations were kept constant since the nitrifying granules had reached the process stability.

Apart from carbon, nitrogen and phosphorus sources, micronutrients were added as (per liter prepared): 25 mg $CaCl_2 \cdot 2H_2O$, 20 mg $MgSO_4 \cdot 7H_2O$, and 10 mg $FeSO_4 \cdot 7H_2O$. The trace elements were (per liter prepared): 0.12 mg $MnCl_2 \cdot 4H_2O$, 0.12 mg $ZnSO_4 \cdot 7H_2O$, 0.03 mg $CuSO_4 \cdot 5H_2O$, 0.05 mg $(NH_4)_6Mo_7O_{24} \cdot 4H_2O$, 0.1 $CoCl_2 \cdot 6H_2O$, 0.1 mg $NiCl_2 \cdot 6H_2O$, 0.05 mg $AlCl_3 \cdot 6H_2O$, and 0.05 mg H_3BO_3. From phase 3 (day 47), both $CaCl_2$ and $MgSO_4$ were removed from feedstock after excessive inorganic precipitation, and accumulation within the granules was observed. The synthetic wastewater pH was not adjusted, but it was around 7.2–7.4.

2.3. Analytical Methods

COD, mixed liquor suspended solids (MLSS), mixed liquor volatile suspended solids (MLVSS), sludge volume index (SVI), and ash content were measured in accordance with standard methods [19]. For the measurement of SVI_5, the settled sludge volume was recorded after 5-min settling in a 100-mL graduated cylinder, as opposed to 30 min for SVI_{30} [20]. After granules formed, the density method used by Beun et al. [21] was employed to measure MLSS due to difficulty in homogenous sampling. Ammonium was measured spectrophotometrically by following the procedure in accordance with that described in the manual BSI [22], using the UV-Visible spectrophotometer Cecil 3000 series (Cecil Instruments Ltd., Cambridge, UK). The pH in the reactors (during aeration), and from the effluent, was daily monitored by using a bench pH-meter (Mettler Toledo, Columbus, OH, USA).

Anions (NO_2^-, NO_3^-, PO_4^{3-}) and cations (Ca^{2+} and Na^+) were measured by the ionic chromatographer 882 Compact IC plus (Metrohm, Switzerland). For the measurement of anions, a Metrosep A Supp 5-150/4.0 column was used with the eluent containing 1 mM $NaHCO_3$ and 3.2 mM Na_2CO_3. For the measurement of cations, a Metrosep C4 250/4.0 column was used with the eluent containing 1.7 mM HNO_3 and 0.7 mM $C_7H_5NO_4$ (dipicolinic acid or Pyridin-2,6-dicarboxylic acid).

The morphologies of granule external surface and internal parts were observed by using the Quanta 250 scanning electron microscopy (SEM) (FEI, Hillsboro, OR, USA). Representative granules were halved and fixed overnight with a solution of 3% glutaraldehyde + 4% formaldehyde in 0.1 M PIPES buffer with a pH of 7.2. Then, the specimens were washed twice with each for 10 min by 0.1 M PIPES buffer with a pH of 7.2, followed by a series of dehydration steps of 10-min washes with ethanol 30%, 50%, 70%, and 95%, respectively. After these, two rinses with absolute ethanol, lasting 20 min each, were conducted. The granules samples were then dried by a Balzers CPD 030 Critical Point Drier, before being carbon coated by a high-resolution sputter and high vacuum carbon coater (Quorum, Q150T ES. West Sussex, UK) for morphology observation. In addition, the energy dispersive X-ray spectroscopy (EDX) coupled with SEM (EDAX, Mahwah, NJ, USA) was used to qualitatively analyze the chemical elements present in the granules.

Crystals in granules were analyzed by XRD (Bruker D2 phaser, Bruker AXS Gmgh, Karlsruhe, Germany). The granule samples were first washed thoroughly with deionized water to remove any soluble chemicals from the feedstock, and then it was kept at $-20\ ^\circ C$ overnight before being freeze-dried (lyophilized). Before XRD analysis, the dry granules were ground into a fine powder.

The elemental composition of granules was analyzed quantitatively by using a multi collector inductively coupled plasma (ICP) emission spectrometer (X-SERIES 2 ICP-MS, Thermo Fisher Scientific, Bremen, Germany). Before the analysis, the samples were dried at $105\ ^\circ C$ overnight and then ground into powder. Fifty milligrams of it was digested by adding 2 mL of aqua regia (a 1:3 mixture of 68% nitric acid and 36% hydrochloric acid) and 0.5 mL perchloric acid (68%) in a Teflon digestion vessel on a hotplate ($150\ ^\circ C$) overnight. After the digestion, 10 mL of concentrated HCl (36%) was added to the completely dry samples. After this solution went dry by evaporating the acid, 3% nitric acid was added to the samples for the analysis by ICP-MS. The data were processed using Plasmalab software.

3. Results and Discussion

3.1. Formation of Heterotrophic Granules with Different COD/N Ratios at Different Temperatures

Temperature effects on granulation were studied at two different influent COD/N ratios, i.e., 5 and 20. It was found that temperature effects at COD/N ratio of 5 were the same as that at COD/N ratio of 20. Thus, only the results of temperature effects on granulation at COD/N ratio of 5 (i.e., the comparison between R3 and R4) are presented in Figure 1. It can be seen that SVI_5 increased sharply in both reactors in the first 3 days and then decreased quickly to less than 100 mL g^{-1} after day 7. This phenomenon is in agreement with those reported by Liu et al. [23] and Liu et al. [24], indicating a typical change of SVI_5 for fast granulation. Once SVI reduced to less than 100 mL g^{-1}, the average sludge size started to increase quickly (Figure 1d), along with the decrease in sludge volume percentage with particle size smaller than 200 μm (Figure 1c). Although SVI_5 in both reactors is similar, sludge size increased more quickly in R3 at 21 °C (Figure 1d), resulting in a higher biomass concentration in R3 (Figure 1b). On day 21, MLVSS in R3 at 21 °C reached 4.6 g L^{-1}, while it was 3.0 g L^{-1} in R4 at 31 °C. Based on the definition of granulation by Liu and Tay [25] that sludge volume percentage with particle size smaller than 200 μm below 50% represents the formation of granule dominant sludge, granular sludge formed on day 9 in R3 at 21 °C, while this occurred on day 11 in R4 at 32 °C. Thus, it can be said that granulation speeds at two different temperatures are almost the same, although moderate temperature seems slightly better than high temperature. It was reported that the time needed for total granule formation at 8 °C approximated twice the time that is normally needed at 20 °C [17], while there was no difference when granules were formed at 30 °C and 40 °C [18]. From these literature reports and our own results presented here, it could be reasoned that temperature range from 20 to 40 °C does not affect granulation speed.

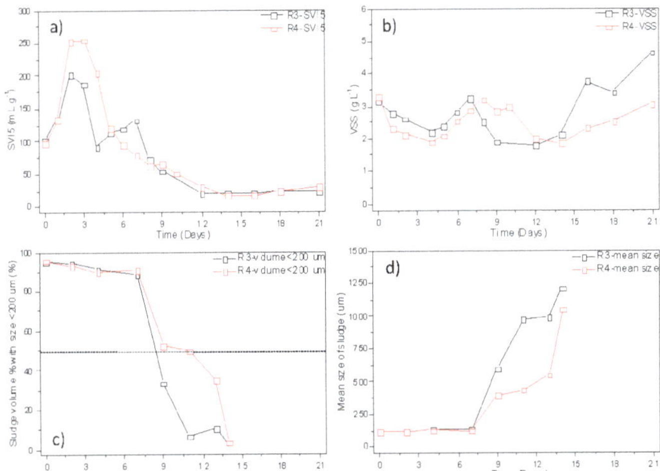

Figure 1. Profiles of SVI_5 (**a**), MLVSS (**b**), sludge volume percentage with a size smaller than 200 μm (**c**), and sludge mean size (**d**) over the time during phase 1 in R3 (21 °C) and R4 (32 °C) at COD/N of 5 with NH_4^+-N concentration of 200 mg L^{-1}.

Figure 2 shows the effects of COD/N ratios on granulation at 32 °C, which is the same with those at 21 °C (data not shown). As in Figure 1, with the effects of temperature on granulation, SVI_5 in both reactors (i.e., R2 and R4) had the same changing patterns (Figure 2a), but sludge size and biomass concentration showed a slight difference after day 13. Based on the definition of granule dominant sludge mentioned above, granular sludge formed on day 11 in R4 with 200 mg NH_4^+-N L^{-1}, while on day 13 in R2 with 50 mg NH_4^+-N L^{-1}.

Because of slightly faster granulation, R4 achieved higher biomass accumulation with a concentration of 3.0 g MLVSS L^{-1} on day 21, whereas it was 1.8 g MLVSS L^{-1} in R2. The results suggested that granulation speed was not affected too much by influent ammonium concentration, although it seems that higher ammonium concentration was slightly better. This result is different from that reported by Yang et al. [26] that initial ammonium concentration of 200 mg L^{-1} inhibited granulation, while granules formed when COD/N ratios were above 6. The key difference between this study and Yang et al., in 2004, is that we employed a short settling time, while 20-min settling time was used by Yang et al. [26]

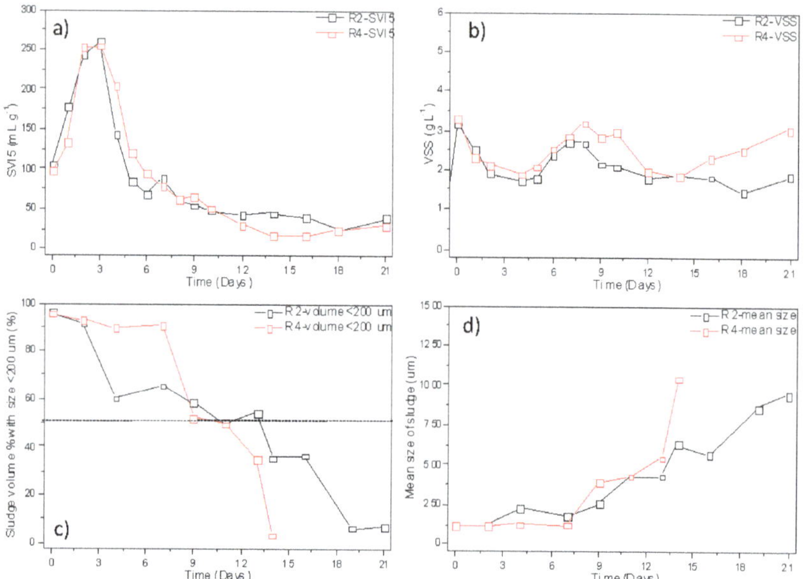

Figure 2. Profiles of SVI_5 (**a**), MLVSS (**b**), sludge volume percentages with a size smaller than 200 μm (**c**), and sludge mean size (**d**) over the time during phase 1 in R2 (with COD/N of 20 and 50 mg NH_4^+-N L^{-1}) and R4 (with COD/N of 5 and 200 mg NH_4^+-N L^{-1}) at the operational temperature of 32 °C.

From Figures 1 and 2, it could be concluded that, as long as the strong hydraulic selective pressure, such as short settling time, was employed to start up granular sludge reactors quickly [27,28], COD/N ratios and temperature have only negligible effects on granulation. This is meaningful for the start-up of the granular sludge reactors under different conditions.

3.2. Mineral Accumulation in Nitrifying Granules Converted from Heterotrophic Granules at Different Temperatures and Minerals Effects on Granules' Long-Term Stability

As shown in Figure 3, the sludge in four reactors showed excellent settleability with SVI_5 lower than 70 mL g^{-1} after 1-week operation, and less than 40 mL g^{-1}, except for R2 after granulation. At the same time, the ash content in sludge increased slightly and slowly from an average value of 18% on day 0 to 19–35% in 4 reactors on day 21, i.e., the end of phase 1 for graduation (Figure 3a), indicating granulation might promote ash accumulation in granules due to the retention of granular sludge. Once granules became dominant with granule percentage higher than 95%, ash contents of granules increased steeply in phase 2, resulting in dense granules with SVI_5 of 14, 5, 23, and 8 mL g^{-1}, respectively, on day 59 (Figure 3b). The comparison between day 21 and day 59 demonstrates that mature granules had much higher ash contents compared with young granules which were just formed. This might be related to granule retention time, and a longer granule retention time allows more mineral accumulation in granules. A similar high ash accumulation within aerobic granules coupled with such low SVI_5 was also described by other research in literature [7].

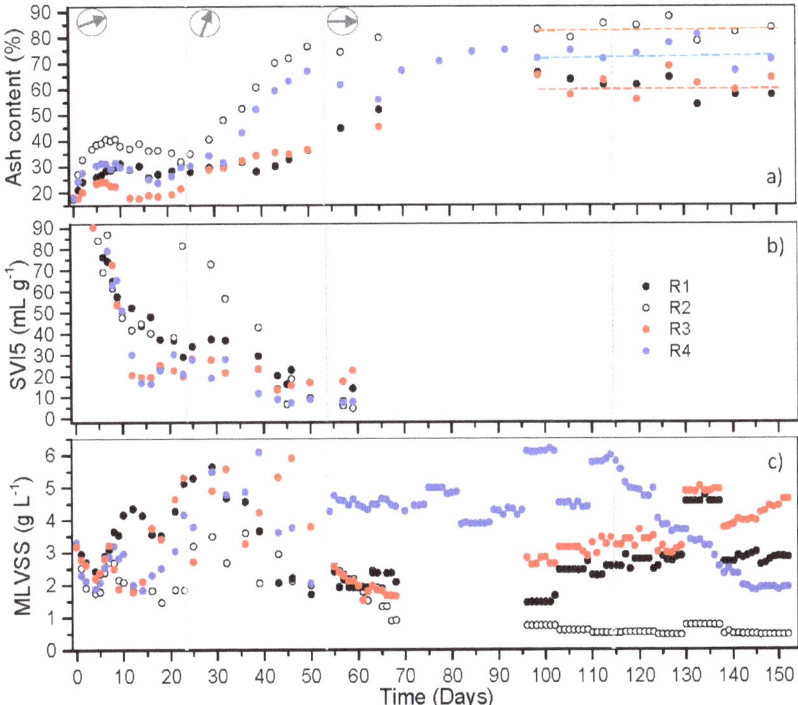

Figure 3. Profiles of ash content (**a**), SVI_5 (**b**), and MLSS-MLVSS (**c**) in R1 and R3 (at 21 °C), R2 and R4 (at 32 °C) over the operation time.

Based on the ash content and its increasing rate during the whole operational period, the ash content curves in Figure 3a can be divided into three distinct stages, from the early days to the stable state with granule ash content increased slightly, dramatically, and negligibly, respectively, as indicated with vertical dotted lines in Figure 3a.

Although ash content in granules showed similar changing trends, it is still very obvious that ash content increasing rates and granule ash contents in 4 reactors were different. Granule ash contents in R2 and R4 at 32 °C increased more dramatically than those in R1 and R3 at 21 °C before day 60 and then stabilized at 80% and 70%, respectively, while it took a longer time for granule ash contents in R1 and R3 at 21 °C to stabilize at 58%. Since the main difference in this phase between R1 and R3, and R2 and R4 is the temperature, it can be believed that the difference in granule ash content increasing rates and stable ash contents were mainly attributed to temperature with higher temperature in R2 and R4 benefiting quicker and more ash accumulation.

During the first 21-days, apart from the temperature difference, there was also a COD/N difference. It can be seen that granule ash content increased more quickly with COD/N of 20 in R1 than COD/N of 5 in R3 at 21 °C. A similar trend was observed with COD/N of 20 in R2 and COD/N of 5 in R4 at 32 °C. Thus, it can be seen that a higher COD/N ratio with little nitrifying activity is favorable to ash accumulation in sludge, which is probably due to a slight pH difference caused by nitrification at low COD/N ratios.

Due to a steep ash content increase in R2 at 32 °C, granule ash content reached 78% on day 50. With such high granule ash content, granule disintegration was observed and granules broke into pieces, which resulted in a quick biomass wash-out and a very low biomass concentration (less than $0.5\ \text{g L}^{-1}$), as shown in Figure 3c. The influent ammonium-nitrogen concentration had to be dropped to $65\ \text{mg L}^{-1}$ to respond to significantly reduced nitrification capacity and performance, as shown in Figure 4. R2 operation basically collapsed, although we still maintained its operation. At 32 °C, since granule ash content increasing rate and ash content at stable state in R4 was slightly lower than R2, probably due to the low COD/N ratio in Phase 1, R4 maintained stable operation for more than 110 days with stable biomass concentration at $5\text{–}6\ \text{g L}^{-1}$; however, granule disintegration was observed after day 110, which led to the biomass washout and reduced biomass concentration, as shown in Figure 3c. Both R2 and R4 at 32 °C experienced granule disintegration, which should not be a coincidence. For R1 and R3 at 21 °C, nitrifying biomass concentrations were stable until the end of operation after 153 days, and ash contents were at around 58%. By comparing two different temperatures and ash contents, it can be seen that higher temperatures resulted in higher ash contents, which might lead to the instability of nitrifying granules in the short term or long term. The stability of granules with high ash contents is seldom studied. However, Xue et al. [29] observed the deteriorated settling velocity of anammox-HAP granules, sludge bulking, and flotation when ash content was as high as 80%, an ash content close to those in R2 and R4 at 32 °C with instability in this study.

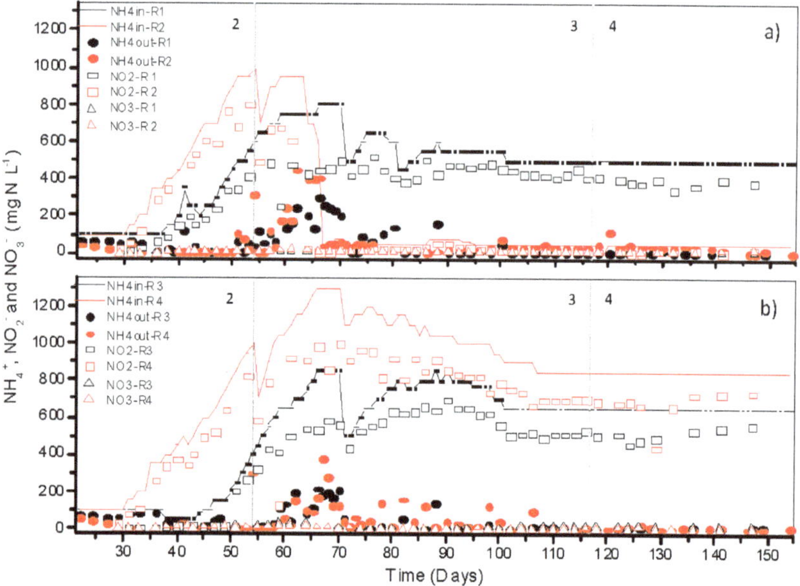

Figure 4. Profiles of nitrogen species over the time during phases 2–4, with COD reduced quickly to 0, and NH$_4^+$-N increased to certain values in R1 and R3 at 21 °C (**a**), and R2 and R4 at 32 °C (**b**).

3.3. Performance of Nitrifying Granules at Different Temperatures and Their Specific Ammonium Oxidization Activities

On day 21, all sludge in 4 reactors was almost pure granules. From day 22, the influent NH$_4^+$-N concentration was increased step-wise, alongside the concurrent decrease in COD concentration to transform newly formed heterotrophic granules to nitrifying granules, as shown in Figure 4. By the end of phase 2 (Table 1), heterotrophic granules were completely transformed into autotrophic nitrifying granules with the higher influent ammonium increasing rates in R2 and R4 at 32 °C than R1 and R3 at 21 °C. Meanwhile, by comparing R1 and R3, and R2 and R4 with two different COD/N ratios at the same temperatures, it was found that initial low COD/N ratio and high ammonium concentration did not help the enrichment rate of nitrifying bacteria in granules. It is well known that the nitrifying bacteria growth rate is higher at a higher temperature than a lower temperature. From the results in this study, it looks as if the enrichment of nitrifying bacteria in heterotrophic granules was highly dependent on temperature instead of initial ammonium concentrations in the influent. However, the maximum influent ammonium concentration reached 1300 mg N L^{-1} in R4 with the initial COD/N ratio of 5 in phase 1, higher than 1000 mg N L^{-1} in R2 with the initial COD/N ratio of 20 in phase 1 at 32 °C. Similar results were obtained at 21 °C in R3 (850 mg N L^{-1}) as shown in Figure 4b and R1 (800 mg N L^{-1}) as shown in Figure 4a. These results suggest that higher initial ammonium concentration might be conducive to the enrichment of nitrifying bacteria that are capable of withstanding higher free ammonia concentrations and, thus, can treat higher influent ammonium concentration. During the whole operation period at 21 °C, the influent ammonium concentration in R3 was always higher than that in R1, indicating further that higher initial ammonium concentration in phase 1 was beneficial for the tolerance of nitrifying granules to high ammonium concentration in the long run.

It was observed, however, that ammonium removal efficiencies in all reactors dropped after they reached the maximum influent ammonium concentrations which had to be reduced to ensure more than 90% ammonium removal efficiencies. R2 collapsed completely due to granule disintegration and biomass wash-out. The free ammonia (FA) concentrations, as shown in Figure S1, increased quickly and reached close to 180 mg N L^{-1} in R2 and R4 at 32 °C and 80 mg N L^{-1} in R1 and R3 at 21 °C, which were higher or close to the maximum FA concentration that AOB could tolerate [30]. The inhibition from FA might be one of the main reasons that nitrifying granules could not deal with the maximum influent ammonium concentration applied to reactors, i.e., 800, 1000, 850, 1300 mg N L^{-1}, for the long run.

Furthermore, it was found from Figure 5 that specific ammonium oxidization rates of nitrifying granules in all reactors dropped from day 58 to day 121, suggesting reduced nitrifying activities. During this period, ash content increased in nitrifying granules, as shown in Figure 3. This phenomenon in nitrifying granules is similar to that reported by Liu et al. [7] that the increased ash content in heterotrophic granules led to reduced microbial activities. Meanwhile, it could be seen from Figure 5 that the specific ammonium oxidization rates of nitrifying granules in R4 at 32 °C were lower than those in R1 and R3 at 21 °C, which is contradictory to the common report that nitrifying bacteria had a higher nitrification rate at a higher temperature. For example, it was reported that a temperature increment at 20 °C resulted in a nitrification rate increase of 4.275% per °C by nitrifying biofilm [31]. Therefore, the most reasonable explanation is that the higher ash content in R4 at 32 °C resulted in the reduced specific ammonium oxidization rate. The reduced microbial activities with the accumulation of inorganics in granules were reported in other studies with other types of granules, as well [15,16,32].

In all 4 reactors, ammonium was mainly oxidized to nitrite with negligible nitrate production (less than 10 mg L^{-1}), and the stable nitritation was maintained during the whole operation period without special controls. High influent ammonium concentration to around 1000 mg L^{-1} might not be the reason for partial nitrification because Chen et al. [28] reported complete nitrification with similar influent ammonium concentration and operational conditions, except for lower calcium concentration and lower ash content. Poot et al. [33] found that the control of the residual ammonium concentration had proven to be effective for repression of *Nitrospira* spp. at 20 °C, and the stratification of an outer AOB layer in the granule structure was found to be highly important to maintain stable partial nitritation in the long term. Thus, it is speculated that relatively high residual ammonium concentration (10 mg N L^{-1}) and high ash content in granules contributed to the maintenance of stable nitritation by increasing mass transfer resistance in granules with ammonium oxidizing bacteria grown in the outer layer of granules.

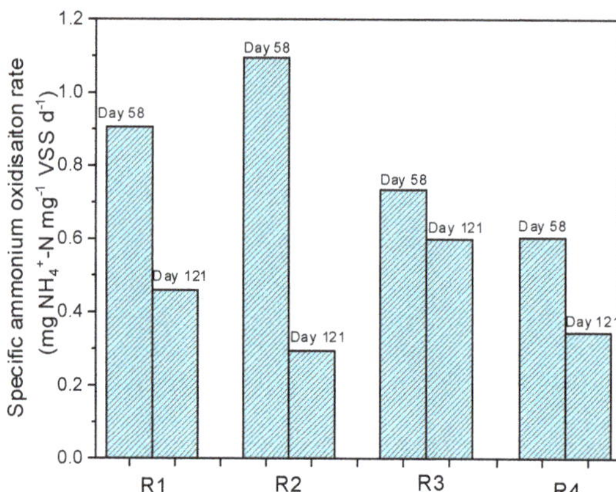

Figure 5. Specific ammonium oxidizing rates of nitrifying granules in 4 reactors on operational days of 58 and 121.

Within 37 days, heterotrophic granules were completely transformed into partial nitrifying granules to treat nitrogen-rich wastewater with influent NH_4^+-N concentration as high as 950 mg L^{-1} at 32 °C and 650 mg L^{-1} at 21 °C. Similar results were also reported by Chen et al. [5], where nitrifying granules could treat 1000 mg NH_4^+-N L^{-1} in less than 55 days from reactor start-up by activated sludge. However, nitrifying sludge was inoculated in their reactors, while activated sludge without nitrification capability was inoculated in our study. The strategy used in this study to form heterotrophic granules rapidly first and then to transform heterotrophic granules to nitrifying granules resulted in the quicker formation of partial nitrifying granules, compared with 70 and 146 days reported in other studies [34–36].

3.4. Calcium Precipitation and Accumulation in Partial Nitrifying Granules at Moderate and High Temperatures

To understand the mineral accumulation in nitrifying granules, metal elements and phosphorus in granules from R3 at 21 °C and R4 at 32 °C were analyzed and shown in Table 2. It can be seen that the inorganic elements contributing to accumulated ash in granules are mainly phosphorus and calcium, with calcium contents up to 17–25% and phosphorus contents up to 4.1–7.7%. According to Table 2, calcium concentration in tap water was 102 mg L^{-1} due to chalk aquifer in Southampton, UK. This hard water is one of the main reasons for the excessive precipitation and accumulation of calcium precipitates in nitrifying granules. Ca/P molar ratio in granules from R3 was 1.74, which is very close to the Ca/P ratio of pure hydroxyapatite (i.e., 1.67). XRD analysis results shown in Figure S2 proved that the main inorganic precipitate in granules from R3 was hydroxyapatite. Based on phosphorus content in Table 2, the calculated hydroxyapatite content of granules in R3 would be 42%, close to the actual ash content of 55%, as shown in Figure 3. Ca/P molar ratio in granules from R4, however, was much higher than 1.67, which reached 4.83. XRD analysis showed the presence of both hydroxyapatite and calcite. Based on the phosphorus content of granules in R4, the calculated hydroxyapatite content was 22%, and, based on calcium content after deducting calcium in hydroxyapatite, the calculated calcite was 42%, resulting in a total ash content of 64%, close to the actual ash content of 73% in Figure 3. The small percentage difference between calculated ash content and actual ash content might be due to a small percentage of other types of amorphous calcium precipitates, which might not yet convert to crystals. It is very interesting to note that higher temperature did not only promote more mineral precipitation and accumulation in granules but also stimulated co-precipitation and accumulation of hydroxyapatite and calcite.

Table 2. The contents of main metal elements and phosphorus in granules from R3 at 21 °C and R4 at 32 °C on day 107 at the end of phase 3.

	Unit	Na	Mg	P	K	Ca	Fe	Ca/P
R3	mg g^{-1} TSS	12.79	0.59	76.80	0.94	172.30	2.94	1.74
R4	mg g^{-1} TSS	11.42	0.91	40.86	0.36	254.40	3.05	4.82
Tap water	mg L^{-1}	13.95	2.18	0.10	1.70	102.70	0.027	
Feedstock elements	mg L^{-1}	7210.00–3285.71	4.15	15.20	19.28	110.46	2.00	5.63

A typical P content in activated sludge is usually less than 1% [37], but, in this study, P content in granules from R3 and R4 reached 7.7% and 4.1%, respectively, which are comparable to the phosphorus content in sewage sludge incineration ash ranging between 4 and 9% [38]. Thus, hydroxyapatite precipitation and accumulation in nitrifying granules could be beneficial for phosphorus removal and recovery. From this perspective, the moderate temperature is more beneficial because hydroxyapatite would be more dominant with a higher phosphorus content at moderate temperature compared with high temperature.

EDX was then conducted to observe the distribution of calcium precipitates on the granule surface and cores by halving granules. It can be seen from Figure 6 that, for granules in both R3 and R4, calcium phosphate and other calcium precipitates were mainly located in the cores of granules, while there were no calcium precipitates on the granule surface. Thus, it is less likely for calcium precipitate to form in bulk liquid and then deposit on the granule surface. The similar element distribution was previously described for both heterotrophic aerobic granules [12] and EBPR granules with 30–35% of ash contents by other researchers [13], who reported that external layers of granules were mainly composed of microorganisms with high contents of C, N, and O, whereas inorganic contents, such as Ca and P, were much higher toward the core of the particles.

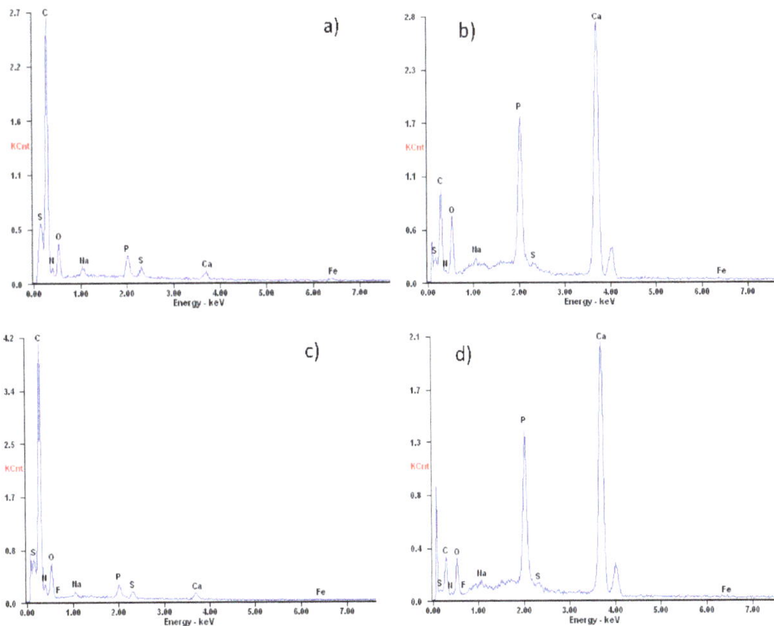

Figure 6. Elemental composition of the (**a**) surface and (**b**) core of granules in R3 and (**c**) surface and (**d**) core of granules in R4 by EDX analysis.

There are two assumptions to explain this phenomenon. One is that inorganic precipitates form in the wastewater first, which then act as carriers for bacteria to attach and form granules [11]. The other possibility is that granule structure and high EPS content in granules could provide favorable micro-environment for biomineralization due to microbial activity, causing biologically induced mineral precipitation in granules [3,13]. It is less likely that mineral precipitation in granules is species-specific, but potential bacteria associated with precipitation of calcium carbonate and calcium phosphate were detected in granular sludge in the co-precipitated calcite and hydroxyapatite [39]. No matter which assumption is tenable, the bacteria tend to grow on the granule surface due to easy access to nutrients and oxygen in water with less mass transfer resistance.

4. Conclusions

Mineral precipitation and accumulation in nitrifying granules were investigated with hard water and two different temperatures, i.e., the moderate temperature at 21 °C and the high temperature at 32 °C. The key findings were summarized as below:

- Initial ammonium concentration and temperature from moderate 21 °C to 32 °C had negligible effects on the speed of heterotrophic granules.
- Mature granules had much higher calcium precipitates accumulated inside compared with newly formed granules, and suspended sludge, indicating the importance of long sludge retention time for the accumulation of calcium precipitates.
- Higher temperature promoted more mineral precipitates in granules, in general. Specifically, high temperature promoted co-precipitation and accumulation of hydroxyapatite and calcite, while moderate temperature was more beneficial for the dominance of hydroxyapatite in granules.
- Increased calcium precipitates in nitrifying granules resulted in reduced microbial activities.
- High mineral content in granules led to the instability of nitrifying granules at a higher temperature in the short or long run.

Overall, mature granules have much higher calcium precipitate contents than young granules due to the long retention time. The temperature has significant effects on calcium precipitation, precipitate species, ash accumulation, ash content, microbial activity, and long-term stability when ash content is high.

Supplementary Materials: The following are available online at https://www.mdpi.com/article/10.3390/pr9101710/s1, Figure S1: Concentrations of free ammonia (FA) (a) and free nitrous acid (FNA) (b) over the operation time with the inreased influent NH4+-N concentration. Figure S2: The inorganic solid phases present within granules from R3 and R4 on day 121 by XRD analysis.

Author Contributions: Conceptualization, Y.-Q.L.; formal analysis, S.C.; investigation, S.C.; writing—original draft preparation, S.C. and Y.-Q.L.; writing—review and editing, Y.-Q.L.; supervision, Y.-Q.L. All authors have read and agreed to the published version of the manuscript.

Funding: This research received no external funding, and the APC was funded by EBNet, BBSRC, UK.

Institutional Review Board Statement: Not applicable.

Informed Consent Statement: Not applicable.

Data Availability Statement: Not applicable.

Conflicts of Interest: The authors declare no conflict of interest.

References

1. Fukuzaki, S.; Nishio, N.; Sakurai, H.; Nagai, S. Characteristics of methanogenic granules grown on propionate in an upflow anaerobic sludge blanket (UASB) reactor. *J. Ferment. Bioeng.* **1991**, *71*, 50–57. [CrossRef]
2. Johansson, S.; Ruscalleda, M.; Colprim, J. Phosphorus recovery through biologically induced precipitation by partial nitritation-anammox granular biomass. *Chem. Eng. J.* **2017**, *327*, 881–888. [CrossRef]

3. de Kreuk, M.; Heijnen, J.; van Loosdrecht, M. Simultaneous COD, nitrogen, and phosphate removal by aerobic granular sludge. *Biotechnol. Bioeng.* **2005**, *90*, 761–769. [CrossRef]

4. Liu, Y.-Q.; Lan, G.-H.; Zeng, P. Size-dependent calcium carbonate precipitation induced microbiologically in aerobic granules. *Chem. Eng. J.* **2016**, *285*, 341–348. [CrossRef]

5. Lin, Y.; Lotti, T.; Sharma, P.; van Loosdrecht, M. Apatite accumulation enhances the mechanical property of anammox granules. *Water Res.* **2013**, *47*, 4556–4566. [CrossRef] [PubMed]

6. Juang, Y.-C.; Adav, S.S.; Lee, D.-J.; Tay, J.-H. Stable aerobic granules for continuous-flow reactors: Precipitating calci-um and iron salts in granular interiors. *Bioresour. Technol.* **2010**, *101*, 8051–8057. [CrossRef] [PubMed]

7. Liu, Y.-Q.; Lan, G.-H.; Zeng, P. Excessive precipitation of CaCO3 as aragonite in a continuous aerobic granular sludge reactor. *Appl. Microbiol. Biotechnol.* **2015**, *99*, 8225–8234. [CrossRef] [PubMed]

8. Jiang, H.-L.; Tay, J.-H.; Liu, Y.; Tay, S.T.-L. Ca^{2+} augmentation for enhancement of aerobically grown microbial gran-ules in sludge blanket reactors. *Biotechnol. Lett.* **2003**, *25*, 95–99. [CrossRef] [PubMed]

9. Wang, S.; Shi, W.; Yu, S.; Yi, X.; Yang, X. Formation of aerobic granules by Mg^{2+} and Al^{3+} augmentation in sequenc-ing batch airlift reactor at low temperature. *Bioprocess Biosyst. Eng.* **2012**, *35*, 1049–1055. [CrossRef] [PubMed]

10. Yilmaz, G.; Bozkurt, U.; Magden, K.A. Effect of iron ions (Fe^{2+}, Fe^{3+}) on the formation and structure of aerobic gran-ular sludge. *Biodegradation* **2017**, *28*, 53–68. [CrossRef]

11. Ma, H.; Xue, Y.; Zhang, Y.; Kobayashi, T.; Kubota, K.; Li, Y.-Y. Simultaneous nitrogen removal and phosphorus recov-ery using an anammox expanded reactor operated at 25° C. *Water Res.* **2020**, *172*, 115510. [CrossRef] [PubMed]

12. Ren, T.-T.; Liu, L.; Sheng, G.-P.; Liu, X.-W.; Yu, H.-Q.; Zhang, M.-C.; Zhu, J.-R. Calcium spatial distribution in aerobic granules and its effects on granule structure, strength and bioactivity. *Water Res.* **2008**, *42*, 3343–3352. [CrossRef] [PubMed]

13. Mañas, A.; Biscans, B.; Sperandio, M. Biologically induced phosphorus precipitation in aerobic granular sludge process. *Water Res.* **2011**, *45*, 3776–3786. [CrossRef]

14. Yu, T.; Tian, L.; You, X.; Wang, L.; Zhao, S.; Kang, D.; Xu, D.; Zeng, Z.; Zhang, M.; Zheng, P. Deactivation mechanism of calcified anaerobic granule: Space occupation and pore blockage. *Water Res.* **2019**, *166*, 115062. [CrossRef] [PubMed]

15. Feldman, H.; Alsina, X.F.; Ramin, P.; Kjellberg, K.; Jeppsson, U.; Batstone, D.J.; Gernaey, K.V. Assessing the effects of intra-granule precipitation in a full-scale industrial anaerobic digester. *Water Sci. Technol.* **2019**, *79*, 1327–1337. [CrossRef] [PubMed]

16. Van Langerak, E.; Ramaekers, H.; Wiechers, J.; Veeken, A.; Hamelers, H.; Lettinga, G. Impact of location of CaCO3 precipitation on the development of intact anaerobic sludge. *Water Res.* **2000**, *34*, 437–446. [CrossRef]

17. de Kreuk, M.; Pronk, M.; van Loosdrecht, M. Formation of aerobic granules and conversion processes in an aerobic granular sludge reactor at moderate and low temperatures. *Water Res.* **2005**, *39*, 4476–4484. [CrossRef]

18. Ab Halim, M.H.; Anuar, A.N.; Jamal, N.S.A.; Azmi, S.I.; Ujang, Z.; Bob, M.M. Influence of high temperature on the performance of aerobic granular sludge in biological treatment of wastewater. *J. Environ. Manag.* **2016**, *184*, 271–280. [CrossRef]

19. APHA. *Standard Methods of Water and Wastewater*, 22nd ed.; American Public Health Association, American Water Works Association, Water Environment Federation: Washington, DC, USA, 2012.

20. Liu, Y.Q.; Kong, Y.H.; Zhang, R.; Zhang, X.; Wong, F.S.; Tay, J.H.; Zhu, J.R.; Jiang, W.J.; Liu, W.-T. Microbial population dynamics of granular aerobic sequencing batch reactors during start-up and steady state periods. *Water Sci. Technol.* **2010**, *62*, 1281–1287. [CrossRef]

21. Beun, J.; Hendriks, A.; van Loosdrecht, M.; Morgenroth, E.; Wilderer, P.; Heijnen, J. Aerobic granulation in a sequencing batch reactor. *Water Res.* **1999**, *33*, 2283–2290. [CrossRef]

22. BSI. *BS 6068-2.11:1984, ISO 7150-1:1984 Water Quality. Physical, Chemical and Biochemical Methods. Determination of Ammonium: Manual Spectrometric Method*; British Standards Institution: London, UK, 1984.

23. Liu, Y.-Q.; Moy, B.; Kong, Y.-H.; Tay, J.-H. Formation, physical characteristics and microbial community structure of aerobic granules in a pilot-scale sequencing batch reactor for real wastewater treatment. *Enzym. Microb. Technol.* **2010**, *46*, 520–525. [CrossRef] [PubMed]

24. Liu, Y.-Q.; Kong, Y.; Tay, J.-H.; Zhu, J. Enhancement of start-up of pilot-scale granular SBR fed with real wastewater. *Sep. Purif. Technol.* **2011**, *82*, 190–196. [CrossRef]

25. Liu, Y.-Q.; Tay, J.-H. Characteristics and stability of aerobic granules cultivated with different starvation time. *Appl. Microbiol. Biotechnol.* **2007**, *75*, 205–210. [CrossRef] [PubMed]

26. Yang, S.-F.; Tay, J.-H.; Liu, Y. Inhibition of free ammonia to the formation of aerobic granules. *Biochem. Eng. J.* **2004**, *17*, 41–48. [CrossRef]

27. Liu, Y.-Q.; Tay, J.-H. Fast formation of aerobic granules by combining strong hydraulic selection pressure with over-stressed organic loading rate. *Water Res.* **2015**, *80*, 256–266. [CrossRef]

28. Chen, F.-Y.; Liu, Y.-Q.; Tay, J.-H.; Ning, P. Rapid formation of nitrifying granules treating high-strength ammonium wastewater in a sequencing batch reactor. *Appl. Microbiol. Biotechnol.* **2015**, *99*, 4445–4452. [CrossRef]

29. Xue, Y.; Ma, H.; Kong, Z.; Guo, Y.; Li, Y.-Y. Bulking and floatation of the anammox-HAP granule caused by low phos-phate concentration in the anammox reactor of expanded granular sludge bed (EGSB). *Bioresour. Technol.* **2020**, *310*, 123421. [CrossRef]

30. Anthonisen, A.C.; Loehr, R.C.; Prakasam, T.B.S.; Srinath, E.G. Inhibition of Nitrification by Ammonia and Nitrous Acid. *J. Water Pollut. Control. Fed.* **1976**, *48*, 835–852.

31. Zhu, S.; Chen, S. The impact of temperature on nitrification rate in fixed film biofilters. *Aquac. Eng.* **2002**, *26*, 221–237. [CrossRef]

32. Xing, B.-S.; Guo, Q.; Yang, G.-F.; Zhang, Z.-Z.; Li, P.; Guo, L.-X.; Jin, R.-C. The properties of anaerobic ammonium oxi-dation (anammox) granules: Roles of ambient temperature, salinity and calcium concentration. *Sep. Purif. Technol.* **2015**, *147*, 311–318. [CrossRef]
33. Poot, V.; Hoekstra, M.; Geleijnse, M.A.; van Loosdrecht, M.C.; Pérez, J. Effects of the residual ammonium concentra-tion on NOB repression during partial nitritation with granular sludge. *Water Res.* **2016**, *106*, 518–530. [CrossRef]
34. Kim, D.-J.; Seo, D. Selective enrichment and granulation of ammonia oxidizers in a sequencing batch airlift reactor. *Process. Biochem.* **2006**, *41*, 1055–1062. [CrossRef]
35. Song, Y.; Ishii, S.; Rathnayak, L.; Ito, T.; Satoh, H.; Okabe, S. Development and characterization of the partial nitrifica-tion aerobic granules in a sequencing batch airlift reactor. *Bioresour. Technol.* **2013**, *139*, 285–291. [CrossRef]
36. Wei, D.; Xue, X.; Yan, L.-G.; Sun, M.; Zhang, G.; Shi, L.; Du, B. Effect of influent ammonium concentration on the shift of full nitritation to partial nitrification in a sequencing batch reactor at ambient temperature. *Chem. Eng. J.* **2014**, *235*, 19–26. [CrossRef]
37. Metcalf, E. *Wastewater Engineering: Treatment, Disposal and Reuse*; McGraw-Hill Inc.: New York, NY, USA, 2003.
38. Franz, M. Phosphate fertilizer from sewage sludge ash (SSA). *Waste Manag.* **2008**, *28*, 1809–1818. [CrossRef] [PubMed]
39. Gonzalez-Martinez, A.; Rodriguez-Sanchez, A.; Rivadeneyra, M.A.; Rivadeneyra, A.; Martin-Ramos, D.; Vahala, R.; Gonzalez-Lopez, J. 16S rRNA gene-based characterization of bacteria potentially associated with phosphate and carbonate precipitation from a granular autotrophic nitrogen removal bioreactor. *Appl. Microbiol. Biotechnol.* **2016**, *101*, 817–829. [CrossRef] [PubMed]

Article

Effect of an Increased Particulate COD Load on the Aerobic Granular Sludge Process: A Full Scale Study

Sara Toja Ortega [1,*], Mario Pronk [2,3] and Merle K. de Kreuk [1]

[1] Section Sanitary Engineering, Department of Water Management, Delft University of Technology, Stevinweg 1, 2628 CN Delft, The Netherlands; m.k.dekreuk@tudelft.nl
[2] Department of Biotechnology, Delft University of Technology, Van der Maasweg 9, 2629 HZ Delft, The Netherlands; m.pronk@tudelft.nl
[3] Royal HaskoningDHV, Laan 1914 35, 3800 AL Amersfoort, The Netherlands
[*] Correspondence: s.tojaortega@tudelft.nl

Abstract: High concentrations of particulate COD (pCOD) in the influent of aerobic granular sludge (AGS) systems are often associated to small granule diameter and a large fraction of flocculent sludge. At high particulate concentrations even granule stability and process performance might be compromised. However, pilot- or full-scale studies focusing on the effect of real wastewater particulates on AGS are scarce. This study describes a 3-month period of increased particulate loading at a municipal AGS wastewater treatment plant. The pCOD concentration of the influent increased from 0.5 g COD/L to 1.3 g COD/L, by adding an untreated slaughterhouse wastewater source to the influent. Sludge concentration, waste sludge production and COD and nutrient removal performance were monitored. Furthermore, to investigate how the sludge acclimatises to a higher influent particulate content, lipase and protease hydrolytic activities were studied, as well as the microbial community composition of the sludge. The composition of the granule bed and nutrient removal efficiency did not change considerably by the increased pCOD. Interestingly, the biomass-specific hydrolytic activities of the sludge did not increase during the test period either. However, already during normal operation the aerobic granules and flocs exhibited a hydrolytic potential that exceeded the influent concentrations of proteins and lipids. Microbial community analysis also revealed a high proportion of putative hydrolysing and fermenting organisms in the sludge, both during normal operation and during the test period. The results of this study highlight the robustness of the full-scale AGS process, which can bear a substantial increase in the influent pCOD concentration during an extended period.

Keywords: aerobic granular sludge; particulate COD; full-scale wastewater treatment; nutrient removal; granule stability

Citation: Toja Ortega, S.; Pronk, M.; de Kreuk, M.K. Effect of an Increased Particulate COD Load on the Aerobic Granular Sludge Process: A Full Scale Study. *Processes* **2021**, *9*, 1472. https://doi.org/10.3390/pr9081472

Academic Editor: Yongqiang Liu

Received: 18 July 2021
Accepted: 20 August 2021
Published: 23 August 2021

1. Introduction

Aerobic granular sludge (AGS) technology is currently applied in more than 80 plants over the world treating domestic and industrial wastewater, under the tradename Nereda® (Royal HaskoningDHV, Amersfoort, The Netherlands) [1]. AGS offers various advantages in comparison to conventional activated sludge systems, including lower space and energy requirements [2]. Furthermore, the granular sludge offers a number of resource recovery options [3]. The technology is operated as a sequencing batch reactor (SBR) consisting of simultaneous anaerobic plug-flow feeding and effluent withdrawal, followed by aeration and settling [4]. The anaerobic feeding regime ensures anaerobic uptake of the readily biodegradable COD (rbCOD) by slow-growing organisms, which store this COD intracellularly and oxidise it in the subsequent aerobic phase [5]. This way, slow-growing organisms such as polyphosphate accumulating organisms (PAO) and glycogen accumulating organisms (GAO) are favoured over ordinary heterotrophic organisms (OHO), which results in smooth granule growth [6].

In addition to rbCOD, domestic wastewater and many industrial wastewaters contain a large fraction of slowly biodegradable, particulate COD (pCOD) [7–9]. This COD fraction may provide additional substrate for nutrient removal and granular sludge production, from which resources could be recovered ultimately. On the other hand, it could challenge the stability of the AGS process, due to the inability of PAO and GAO to consume more complex COD directly; pCOD needs to be extracellularly hydrolysed to rbCOD, after which it can be assimilated by the microorganisms in the sludge [10]. Hydrolysis is often regarded as rate-limiting in the degradation of particulates [11,12]. It relies on contact between the substrate and hydrolytic enzymes, which are mainly reported to be biomass-bound [13–15]. In activated sludge, the open floc structure allows good contact between sludge flocs and pCOD, rendering a continuous and homogeneous release of rbCOD throughout the floc [16]. In contrast, the dense biofilm structure of AGS limits the accessibility of pCOD to hydrolytic enzymes in the sludge. Such a mass-transfer limitation has induced irregular granule growth and deterioration of granule stability in lab-scale reactors fed with high influent particulates [17–19]. This is attributed to hydrolysis of pCOD into rbCOD during the aeration phase on the granule surface, leading to substrate concentration gradients and aerobic rbCOD consumption by fast-growing OHO [20]. Nevertheless, AGS has been cultivated successfully on domestic wastewater [21,22], high-strength industrial wastewaters [23–25] and faecal sludge-containing wastewater [26]. Even though the influent to these reactors contained a high concentration of particulates, no filamentous outgrowth was observed. However, these studies reported much higher effluent suspended solids than generally achieved in the effluent of full-scale Nereda® reactors or found a flocculent sludge fraction accompanying the AGS. Overall, the sludge bed consisted of smaller granules compared to lab reactors fed with synthetic media containing solely rbCOD in all reported lab- and pilot-scale experiments. Layer et al. [27] hypothesised that the flocculent sludge fraction benefitted the AGS process, by capturing influent pCOD and avoiding its aerobic degradation at the granule surface. Haaksman et al. [28], similarly, proposed that rbCOD leakage to the aerobic phase mainly results in flocculent sludge growth, which has to be discharged regularly in order to maintain a stable granule bed. These studies suggest that AGS can adapt well to high concentrations of pCOD in the wastewater when suitable operating conditions are employed. Moreover, they suggest that granule stability is supported by an equilibrium between different biomass fractions within the AGS reactor. However, reports on full-scale AGS treating particulate-rich wastewater are scarce and they do not focus on the removal of pCOD. Thus, the effect of pCOD on granule stability and sludge bed composition at full-scale remains elusive. Furthermore, little is known about the enzymatic activity and microbial community of full-scale AGS cultivated on high influent pCOD concentrations.

To study the effect of particulates on AGS stability, a full-scale study was conducted at wastewater treatment plant (WWTP) Epe, the Netherlands, equipped with three 4500 m^3 reactors. The Nereda® reactors were operated with an increased influent COD load during a 4-month period (4 March–11 July 2019). The extra influent COD originated from the temporary discharge of untreated wastewater of a nearby slaughterhouse and consisted mainly of pCOD (71–86% of the total COD). During the experimental period, sludge growth, granule size and morphology and plant performance were monitored. Furthermore, the hydrolytic enzyme activity of the sludge was monitored to study its location (granular or flocculent sludge) and to compare the hydrolysis rates during operation with and without this additional untreated slaughterhouse wastewater. In this case, 16S rRNA gene amplicon sequencing was used to identify the main organisms in the sludge, and to detect shifts in the microbial community during increased pCOD loading. In this way, the present study aimed to assess the robustness of full-scale AGS reactors faced with an increase in influent particulate content, while exploring the involvement of different sludge fractions in the removal of particulates.

2. Materials and Methods

2.1. Description of the Plant and Additional Influent Dosing

WWTP Epe, operated by the Water Authority Vallei en Veluwe, treats the wastewater of the municipality of Epe (The Netherlands) and its surroundings. The wastewater is pre-treated with 3 mm screens and grit removal and treated in three 4500 m^3 AGS reactors with a diameter of 25 m and a depth of 9.2 m, designed and built by Royal HaskoningDHV. The reactors are operated in SBR mode with simultaneous anaerobic feeding/effluent discharge, aeration and settling. The wastewater is fed from the bottom of the reactor in a plug-flow regime, at an average upflow velocity of approximately 1 m h^{-1}. At the end of the settling phase, the slowest settling sludge is removed selectively. The influent of WWTP Epe is a mixture of domestic (70%), and industrial (30%) wastewater mainly originating from slaughterhouses. Wastewater composition is summarised in Table 1. The wastewater is treated to reach the effluent consents summarised in Table 2. The plant was designed to treat an average flow of 8000 m^3 d^{-1}, and a load of 59,000 population equivalents (PE). Nevertheless, the conditions at the time of the study have not reached the design values: the average flow is approximately 5000 m^3 d^{-1} and the load is 35,000 PE.

In the period 4 March–11 July 2019, the wastewater of a nearby slaughterhouse was discharged untreated to the municipal sewer network leading to WWTP Epe. The additional influent increased the substrate loading to the Nereda® reactors, approaching the design values. The objective was to enhance the growth of granular sludge while maintaining good pollutant removal performance. In a regular situation, the slaughterhouse treats its own wastewater using a dissolved air flotation (DAF) system. The treatment involves addition of coagulants and removal of solids by flotation, removing 80–90% of the COD of the wastewater. The treated wastewater, mostly composed of soluble COD, is discharged to the municipal sewer network and arrives at WWTP Epe, at approximately 500 m distance. During the test period, in-house treatment was stopped and concentrated raw wastewater was discharged, with a high content of suspended solids (see Table 1). The wastewater was continuously discharged during 12 h per day, with an average flow of 300 m^3 d^{-1} and a maximum flow of 50 m^3 h^{-1}.

Table 1. Influent composition of WWTP Epe during normal operation and during the test period. Values are expressed as average ± standard deviation. Routine measurements: normal operation $n = 79$; test period $n = 21$. Occasional miscellaneous analysis: normal operation $n = 6$; test period $n = 3$. Asterisks denote significant changes (* = $p < 0.05$; *** = $p < 0.001$).

	Normal Operation	Test Period	Significance
Routine measurements			
tCOD [g m^{-3}]	840 ± 254	1456 ± 692	***
TSS [g m^{-3}]	317 ± 123	633 ± 361	***
BOD$_5$ [g m^{-3}]	341 ± 109	537 ± 249	***
TN [g m^{-3}]	77 ± 21	101 ± 39	***
TP [g m^{-3}]	8 ± 2	13 ± 5	***
Q [m^3/d]	4696 ± 2099	5229 ± 3418	
COD/N [g/g]	11 ± 3	14 ± 4	***
COD/P [g/g]	101 ± 15	117 ± 24	***
Occasional miscellaneous analysis			
tCOD [g m^{-3}]	864 ± 274	1713 ± 572	*
sCOD [g m^{-3}]	339 ± 129	386 ± 49	
Acetate [g COD m^{-3}]	82 ± 59	43 ± 37	
Propionate [g COD m^{-3}]	14 ± 12	16 ± 17	
Lipids [g COD m^{-3}]	45 ± 20	290 ± 77	*
Total carbohydrates [g COD m^{-3}]	274 ± 155	538 ± 125	
Total proteins [g COD m^{-3}]	90 ± 11	162 ± 53	
Soluble carbohydrates [g COD m^{-3}]	18 ± 1	21 ± 2	
Soluble proteins [g COD m^{-3}]	20 ± 7	33 ± 23	

Table 2. Average effluent concentration of the main wastewater pollutants, in the months before the test period and during the test period. Standard deviation values are given between parentheses.

	COD $[\mathrm{g\ m^{-3}}]$	BOD$_5$ $[\mathrm{g\ m^{-3}}]$	TN $[\mathrm{g\ m^{-3}}]$	NH$_4^+$-N $[\mathrm{g\ m^{-3}}]$	NO$_X$-N $[\mathrm{g\ m^{-3}}]$	PO$_4$-P $[\mathrm{g\ m^{-3}}]$	TP $[\mathrm{g\ m^{-3}}]$	TSS $[\mathrm{g\ m^{-3}}]$
Effluent consent		7	5				0.3	30
Before test period	26 ($\pm$6)	1.5 ($\pm$0.5)	4.0 ($\pm$2.0)	0.3 ($\pm$0.2)	2.4 ($\pm$1.8)	0.04 ($\pm$0.01)	0.14 ($\pm$0.04)	5.2 ($\pm$1.6)
During test period	32 ($\pm$9)	2.4 ($\pm$1.3)	3.5 ($\pm$1.4)	0.3 ($\pm$0.3)	1.5 ($\pm$0.9)	0.07 ($\pm$0.09)	0.23 ($\pm$0.12)	5.5 ($\pm$1.2)

2.2. Sampling and Sample Handling

For this study, reactor mixed liquor samples and influent wastewater samples were collected. The samples were collected during two phases: before the test period (in a stretch of 5 months), and during the final week of the test period.

2.2.1. Influent Wastewater

Influent wastewater was collected using a flow-proportional 24-hour sampler, after screening and grit removal. The samples were stored at 4 °C in the sampler chamber, and then transported to the lab for analysis. Part of the wastewater was filtered using a vacuum filter device to characterise the particulate size-fractions. Sequential filtering was performed using filters with the following pore sizes: 100 μm, 10 μm, 1 μm and 0.1 μm (Product details: 100 μm: stainless steel mesh (Anglo Staaal, Borne, The Netherlands); 10 μm: Cyclopore®, polycarbonate [PC], 1 μm: Whatman® GF/B, glass fibre; 0.1 μm: Cyclopore®, PC. Whatman, Buckinghamshire, UK). The filtrate of each step was collected and stored for analysis. A small sample was also filtered through 0.45 μm for soluble COD (sCOD) analysis (Durapore® filters, PVDF. Merck, Darmstadt, Germany). Samples were preserved at 4 °C for short-term storage, and −20 °C for long-term storage.

2.2.2. Reactor Mixed Liquor

Reactor mixed liquor was sampled to study hydrolytic activity and sludge characteristics. Mixed liquor samples were collected during the aeration phase, at least 40 min after the beginning of aeration to ensure a completely mixed sample. Sieve fractions of the mixed liquor were obtained by pouring the mixed liquor through a stack of sieves of 0.045, 0.2 and 1 mm, and gently rinsing with tap water. The fraction smaller than 0.045 mm was centrifuged and the supernatant was kept. The following fractions were collected: mixed sludge; bulk (<0.045 mm fraction, centrifuged); flocs (0.045–0.2 mm); and granules (>1 mm). The fraction between 0.2 and 1 mm was excluded from analysis due to its high amount of debris and heterogeneous composition.

2.3. Analytical Methods

Several measurements were performed on the influent and its size fractions, as well as the two samples of slaughterhouse wastewater. TSS and VSS were measured according to the Standard Methods [29]. Volatile fatty acids (VFA) acetate and propionate were quantified using high performance liquid chromatography (HPLC) with an Aminex HPX-87H column from Bio-Rad (Hercules, CA, USA). tCOD was measured using photochemical test kits from Hach (Dusseldorf, Germany). sCOD was determined by measuring COD in the wastewater filtered through 0.45 μm. Particulate COD (pCOD) was calculated as tCOD − sCOD. Proteins were quantified using the modified Lowry method [13]. Carbohydrates were quantified using the anthrone-sulfuric acid method [30]. Lipids were measured by Merieux Nutrisciences (Resana, Italy) using the gravimetric method.

Sludge was inspected using a Keyence VHX-700F digital microscope (Mechelen, Belgium). The TS and VS of the sludge used in activity assays were determined according to the standard Methods [29].

2.4. Hydrolytic Enzyme Activity Tests

Lipase and protease enzyme activities of the sludge were assessed in the mixed sludge, granules, flocs and bulk liquid. The enzyme assays were performed within 8 h after sampling. The procedure of the assays is described in detail in Toja Ortega et al. (in press) [31]. In short, the sludge was incubated with chromogenic substrates, in anaerobic vials with a sampling port. The assay was conducted at 20 °C, pH 7.5 and at fully mixed conditions, using a Fisherbrand Seastar orbital shaker (Thermo Fisher Scientific, Waltham, USA) running at 120 rpm. The vials were flushed with N_2 gas for 2 min to provide anaerobic conditions. The substrates used were azocasein for the protease assays, and *p*-nitrophenyl-palmitate (*p*NP-palmitate) for the lipase assays (Sigma-Aldrich, Darmstadt, Germany). The assay vials were sampled every 15–30 min and the samples were filtered through 0.45 μm with a syringe filter to remove biomass. 1 mL sample was mixed with 1 mL TCA to stop enzyme activity. Samples were stored at −20 °C until analysed, and then thawed, centrifuged and filtered through 0.45 μm. The filtrate was mixed on a 1:1 proportion with NaOH 2M and its absorbance was measured, at 410 nm in the lipase assay and 440 nm in the protease assay. The increase in absorbance over time was translated to substrate hydrolysis rate, as described in Toja Ortega et al. [31]. The sludge-specific activity was calculated considering the sludge concentration in the vials. Finally, the total activity contained in the reactor was calculated by multiplying the specific activity of each sludge fraction by its abundance in the reactor.

2.5. Routine Measurements

Routine influent and effluent measurements were conducted by a certified lab and provided by Water Authority Vallei en Veluwe. These included: COD, biological oxygen demand (BOD_5), total nitrogen (TN), total phosphorus (TP), total suspended solids (TSS), ammonia nitrogen (NH_4-N), nitrate/nitrite nitrogen (NO_X-N) and flow. As were the excess sludge production, long-term reactor sludge concentration and granule size distribution measurements.

In addition to influent and effluent composition, key performance indicators (KPI) were used to follow reactor performance during the SBR cycle. Online samplers measure nutrient profiles during the AGS cycle (NH_4-N, NO_3-N, PO_4-P, dissolved oxygen [DO]). Measurement data were retrieved from the Aquasuite Nereda® controllers ® (Royal HaskoningDHV, Amersfoort, The Netherlands) and provided by Royal Haskoning DHV.

2.6. Microbial Population Analysis

2.6.1. Sludge Processing and DNA Extraction

The mixed sludge and sludge sieve fractions were homogenised using a Potter-Elvehjem tissue grinder to ensure a representative sludge sample for DNA extraction. The homogenised sludge was transferred to 1.5 mL Eppendorf tubes (Eppendorf, Hamburg, Germany) and centrifuged in a microcentrifuge (Eppendorf, Hamburg, Germany) at 14,000 g for 5 min. Around 30 mg of pellet were added to extraction tubes from the FastDNA spin kit for soil (MP Biomedicals, Irvine, CA, USA). The DNA was extracted following the protocol optimised by Albertsen et al. for activated sludge samples [32]. Each sample was extracted three times, to improve the recovery of the DNA of all microorganisms in the samples [33,34]. The concentration of the extracted DNA was measured using a Qubit dsDNA HS assay kit (Thermo Fisher Scientific, Waltham, MA, USA).

2.6.2. 16S rRNA Gene Amplicon Sequencing and Data Analysis

The 16S rRNA gene was amplified and paired-end sequenced in an llumina NovaSeq 6000 platform by Novogene (Beijing, China). The hypervariable regions V3–V4 were amplified and sequenced, using the primer set 341F [5′–CCTAYGGGRBGCASCAG–3′] and 806R [5′–GGACTACNNGGGTATCTAAT–3′]. The raw reads were deposited in the National Center for Biotechnology Information (NCBI) Sequence Read Archive (SRA) on BioProject PRJNA746138.

The trimmed and merged sequences provided by Novogene were processed using QIIME2, version 2020.2 [35]. The sequences were quality-filtered using Deblur [36], trimming the sequences the 3′ end at position 403 (parameter p-trunc-len). The remaining sequences were assembled into a phylogenetic tree to perform diversity analyses, using the q2-phylogeny plugin. Beta diversity metrics (Bray–Curtis and Unweighted Unifrac) were generated, and differences in beta diversity between sludge types and experimental conditions were analysed using PERMANOVA [37]. A p-value of 0.05 was used as cut-off for significance. Finally, taxonomic affiliation of the sequences was determined by aligning the sequences to the MiDAS 3.6 database [38]. Sample subsetting, visualisation and further statistical analysis was performed in R, using the Phyloseq and Ampvis2 packages [39,40]. Abundant taxa were defined as taxa with 1% abundance of more. Per sludge type, significant differences in abundant taxa before and after the test period were determined through t-tests. A Bonferroni-corrected p-value was used, of 0.01 divided by the number of abundant taxa per sludge type.

3. Results

3.1. Additional Influent Dosing and Changes in Influent Composition

During the test period, the influent COD concentration almost doubled (from 840 to 1460 g m^{-3}). The pCOD, in particular, increased from 525 to 1327 g m^{-3}, along with the TSS content, which increased from 317 to 633 g m^{-3}. Table 1 summarises the characteristics of the influent to the wastewater treatment plant of Epe, before and during the addition of the untreated slaughterhouse wastewater. The VFA content in the influent was very variable, and the average VFA concentrations before and after the test period did not significantly change ($p = 0.39$). The sCOD measurements did not indicate an increase in soluble compounds in the wastewater either. Furthermore, particle size measurements of the influent showed that the extra COD during the test period came in the fractions between 1 and 100 μm, especially in the size fraction of 10 to 100 μm (Supplementary Information S1). The smallest COD fraction (<0.1 μm) had an average concentration of 260 ± 92 g m^{-3}, before the test period, and 331 ± 61 g m^{-3} during the test period. However, the differences on this size fraction between both situations are not statistically significant ($p = 0.11$). From the macromolecules quantified (proteins, carbohydrates and lipids), lipid concentration increased most, from 45 to 290 g m^{-3}.

In short, there was a significant increase in the influent concentration of tCOD ($p < 0.001$) and of TSS ($p < 0.001$). sCOD and VFA concentrations, in contrast, were not significantly affected by the addition of the untreated slaughterhouse wastewater, since the location is only 500 m from the WWTP. Apparently, there was not enough residence time in the sewer for fermentation of this wastewater, despite the warm spring season. Even though sCOD did not increase, the BOD$_5$ was higher during the test period, indicating the presence of particulate substrates with high biodegradability in the added untreated slaughterhouse wastewater.

3.2. Sludge Production

The biomass concentration of the reactors increased during the 4 months of changed influent. On the year before the test, the average TS concentration of the reactors was 5.8 ± 1.7 g TS/L. At the end of the test period, the average sludge concentration was 8.3 ± 0.6 g TS/L. However, it should be noted that already the months before the test period the sludge concentration had started to increase (Supplementary Information S2). The granule bed was composed of different sludge morphologies (size fractions). The measured percentage of each sludge size fraction over the total sludge oscillated during the experimental period. Overall, the trend showed an increased large granule fraction (from 56 to 67%) and a decreased smaller granule fraction (from 23 to 18%) and floc fraction (from 20 to 15%). However, such tendency could not be attributed to the test period alone, because the change in sludge bed composition occurred during a longer time period and not only during the test itself (Supplementary Information S2). Furthermore, protozoa growth was observed on granule surfaces (Figure 1), but this did not affect the settleability of the sludge; SVI_5 was maintained around 40–45 mL/g VS.

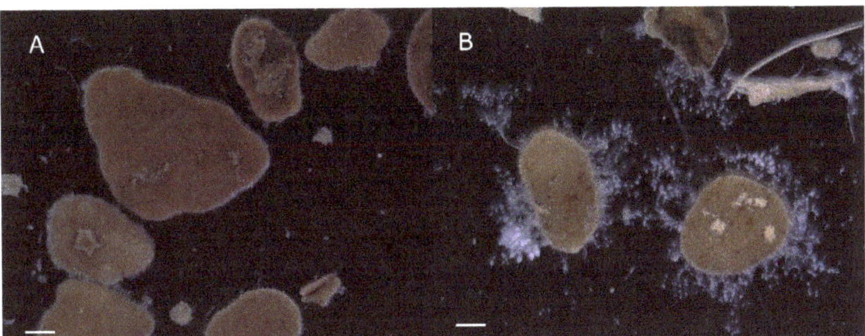

Figure 1. Micrographs of large granules (>1 mm): (**A**) on 23 January 2019 (normal operation); and (**B**) on 9 July 2019 (at the end of the test period). Size bar: 0.5 mm.

During the test period, the volume of the wasted sludge increased (Figure 2). The amount of wet sludge withdrawn doubled during the test period, from 870 to 1640 tonnes per month. The TS measurements of the waste sludge were lower during the test period than during normal operation (40 ± 10 g TS/L versus 31 ± 8 g TS/L). Thus, in terms of dry weight the increase in wasted sludge was less pronounced: 48.8 tonnes TS per month instead of the normal 34.7 tonnes TS per month, or around 13 tonnes TS per month higher. These data show that the additional influent particulates were at least partly degraded during the process. If the particulates were removed non-degraded with the excess sludge, the sludge production would have been considerably higher during the test period. Considering the increase of 0.3 g TSS/L in the influent, sludge wasting would increase 50 tonnes TS/month instead of the observed 13 tonnes TS/month. The sludge production data should be interpreted with caution, though. The measurement of the dry solids content of the excess sludge was not very accurate, due to difficulties associated with full-scale monitoring. The operation of the sludge thickening unit was fluctuating, resulting on a variable spill concentration. Therefore, the TS measurements on weekly grab samples of the thickened sludge might not be fully representative of the TS concentration of the spill over the whole period. Nevertheless, such a large gap between the expected and observed spill production still indicates that a fraction of particulate COD was likely to be consumed by the sludge. For a more detailed mass balance, check Supplementary Information S5.

3.3. Hydrolytic Enzyme Activity Tests

Hydrolytic enzyme activity assays showed that both granules and flocs have the ability to hydrolyse proteins and lipids (Figure 3). No protease and lipase activity was detected in the bulk liquid.

Before adding the particulate-rich influent stream, the specific protease activity (i.e., per gram of VS) of granules was about half of that of flocs. Nevertheless, due to the higher percentage of granular than flocculent sludge in the reactor, the total contribution to protease activity of the granule fraction in the reactor was as high as that of the flocculent fraction (Figure 3B). It is noteworthy that the large granule and floc activities do not add up to the total activity in the mixed sludge. This can, at least partly, be explained by the activity of the small granule fraction (0.2–1 mm) which was not measured individually but is also part of the mixed liquor sample [31]. Increasing the influent pCOD did not significantly affect the biomass-specific protease activity of the mixed liquor and the granule fraction, and flocculent sludge activity decreased (Figure 3A). The total activity at reactor scale seemed to increase due to higher sludge concentration after the test period, although the large variation in activity in the mixed sludge fraction makes the difference statistically just not significant ($p = 0.056$) (Figure 3B). The total protease activity contributed by large granules did increase significantly ($p = 0.03$).

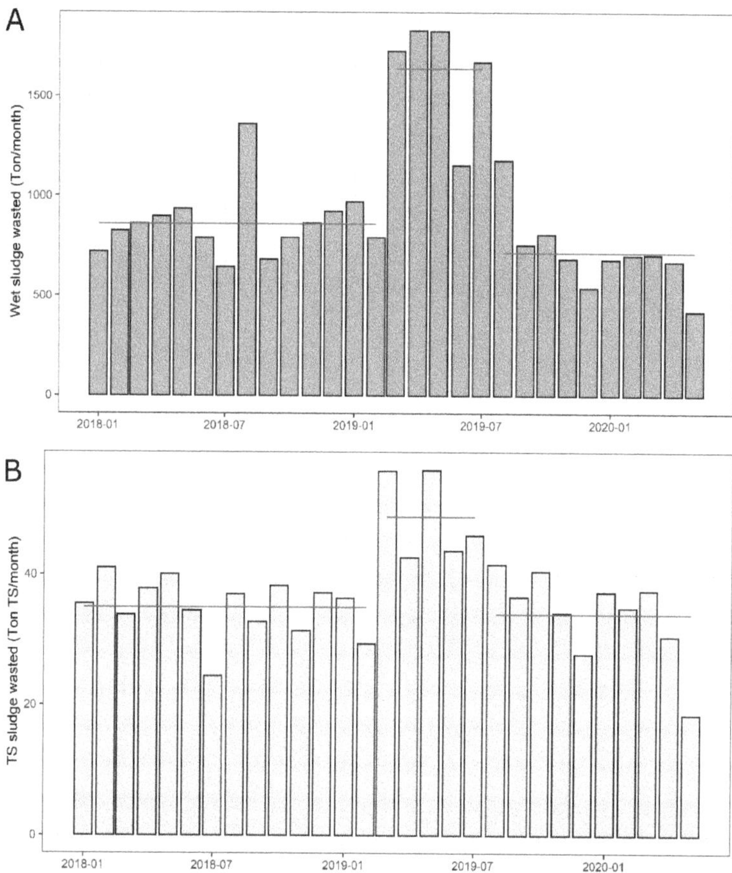

Figure 2. Monthly waste sludge production. (**A**) Tons of wet sludge produced (**B**) Tons of TS produced, based on the measured TS content of the waste sludge (40 g/L during normal operation and 31 g/L during the test period). The blue lines indicate the average sludge production in the periods before, during and after the test period.

Regarding the hydrolysis of lipids, before the addition of the extra untreated slaughterhouse effluent the lipase activity was considerably higher in flocs than in granules (approximately 9-fold) (Figure 3A). Due to such a large difference in specific activity, also at a reactor level, the flocs contributed more to lipid hydrolysis than granules (three-fold). After the influent change, the specific lipase activity in flocculent sludge increased ($p = 0.02$) and was approximately 10-fold of the specific lipase activity of granules. The biomass-specific lipase activity of granules, as with protease, did not change. On reactor-scale, the total mixed liquor lipase activity increased significantly due to the increased biomass concentration in the reactor ($p = 0.009$). Regarding the contribution of each biomass fraction, the total lipase activity of granules increased due to the increased granular sludge concentration ($p = 0.007$), so the activity of flocs became only twice as high as that of large granules at reactor scale. In any case, flocs seemed to be strongly involved in lipid hydrolysis, although granules also revealed measurable activity.

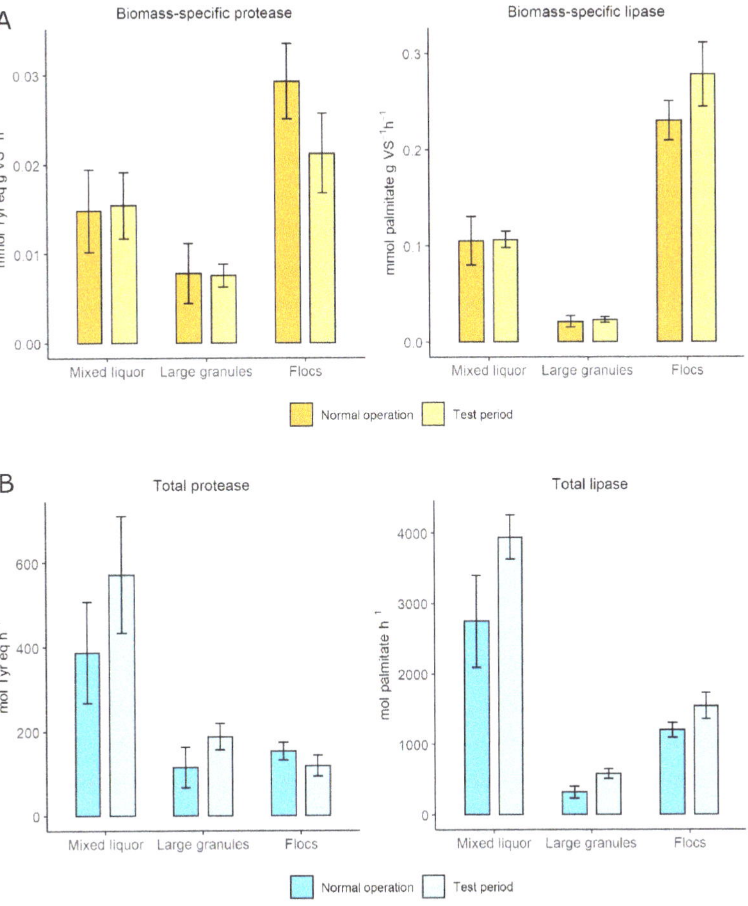

Figure 3. Hydrolytic activity of mixed sludge, granules and flocs. (**A**) Specific hydrolytic activities of the mixed liquor (Mix), and the sieve fractions large granules (>1 mm) and flocs (<0.2 mm); (**B**) Total hydrolytic activities at reactor-scale in the mixed liquor, and the total activity contributed by large granules and flocs. The fraction in the range 0.2 to 1 mm was not analysed but can be estimated by the difference between the mixed liquor and the sum of the other two fractions. Each bar represents the average of: 2 reactor samples and duplicate vials before the influent change ($n = 4$); 3 reactor samples and duplicate vials at the end of the test period ($n = 6$).

3.4. Reactor Performance and Nutrient Cycles

Effluent quality did not deteriorate during the test period (Table 2). Regarding reactor performance, several key performance indicators (KPI) were higher during the test period, including phosphate uptake rate, total phosphate release and ammonia uptake rate (nitrification) (Supplementary Information S3). However, a similar trend was observed in the spring–summer period (April–July) of the years 2018 and 2020. Furthermore, the changes in the different KPI rate values can be theoretically coupled to temperature increase, so the increase was more likely attributed to seasonal variations than to the influent composition. Denitrification rate seems to be an exception. Denitrification rate was higher during the test period (Figure 4), and no similar increase was observed in the previous summer, except in reactor 2. In the following summer (2020) denitrification rate was higher than in the winter, but the change was not as pronounced as during the test period. Furthermore, changes in denitrification rate in previous years clearly followed increases in temperature, while during the test period the increase in denitrification rate was uncoupled from the increase in temperature. Nitrate uptake rates were determined during post-denitrification phases in all cycles. Simultaneous nitrification denitrification (SND) was not monitored through the KPI, because intermittent aeration was used, which did not allow a straightforward comparison of the produced and consumed nitrate.

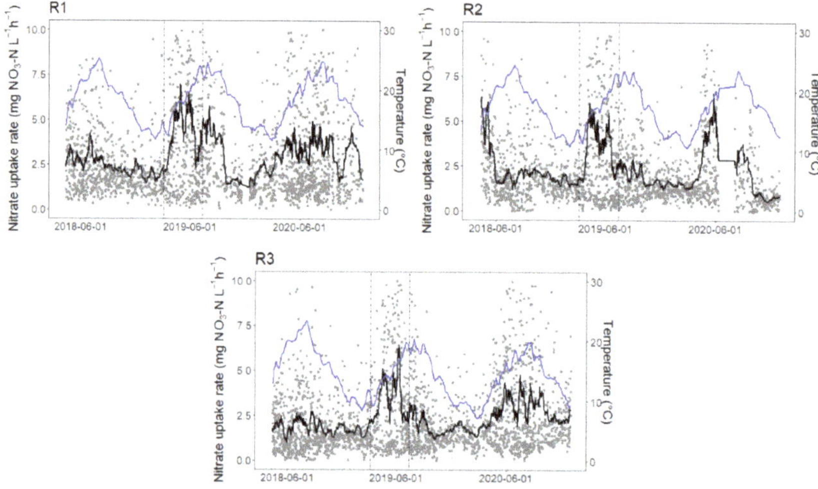

Figure 4. Denitrification rates measured during the post-denitrification period. R1, R2 and R3 indicate Nereda® reactors R1, R2 and R3 of WWTP Epe. Each dot represents the average rate in one SBR cycle. The continuous black line represents the moving average, with a sliding window of 7 days. The blue line indicates the temperature in the reactors. The test period is delimited by dashes.

3.5. Microbial Community Composition: 16S rRNA Sequencing Results

The microbial community of the mixed liquor, large granules and flocs differed significantly, in terms of beta-diversity ($p = 0.001$). Significant differences were observed too between the samples taken during normal operation and during the test period. The shift in microbial composition during the test period was most pronounced in flocculent sludge, and less in large granules. The distances between the analysed samples are presented in Figure 5A.

The 20 most abundant genera in the sludge are depicted in Figure 5B. Different genera of putative PAO and GAO were abundant in the sludge, such as Ca. *Accumulibacter*, *Tetrasphaera*, *Propionivibrio*, *Dechloromonas* and Ca. *Competibacter*. In the mixed sludge, a large proportion of the sequenced reads belonged to Ca. *Competibacter*, and the proportion increased during the test period from 7.5 to 10% of the reads. The genus *Tetrasphaera* was also abundant but did not significantly change over the test period in the mixed sludge. This putative PAO genus was predominant in the flocculent sludge fraction, where its proportion significantly increased during the test period. Large granules also had a considerably high percentage of *Tetrasphaera* (5.3–6.8%) and Ca. *Accumulibacter* (1.7–2.7%).

The genus *Nitrospira* was also at high relative abundance in the sludge. Only a marginal percentage of *Nitrotoga* (0.002%) was detected, making *Nitrospira* the dominant nitrite oxidising bacteria (NOB). *Nitrospira* was mainly associated to large granules, where it also became more abundant during the test period. The AOB *Nitrosomonas* was detected but at low relative abundance (0.015%).

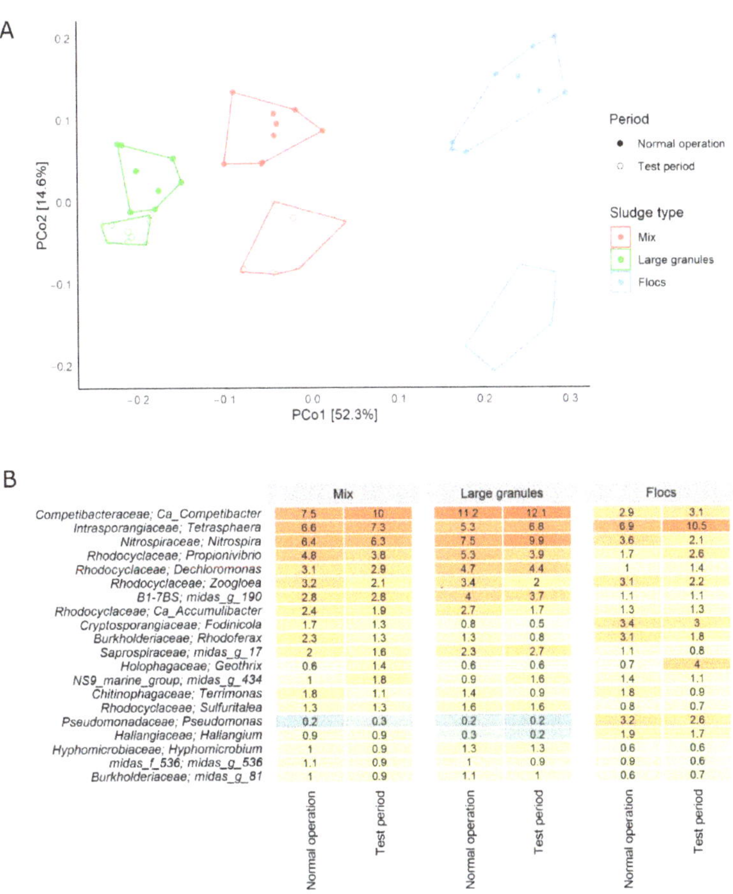

	Mix		Large granules		Flocs	
	Normal operation	Test period	Normal operation	Test period	Normal operation	Test period
Competibacteraceae; Ca_Competibacter	7.5	10	11.2	12.1	2.9	3.1
Intrasporangiaceae; Tetrasphaera	6.6	7.3	5.3	6.8	6.9	10.5
Nitrospiraceae; Nitrospira	6.4	6.3	7.5	9.9	3.6	2.1
Rhodocyclaceae; Propionivibrio	4.8	3.8	5.3	3.9	1.7	2.6
Rhodocyclaceae; Dechloromonas	3.1	2.9	4.7	4.4	1	1.4
Rhodocyclaceae; Zoogloea	3.2	2.1	3.4	2	3.1	2.2
B1-7BS; midas_g_190	2.8	2.8	4	3.7	1.1	1.1
Rhodocyclaceae; Ca_Accumulibacter	2.4	1.9	2.7	1.7	1.3	1.3
Cryptosporangiaceae; Fodinicola	1.7	1.3	0.8	0.5	3.4	3
Burkholderiaceae; Rhodoferax	2.3	1.3	1.3	0.8	3.1	1.8
Saprospiraceae; midas_g_17	2	1.6	2.3	2.7	1.1	0.8
Holophagaceae; Geothrix	0.6	1.4	0.6	0.6	0.7	4
NS9_marine_group; midas_g_434	1	1.8	0.9	1.6	1.4	1.1
Chitinophagaceae; Terrimonas	1.8	1.1	1.4	0.9	1.8	0.9
Rhodocyclaceae; Sulfuritalea	1.3	1.3	1.6	1.6	0.8	0.7
Pseudomonadaceae; Pseudomonas	0.2	0.3	0.2	0.2	3.2	2.6
Haliangiaceae; Haliangium	0.9	0.9	0.3	0.2	1.9	1.7
Hyphomicrobiaceae; Hyphomicrobium	1	0.9	1.3	1.3	0.6	0.6
midas_f_536; midas_g_536	1.1	0.9	1	0.9	0.9	0.6
Burkholderiaceae; midas_g_81	1	0.9	1.1	1	0.6	0.7

Figure 5. Microbial community composition of the sludge from WWTP Epe derived from 16S rRNA sequencing results. (**A**) Principal coordinates analysis (PCoA) plot based on the Bray-Curtis distance matrix between samples. (**B**) 20 most abundant genera and their families, in the mixed liquor sludge (Mix) and the large granule and floc fractions.

Microorganisms with a central role in hydrolysis and fermentation have been studied less than PAO, GAO and nitrifiers. This is partly due to the large array of organisms that can perform these functions. Several fermenting microorganisms were identified by Layer et al. in AGS reactors fed with complex wastewaters [27]. The families *Saprospiraceae* and *Chitinophagaceae*, and the phylum *Chloroflexi* found in their study were at high relative abundances in Epe too (4.8–5.1%, 2.9–3.8% and 2–3.5%, respectively). Many members of these taxa can perform hydrolysis and fermentation [41]. Between the abundant genera (>1% relative abundance) in Epe, there were several putative fermenters and hydrolysers: *Rhodoferax* (1.3–1.7%), *Fodinicola* (1.3–2.3%), *Terrimonas* (1.2–1.8%) and *Geothrix* (0.7–1.4%). *Rhodoferax* are denitrifying organisms and putative fermenters, which seem to be able to utilise carbohydrates, amino acids and short chain fatty acids. *Fodinicola*, in turn, are aerobic filamentous organisms that can hydrolyse proteins and carbohydrates. *Terrimonas*, too, are likely strictly aerobic hydrolysers. Last, *Geothrix* might play a role in lipid metabolism, as they can metabolise both short-chain and long-chain fatty acids, using nitrate as electron acceptor [42]. However, the metabolism of these genera has not been systematically studied in situ and therefore it is difficult to infer their function in our system.

4. Discussion

4.1. Changes in Hydrolytic Activity of the Sludge

From the enzyme activities measured, none was detected in the bulk liquid. All lipase and protease activity were associated to the biomass, both to granules and to flocs. This observation is in line with previous biofilm and activated sludge studies that found hydrolytic activity to be predominantly associated to the biomass [13,17,43]. Flocs had higher specific lipase and protease activity than granules, as observed in WWTP Garmerwolde [31], most markedly in the case of lipase. The increased particulate load period had an effect on the biomass-specific activities of flocs, with a decrease in protease activity and an increase in lipase activity. This could be related to the increased influent lipid concentration in the test period, whereas the protein concentration did not change as sharply. That only the biomass-specific hydrolytic activities of flocs changed over the test period indicates that floc activity is more rapidly influenced by changes in the influent than granules. Furthermore, it seems from our results that flocculent sludge is actively involved in lipid hydrolysis and can increase the production of enzymes during a higher exposure to lipid substrates.

The change in influent composition did not affect the biomass-specific activity of granules. The total enzyme activity in the reactors mostly increased because of overall granule growth. The newly produced granules had the same biomass-specific activity, so the total hydrolytic activity increased. Under normal operation, the granular sludge already had, theoretically, enough potential to hydrolyse all the proteins and lipids from the more concentrated influent within one SBR cycle (Supplementary Information S4). However, it would be unrealistic to assume that the rates measured in laboratory assays, based on kDa-size substrates with good biodegradability, are directly translatable to full-scale conditions. Nevertheless, the enzymatic activity of the granules does not seem to be a limiting factor for complete hydrolysis of influent pCOD. Similar results were obtained at the AGS reactors at WWTP Garmerwolde [31]. The hydrolysis rate apparently does not depend so much on enzyme content of the granules but on mass-transfer of the substrate into the biofilm (granule). In biofilm reactors, hydrolysis can be described using an areal hydrolysis rate (g TOC/m^2/h), which accounts for the biofilm area in the reactor [44].

4.2. Microbial Community Composition and Shifts

4.2.1. Microorganisms Involved in P and N Removal

In this section the most abundant taxa are discussed under the assumption that the relative abundances in the sequenced 16S rRNA genes reflect the abundance of their respective microorganisms in the sludge. This might not hold true in all cases, due to biases associated to 16S rRNA gene amplicon sequencing, including biases in the extraction and amplification of the DNA and differences in DNA copy number among microbial groups. For example, the relative abundances of Ca. *Accumulibacter* and *Tetrasphaera* are underestimated when determined by 16S sequencing compared to FISH, while *Dechloromonas* seems to be overestimated [45]. However, the data can provide a glimpse of the main microbial components in the sludge. The sequencing results revealed a large proportion (~20% of the reads) of microorganisms related to enhanced biological phosphorus removal (EBPR) and nitrogen removal [45,46]. This is in line with the good nutrient removal performances that were achieved in the plant. The NOB *Nitrospira* was one of the most abundant microorganisms in the sludge, comprising 6.3–6.4% of the reads in the mixed sludge. Numerous studies have reported *Nitrospira* as the main NOB in full-scale EBPR plants, probably due to its high substrate affinity, which allows them to thrive under low oxygen and nitrite concentrations [47,48]. The only detected AOB was *Nitrosomonas*, with low relative abundance (0.015%). Finding AOB at low relative abundances in 16S rRNA amplicon sequencing data does not necessarily imply a low AOB activity in the sludge [49]. That was also the case in this study where no issues with nitrification were identified.

The classical PAO genus Ca. *Accumulibacter* was present in relatively high abundance (1.9–2.4%) in the mixed sludge. In general, at more complex influent composition fermentative PAO and GAO populations seem to be favored [27,50–52], which is in line with the observations for the sludge in Epe. The fermentative PAO *Tetrasphaera* was found in high relative abundance, accounting for 6.6–7.3% of the reads in the mixed sludge samples. *Tetrasphaera* is often found in AS and AGS fed with complex wastewaters. They can use a wider range of substrates compared to Ca. *Accumulibacter*, including sugars and amino acids which they can utilise through a fermentative metabolism [45]. The fermentative *Dechloromonas* was also abundant in the mixed sludge samples. Members of this genus seem to possess a PAO metabolism, but others behave as GAO [50]. Therefore, their contribution to EBPR in our system is unclear. Overall, GAO were abundant in the sludge from Epe, the main genera being Ca. *Competibacter* (7.5–10%) and the fermentative GAO *Propionivibrio* (3.8–4.8%). The high COD/P ratio of the wastewater (>100 mg COD/mg P) likely favored the growth of GAOs. GAOs in most cases are not harmful to the EBPR process and indicate an excess of COD [50]. In our study too, effective EBPR was observed regardless of the high GAO abundance.

Similar to previous studies, slow-growing microorganisms (PAO, GAO and nitrifiers) were enriched in the granular sludge in comparison with flocs, due to the longer solids retention time (SRT) of granules [51,53]. Furthermore, granules have preferential access to influent substrates during the feeding phase, due to their localisation at the bottom of the reactor resulting from their higher settling velocity compared to smaller granules and flocs [54]. This further enhances the access of PAO and GAO to the rbCOD in the influent. Interestingly, *Tetrasphaera* was more concentrated in flocculent sludge, and even increased during the test period, while previous studies reported higher *Tetrasphaera* abundances in granules than in flocs [51]. The reasons for the discrepancy with our study might be related to the differences in substrate utilisation of both PAO. Ca. *Accumulibacter*, in granules, have preferential access to the VFA in the influent. Meanwhile, flocculent sludge has the ability to entrap more particulate substrates throughout the cycle, and also has a higher hydrolytic capacity, making monomers such as amino acids and monosaccharides available to *Tetrasphaera* throughout the cycle. However, this is a mere speculation that should be tested in more AGS plants and using more quantitative methods than 16S rNA sequencing.

4.2.2. Microbial Community Changes during the Test Period

The changes in microbial diversity during the test period were more pronounced in flocculent sludge than in large granules. The shorter SRT of flocculent sludge, and the higher immigration rates to this sludge fraction, can explain this observation [51]. The new influent did not have a strong impact on the microbial communities in the more rigid and older granules, which relates to the stable performance of the plant and the minimal changes in the hydrolytic activity of granules. The larger shift in microbial community of flocs is also in line with the activity changes measured in this fraction.

In terms of changes in individual taxa, the GAO Ca. *Competibacter* proliferated during the test period, possibly due to the increased COD/P ratio. No clear increase was observed among the putative hydrolyser and fermenter taxa, even if the reactors received higher particulate contaminant concentrations. The abundance of *Chloroflexi* and *Geothrix* increased, but *Saprospiraceae* and *Fodinicola* decreased, while *Tetrasphaera*, *Propionivibrio*, *Dechloromonas* and *Rhodoferax* did not change significantly. The sludge contained various organisms capable of metabolising different types of organic matter already during normal operation, and at increased particulate loads there were shifts among these groups but no sharp increase in their relative abundances. The lack of a clear increase aligns with the results from the hydrolytic tests, as the specific hydrolytic activity of the sludge did not change either. Lipid metabolism in flocculent sludge was an exception; the specific lipase activity of flocculent sludge increased. In flocculent sludge, the abundance of *Geothrix* increased from 0.7 to 4.2%, which might be related to the increase in lipase activity. However, the metabolism of *Geothrix* in aerobic wastewater treatment systems has not been studied. Further analysis would be insightful to understand their role in this ecosystem. Filamentous microorganisms related to LCFA metabolism (*Ca. Microthrix*, *Gordonia*) were not abundant in the sludge, and Ca. *Microthrix* even decreased during the test period (from

0.5 to 0.09%). No foaming or bulking was observed in the reactors even if the lipid load increased considerably.

Protozoa were not studied through sequencing, as only the bacterial and archaeal 16S gene was targeted. However, microscopy revealed a protozoa bloom in the sludge during the test period (Figure 3). Protozoa are likely to take up particulates, although their contribution to pCOD removal was not quantified [10,55]. They did not deteriorate the settleability of the sludge and might contribute to granule stability by metabolising particulates at the granule surfaces.

4.3. Impact of Particulates on Wastewater Treatment and Granular Sludge Growth

The treatment performance during the test period remained stable, in terms of COD, P and N removal. Effluent suspended solids did not increase either. Furthermore, during the test period the denitrification rates were higher than usual (on average, 4.0 mg NO_3-N L^{-1} h^{-1} compared to 3.1 mg NO_3-N L^{-1} h^{-1} during normal operation). Previous studies showed that the particulates could significantly contribute to denitrification in AS [56,57], as well as in biofilm reactors [58,59]. In those studies, denitrification relied partly on intracellularly stored substrates, but mostly on particulates that remained entrapped in the sludge and kept hydrolysing during the anoxic phase. In AGS reactors, this would mainly benefit post-denitrification, since simultaneous nitrification-denitrification (SND) requires the anaerobic storage of substrates and subsequently the coexistence of different redox zones during aeration. Particulates would have to be anaerobically hydrolysed and fermented by granules and diffuse below the granule surface in order to be stored as PHA and contribute to SND. This is likely a slow process due to mass-transfer limitation. Even if the anaerobic uptake of particulates by granules would be limited, SND can be enhanced by applying optimised aeration strategies. Layer et al. proposed to apply alternating aeration (intermittent switching on and off) or 2-step aeration (a pulse of high aeration followed by low DO for the rest of the reaction phase) to improve SND when AGS was fed with complex wastewater [60]. This way, with an increased particulate COD load such as the one in the present study it is likely possible to enhance SND, as well as to reduce the post-denitrification phase length and achieve lower effluent N.

During the test period, sludge growth was observed in the reactors at WWTP Epe. The total sludge concentration of all three reactors increased (from 5.7 to 7.6 g TS/L in R1, 5.8 to 8.5 g TS/L in R2 and 5.9 to 8.7 g TS/L in R3), as well as the granule proportion (from 58 to 67%). Thus, granule growth was observed following the increase in influent COD. After the test period, the sludge concentration remained high but did not increase further. Nevertheless, the sludge concentration in the reactors, as well as the proportion of large granules, had started to increase before the test period and there were large oscillations in the measurements (Supplementary Information S2). This makes it difficult to conclude whether granular sludge growth was supported by the added particulates. Part of these particulates were consumed in the reactor, but it is not clear if this was accomplished by the granules or by the flocculent sludge fraction. Nonetheless, the study does show that the sustained increased load of biodegradable influent particulates did not negatively impact granule growth. This holds for an influent particulate concentration of approximately 1.3 g COD/L (77% of the COD). The granule size did not decrease, as opposed to previous studies where at more complex wastewater feed the sludge bed was composed of smaller granule sizes [21,27]. The proportion of flocculent sludge did not increase either, probably due to increased sludge wasting (Figure 2). Furthermore, the higher influent particulate concentration did not result in filamentous outgrowths in the granules or disruption of granule stability as seen in previous studies [18,19]. Only a higher concentration of protozoa was observed on granule surfaces. According to our results, a moderate increase of influent suspended solids does not necessarily deteriorate granule integrity or size distribution. There might be a few reasons for this: (1) The granules can anaerobically hydrolyse part of the pCOD, as indicated by their relatively high hydrolytic activities. The available pCOD during aeration is therefore lowered; (2) protozoa at the granule surfaces might take up a considerable amount of particulates aerobically, limiting the creation of substrate gradients at the granule surfaces [55]; and (3) the flocculent sludge can capture and hydrolyse particulates, as indicated by its high specific hydrolytic activity, an observation in line with the hypothesis of Layer et al. [27]. An adjusted sludge wasting maintains the sludge bed composition avoiding a buildup of the flocculent sludge fraction and particulate material in the reactor.

5. Conclusions

We studied a full-scale Nereda® plant faced with an increased influent pCOD concentration during a 3-month period. Doubling the influent pCOD concentration did not interfere with the smooth granule morphology and did not compromise nutrient removal efficiency. In line with this observation, the sludge exhibited high hydrolytic activity and a high proportion of putative hydrolysing and fermenting organisms. The exact contribution of pCOD to sludge growth and nutrient removal could not be clarified and deserves further research. Nonetheless, the results of this study point towards good performance stability of the AGS process operated with an increased influent particulate concentration. An increased COD could permit the achievement of more efficient N removal and meeting more stringent effluent discharge requirements, without adding external C sources or post-treatment steps. Therefore, adding a COD-rich wastewater stream, even with a high particulate content, can be a good solution when a higher COD load to AGS reactors is desired.

Supplementary Materials: The following are available online at https://www.mdpi.com/article/10.3390/pr9081472/s1, Figure S1: Distribution of the influent COD in different size fractions, during normal operation and during the test period, Figure S2: Mixed liquor suspended solids (MLSS) concentration of the reactors at WWTP Epe, and the concentration of different granule size fractions, Figures S3.1–S3.5: Key performance indicators from reactors 1–3 of the WWTP Epe, Tables S4.1–S4.3: Calculation of total hydrolytic activity during feeding, Table S5: Comparison of estimated and observed sludge production.

Author Contributions: Conceptualisation, M.K.d.K., M.P. and S.T.O.; methodology, S.T.O.; formal analysis, S.T.O.; investigation, S.T.O.; writing—original draft preparation, S.T.O.; writing—review and editing, M.K.d.K. and M.P.; supervision, M.K.d.K. and M.P.; funding acquisition, M.K.d.K. All authors have read and agreed to the published version of the manuscript.

Funding: This research was funded by the Netherlands Organisation for Scientific Research (NWO) (VIDI grant number: 016.168.320).

Institutional Review Board Statement: Not applicable.

Informed Consent Statement: Not applicable.

Data Availability Statement: The data of this study is available upon request to the authors. The 16S rRNA amplicon sequencing data are deposited on BioProject PRJNA746138.

Acknowledgments: The authors thank Water Authority Vallei en Veluwe for the sampling and sharing the plant information. They also would like to thank Royal Haskoning DHV and more specifically Matthijs Oosterhuis for assistance during the research and Edward van Dijk for sharing the reactor monitoring data. The help in the laboratory by Nadia Boulif is much appreciated too.

Conflicts of Interest: The authors declare no conflict of interest.

References

1. Nereda®Plants. Available online: https://www.royalhaskoningdhv.com/en-gb/nereda/nereda-plants-a-to-z (accessed on 20 April 2021).
2. De Bruin, L.M.M.; de Kreuk, M.K.; van der Roest, H.F.R.; Uijterlinde, C.; van Loosdrecht, M.C.M. Aerobic Granular Sludge Technology: An Alternative to Activated Sludge? *Water Sci. Technol.* **2004**, *49*, 1–7. [CrossRef]
3. De Amorim de Carvalho, C.; Ferreira dos Santos, A.; Tavares Ferreira, T.J.; Sousa Aguiar Lira, V.N.; Mendes Barros, A.R.; Bezerra dos Santos, A.B. Resource recovery in aerobic granular sludge systems: Is it feasible or still a long way to go? *Chemosphere* **2021**, *274*, 129881. [CrossRef] [PubMed]

4. Pronk, M.; de Kreuk, M.; de Bruin, B.; Kamminga, P.; Kleerebezem, R.; van Loosdrecht, M. Full scale performance of the aerobic granular sludge process for sewage treatment. *Water Res.* **2015**, *84*, 207–217. [CrossRef]
5. de Kreuk, M.; Heijnen, J.; van Loosdrecht, M. Simultaneous COD, nitrogen, and phosphate removal by aerobic granular sludge. *Biotechnol. Bioeng.* **2005**, *90*, 761–769. [CrossRef]
6. De Kreuk, M.; Van Loosdrecht, M. Selection of slow growing organisms as a means for improving aerobic granular sludge stability. *Water Sci. Technol.* **2004**, *49*, 9–17. [CrossRef]
7. Sophonsiri, C.; Morgenroth, E. Chemical composition associated with different particle size fractions in municipal, industrial, and agricultural wastewaters. *Chemosphere* **2004**, *55*, 691–703. [CrossRef]
8. Karahan, Ö.; Dogruel, S.; Dulekgurgen, E.; Orhon, D. COD fractionation of tannery wastewaters—Particle size distribution, biodegradability and modeling. *Water Res.* **2008**, *42*, 1083–1092. [CrossRef]
9. Ravndal, K.T.; Opsahl, E.; Bagi, A.; Kommedal, R. Wastewater characterisation by combining size fractionation, chemical composition and biodegradability. *Water Res.* **2017**, *131*, 151–160. [CrossRef]
10. Morgenroth, E.; Kommedal, R.; Harremoës, P. Processes and Modelling of Hydrolysis of Particulate Organic Matter in Aerobic Wastewater Treatment—A Review. *Water Sci. Technol.* **2002**, *45*, 25–40. [CrossRef]
11. Henze, M.; Grady, C.P.L.; Gujer, W.; van Marais, G.R.; Matsuo, T. *Activated Sludge Model No.1*; IWA Publishing: London, UK, 1986.
12. Gujer, W.; Henze, M.; Mino, T.; van Loosdrecht, M. Activated Sludge Model No. 3. *Water Sci. Technol.* **1999**, *39*, 183–193. [CrossRef]
13. Fr/olund, B.; Griebe, T.; Nielsen, P.H. Enzymatic activity in the activated-sludge floc matrix. *Appl. Microbiol. Biotechnol.* **1995**, *43*, 755–761. [CrossRef]
14. Goel, R.; Mino, T.; Satoh, H.; Matsuo, T. Comparison of hydrolytic enzyme systems in pure culture and activated sludge under different electron acceptor conditions. *Water Sci. Technol.* **1998**, *37*, 335–343. [CrossRef]
15. Karahan, Ö.; Martins, A.; Orhon, D.; Van Loosdrecht, M.C.; Özgün, O.K. Experimental evaluation of starch utilization mechanism by activated sludge. *Biotechnol. Bioeng.* **2006**, *93*, 964–970. [CrossRef]
16. Martins, A.M.; Karahan, Ö.; van Loosdrecht, M.C. Effect of polymeric substrate on sludge settleability. *Water Res.* **2011**, *45*, 263–273. [CrossRef]
17. Mosquera-Corral, A.; Montràs, A.; Heijnen, J.; van Loosdrecht, M. Degradation of polymers in a biofilm airlift suspension reactor. *Water Res.* **2003**, *37*, 485–492. [CrossRef]
18. Schwarzenbeck, N.; Borges, J.M.; Wilderer, P.A. Treatment of dairy effluents in an aerobic granular sludge sequencing batch reactor. *Appl. Microbiol. Biotechnol.* **2005**, *66*, 711–718. [CrossRef]
19. de Kreuk, M.; Kishida, N.; Tsuneda, S.; van Loosdrecht, M. Behavior of polymeric substrates in an aerobic granular sludge system. *Water Res.* **2010**, *44*, 5929–5938. [CrossRef]
20. Picioreanu, C.; van Loosdrecht, M.C.; Heijnen, J.J. Effect of diffusive and convective substrate transport on biofilm structure formation: A two-dimensional modeling study. *Biotechnol. Bioeng.* **2000**, *69*, 504–515. [CrossRef]
21. Derlon, N.; Wagner, J.; da Costa, R.H.R.; Morgenroth, E. Formation of aerobic granules for the treatment of real and low-strength municipal wastewater using a sequencing batch reactor operated at constant volume. *Water Res.* **2016**, *105*, 341–350. [CrossRef]
22. Cetin, E.; Karakas, E.; Dulekgurgen, E.; Ovez, S.; Kolukirik, M.; Yilmaz, G. Effects of high-concentration influent suspended solids on aerobic granulation in pilot-scale sequencing batch reactors treating real domestic wastewater. *Water Res.* **2018**, *131*, 74–89. [CrossRef]
23. Yilmaz, G.; Lemaire, R.; Keller, J.; Yuan, Z. Simultaneous nitrification, denitrification, and phosphorus removal from nutrient-rich industrial wastewater using granular sludge. *Biotechnol. Bioeng.* **2008**, *100*, 529–541. [CrossRef]
24. Morales, N.; Figueroa, M.; Fra-Vázquez, A.; del Río, A.V.; Campos, J.; Mosquera-Corral, A.; Méndez, R. Operation of an aerobic granular pilot scale SBR plant to treat swine slurry. *Process. Biochem.* **2013**, *48*, 1216–1221. [CrossRef]
25. Stes, H.; Caluwé, M.; Dockx, L.; Cornelissen, R.; De Langhe, P.; Smets, I.; Dries, J. Cultivation of aerobic granular sludge for the treatment of food-processing wastewater and the impact on membrane filtration properties. *Water Sci. Technol.* **2021**, *83*, 39–51. [CrossRef]
26. Barrios-Hernández, M.L.; Buenaño-Vargas, C.; García, H.; Brdjanovic, D.; van Loosdrecht, M.C.; Hooijmans, C.M. Effect of the co-treatment of synthetic faecal sludge and wastewater in an aerobic granular sludge system. *Sci. Total. Environ.* **2020**, *741*, 140480. [CrossRef]
27. Layer, M.; Adler, A.; Reynaert, E.; Hernandez, A.; Pagni, M.; Morgenroth, E.; Holliger, C.; Derlon, N. Organic substrate diffusibility governs microbial community composition, nutrient removal performance and kinetics of granulation of aerobic granular sludge. *Water Res. X* **2019**, *4*, 100033. [CrossRef]
28. Haaksman, V.; Mirghorayshi, M.; van Loosdrecht, M.; Pronk, M. Impact of aerobic availability of readily biodegradable COD on morphological stability of aerobic granular sludge. *Water Res.* **2020**, *187*, 116402. [CrossRef]
29. APHA. *Standard Methods for the Examination of Water and Wastewater*, 21st ed.; American Public Health Association: Washington, DC, USA, 2005.
30. Dubois, M.; Gilles, K.A.; Hamilton, J.K.; Rebers, P.A.; Smith, F. Colorimetric Method for Determination of Sugars and Related Substances. *Anal. Chem.* **1956**, *28*, 350–356. [CrossRef]
31. Ortega, S.T.; Pronk, M.; de Kreuk, M.K. Anaerobic hydrolysis of complex substrates in full-scale aerobic granular sludge: Enzymatic activity determined in different sludge fractions. *Appl. Microbiol. Biotechnol.* **2021**, 1–14. [CrossRef]

32. Albertsen, M.; Karst, S.M.; Ziegler, A.S.; Kirkegaard, R.H.; Nielsen, P.H. Back to Basics—The Influence of DNA Extraction and Primer Choice on Phylogenetic Analysis of Activated Sludge Communities. *PLoS ONE* **2015**, *10*, e0132783. [CrossRef]
33. Feinstein, L.M.; Sul, W.J.; Blackwood, C.B. Assessment of Bias Associated with Incomplete Extraction of Microbial DNA from Soil. *Appl. Environ. Microbiol.* **2009**, *75*, 5428–5433. [CrossRef]
34. Jones, M.D.; Singleton, D.R.; Sun, W.; Aitken, M.D. Multiple DNA Extractions Coupled with Stable-Isotope Probing of Anthracene-Degrading Bacteria in Contaminated Soil. *Appl. Environ. Microbiol.* **2011**, *77*, 2984–2991. [CrossRef] [PubMed]
35. Bolyen, E.; Rideout, J.R.; Dillon, M.R.; Bokulich, N.A.; Abnet, C.C.; Al-Ghalith, G.A.; Alexander, H.; Alm, E.J.; Arumugam, M.; Asnicar, F.; et al. Reproducible, interactive, scalable and extensible microbiome data science using QIIME 2. *Nat. Biotechnol.* **2019**, *37*, 852–857. [CrossRef]
36. Amir, A.; McDonald, D.; Navas-Molina, J.A.; Kopylova, E.; Morton, J.T.; Xu, Z.Z.; Kightley, E.P.; Thompson, L.R.; Hyde, E.R.; Gonzalez, A.; et al. Deblur Rapidly Resolves Single-Nucleotide Community Sequence Patterns. *mSystems* **2017**, *2*, e00191-16. [CrossRef]
37. Anderson, M.J. A new method for non-parametric multivariate analysis of variance. *Austral Ecol.* **2001**, *26*, 32–46. [CrossRef]
38. Nierychlo, M.; Andersen, K.S.; Xu, Y.; Green, N.; Jiang, C.; Albertsen, M.; Dueholm, M.S.; Nielsen, P.H. MiDAS 3: An ecosystem-specific reference database, taxonomy and knowledge platform for activated sludge and anaerobic digesters reveals species-level microbiome composition of activated sludge. *Water Res.* **2020**, *182*, 115955. [CrossRef] [PubMed]
39. McMurdie, P.J.; Holmes, S. phyloseq: An R Package for Reproducible Interactive Analysis and Graphics of Microbiome Census Data. *PLoS ONE* **2013**, *8*, e61217. [CrossRef] [PubMed]
40. Andersen, K.S.; Kirkegaard, R.H.; Karst, S.M.; Albertsen, M. Ampvis2: An R Package to Analyse and Visualise 16S RRNA Amplicon Data. *bioRxiv* **2018**, 299537. [CrossRef]
41. Xia, Y.; Kong, Y.; Nielsen, P.H. In situ detection of protein-hydrolysing microorganisms in activated sludge. *FEMS Microbiol. Ecol.* **2007**, *60*, 156–165. [CrossRef]
42. Coates, J.D.; Ellis, D.J.; Gaw, C.V.; Lovley, D.R. Geothrix fermentans gen. nov., sp. nov., a novel Fe(III)-reducing bacterium from a hydrocarbon-contaminated aquifer. *Int. J. Syst. Evol. Microbiol.* **1999**, *49*, 1615–1622. [CrossRef]
43. Confer, D.R.; Logan, B.E. Location of protein and polysaccharide hydrolytic activity in suspended and biofilm wastewater cultures. *Water Res.* **1998**, *32*, 31–38. [CrossRef]
44. Kommedal, R.; Milferstedt, K.; Bakke, R.; Morgenroth, E. Effects of initial molecular weight on removal rate of dextran in biofilms. *Water Res.* **2006**, *40*, 1795–1804. [CrossRef] [PubMed]
45. Stokholm-Bjerregaard, M.; McIlroy, S.J.; Nierychlo, M.; Karst, S.M.; Albertsen, M.; Nielsen, P.H. A Critical Assessment of the Microorganisms Proposed to be Important to Enhanced Biological Phosphorus Removal in Full-Scale Wastewater Treatment Systems. *Front. Microbiol.* **2017**, *8*, 718. [CrossRef] [PubMed]
46. Saunders, A.; Albertsen, M.; Vollertsen, J.; Nielsen, P.H. The activated sludge ecosystem contains a core community of abundant organisms. *ISME J.* **2016**, *10*, 11–20. [CrossRef] [PubMed]
47. Nielsen, P.H.; Mielczarek, A.T.; Kragelund, C.; Nielsen, J.L.; Saunders, A.; Kong, Y.; Hansen, A.A.; Vollertsen, J. A conceptual ecosystem model of microbial communities in enhanced biological phosphorus removal plants. *Water Res.* **2010**, *44*, 5070–5088. [CrossRef] [PubMed]
48. Daims, H.; Nielsen, J.L.; Nielsen, P.H.; Schleifer, K.-H.; Wagner, M. In Situ Characterization of Nitrospira -Like Nitrite-Oxidizing Bacteria Active in Wastewater Treatment Plants. *Appl. Environ. Microbiol.* **2001**, *67*, 5273–5284. [CrossRef]
49. Weissbrodt, D.G.; Neu, T.R.; Kuhlicke, U.; Rappaz, Y.; Holliger, C. Assessment of bacterial and structural dynamics in aerobic granular biofilms. *Front. Microbiol.* **2013**, *4*, 175. [CrossRef]
50. Nielsen, P.H.; McIlroy, S.J.; Albertsen, M.; Nierychlo, M. Re-evaluating the microbiology of the enhanced biological phosphorus removal process. *Curr. Opin. Biotechnol.* **2019**, *57*, 111–118. [CrossRef]
51. Ali, M.; Wang, Z.; Salam, K.W.; Hari, A.R.; Pronk, M.; Van Loosdrecht, M.C.M.; Saikaly, P.E. Importance of Species Sorting and Immigration on the Bacterial Assembly of Different-Sized Aggregates in a Full-Scale Aerobic Granular Sludge Plant. *Environ. Sci. Technol.* **2019**, *53*, 8291–8301. [CrossRef]
52. Campo, R.; Sguanci, S.; Caffaz, S.; Mazzoli, L.; Ramazzotti, M.; Lubello, C.; Lotti, T. Efficient carbon, nitrogen and phosphorus removal from low C/N real domestic wastewater with aerobic granular sludge. *Bioresour. Technol.* **2020**, *305*, 122961. [CrossRef]
53. Winkler, M.K.; Kleerebezem, R.; Khunjar, W.O.; de Bruin, B.; van Loosdrecht, M.C. Evaluating the solid retention time of bacteria in flocculent and granular sludge. *Water Res.* **2012**, *46*, 4973–4980. [CrossRef]
54. van Dijk, E.J.; Pronk, M.; van Loosdrecht, M.C. A settling model for full-scale aerobic granular sludge. *Water Res.* **2020**, *186*, 116135. [CrossRef] [PubMed]
55. Schwarzenbeck, N.; Erley, R.; Mc Swain, B.S.; Wilderer, P.A.; Irvine, R.L. Treatment of Malting Wastewater in a Granular Sludge Sequencing Batch Reactor (SBR). *Acta Hydrochim. Hydrobiol.* **2004**, *32*, 16–24. [CrossRef]
56. Drewnowski, J.; Makinia, J. The role of biodegradable particulate and colloidal organic compounds in biological nutrient removal activated sludge systems. *Int. J. Environ. Sci. Technol.* **2013**, *11*, 1973–1988. [CrossRef]
57. How, S.W.; Sin, J.H.; Wong, S.Y.Y.; Lim, P.B.; Aris, A.M.; Ngoh, G.C.; Shoji, T.; Curtis, T.P.; Chua, A.S.M. Characterization of slowly-biodegradable organic compounds and hydrolysis kinetics in tropical wastewater for biological nitrogen removal. *Water Sci. Technol.* **2020**, *81*, 71–80. [CrossRef]

58. Krasnits, E.; Beliavski, M.; Tarre, S.; Green, M. The contribution of suspended solids to municipal wastewater PHA-based denitrification. *Environ. Technol.* **2013**, *35*, 313–321. [CrossRef]
59. Krasnits, E.; Beliavsky, M.; Tarre, S.; Green, M. PHA based denitrification: Municipal wastewater vs. acetate. *Bioresour. Technol.* **2013**, *132*, 28–37. [CrossRef]
60. Layer, M.; Villodres, M.G.; Hernandez, A.; Reynaert, E.; Morgenroth, E.; Derlon, N. Limited simultaneous nitrification-denitrification (SND) in aerobic granular sludge systems treating municipal wastewater: Mechanisms and practical implications. *Water Res. X* **2020**, *7*, 100048. [CrossRef] [PubMed]

Article

Effect of Salinity on Cr(VI) Bioremediation by Algal-Bacterial Aerobic Granular Sludge Treating Synthetic Wastewater

Bach Van Nguyen [1,2,†], Xiaojing Yang [1,3,†], Shota Hirayama [1], Jixiang Wang [1], Ziwen Zhao [1], Zhongfang Lei [1,*], Kazuya Shimizu [1], Zhenya Zhang [1] and Sinh Xuan Le [2]

[1]   Graduate School of Life and Environmental Sciences, University of Tsukuba, 1-1-1 Tennodai, Tsukuba 305-8572, Japan; s1926055@s.tsukuba.ac.jp (B.V.N.); yangxj-tsukuba@outlook.com (X.Y.); s2021164@s.tsukuba.ac.jp (S.H.); s1836028@s.tsukuba.ac.jp (J.W.); tsukubazhaoziwen@outlook.com (Z.Z.); shimizu.kazuya.fn@u.tsukuba.ac.jp (K.S.); zhang.zhenya.fu@u.tsukuba.ac.jp (Z.Z.)

[2]   Institute of Marine Environment and Resources, Vietnam Academy of Science and Technology, Haiphong 180000, Vietnam; sinhlx@imer.vast.vn

[3]   School of Environmental Science and Engineering, Sun Yat-Sen University, Guangzhou 510275, China

*   Correspondence: lei.zhongfang.gu@u.tsukuba.ac.jp

†   These authors contributed equally to this work.

Abstract: Heavy metal-containing wastewater with high salinity challenges wastewater treatment plants (WWTPs) where the conventional activated sludge process is widely applied. Bioremediation has been proven to be an effective, economical, and eco-friendly technique to remove heavy metals from various wastewaters. The newly developed algal-bacterial aerobic granular sludge (AGS) has emerged as a promising biosorbent for treating wastewater containing heavy metals, especially Cr(VI). In this study, two identical cylindrical sequencing batch reactors (SBRs), i.e., R1 (Control) and R2 (with 1% additional salinity), were used to cultivate algal-bacterial AGS and then to evaluate the effect of salinity on the performance of the two SBRs. The results reflected that less filamentation and a rougher surface could be observed on algal-bacterial AGS when exposed to 1% salinity, which showed little influence on organics removal. However, the removals of total inorganic nitrogen (TIN) and total phosphorus (TP) were noticeably impacted at the 1% salinity condition, and were further decreased with the co-existence of 2 mg/L Cr(VI). The Cr(VI) removal efficiency, on the other hand, was 31–51% by R1 and 28–48% by R2, respectively, indicating that salinity exposure may slightly influence Cr(VI) bioremediation. In addition, salinity exposure stimulated more polysaccharides excretion from algal-bacterial AGS while Cr(VI) exposure promoted proteins excretion.

Keywords: algal-bacterial aerobic granular sludge; bioremediation; hexavalent chromium; salinity; wastewater treatment

Citation: Nguyen, B.V.; Yang, X.; Hirayama, S.; Wang, J.; Zhao, Z.; Lei, Z.; Shimizu, K.; Zhang, Z.; Le, S.X. Effect of Salinity on Cr(VI) Bioremediation by Algal-Bacterial Aerobic Granular Sludge Treating Synthetic Wastewater. *Processes* **2021**, *9*, 1400. https://doi.org/10.3390/pr9081400

Academic Editor: Enrico Marsili

Received: 2 July 2021
Accepted: 10 August 2021
Published: 13 August 2021

Publisher's Note: MDPI stays neutral with regard to jurisdictional claims in published maps and institutional affiliations.

1. Introduction

The rapidly increasing global population, along with the development of industrialization, has triggered various environmental issues in many countries worldwide. Heavy metal (HM) in wastewater has emerged as a global environmental concern because of its adverse effects on human beings and ecosystems [1]. Among the HMs commonly found in wastewater, hexavalent chromium (Cr(VI)) is considered as one of the most harmful and toxic metal forms because of its high carcinogenic and mutagenic properties [2,3]. More specifically, exposure to Cr(VI) through inhalation, digestion, or direct contact can negatively affect human health, leading to liver and kidney damage, internal hemorrhage, and respiratory disorders [4]. Furthermore, Cr(VI)-contaminated wastewater is gaining more attention due to its diverse applications and higher concentrations in water contributed by various industrial processes such as metallurgy, refractory materials, textile dyeing, leather tanning, electroplating, and chemicals production [5,6].

To date, many techniques have been developed and applied to remove Cr(VI) from wastewater, such as chemical precipitation, reverse osmosis, ion exchange, and adsorption [4,7]. In general, physicochemical and biological methods are the common solutions to Cr(VI)-containing wastewater treatment [8]. However, physicochemical methods (such as electrolysis, membrane separation, and flocculation) often consume energy, with high costs in the long-term operation [9]. In addition, these techniques usually generate various secondary pollutants [10]. In contrast, biological treatment processes are cost-effective with low risk in respect of secondary pollutants. Among the biological processes, bioremediation has been extensively examined to be effective, economical, and environmentally friendly for HMs, especially Cr(VI) removal from wastewater [11,12]. The term of bioremediation commonly refers to the process of degrading contaminants from soil, water, and other environments using living organisms, such as microbes, bacteria, fungi, and plants [13].

Based on previous studies [14,15], the newly developed algal-bacterial aerobic granular sludge (AGS) is a promising biosorbent for Cr(VI) which can be applied to treat Cr(VI) contaminated wastewater. The algal-bacterial AGS has been reported to possess the merits of both algae (high surface area to volume ratio, different functional groups, and several algal cell wall components) and AGS (compact structure, superior settleability, and dense microbial structure). Nevertheless, when taking its practical application into consideration, the co-existence of Cr(VI) and salt ions is inevitable. For example, tannery wastewater may exert toxic effects on the microbial community in the wastewater treatment systems [16], thus adversely impacting the performance of wastewater treatment plants (WWTPs). Besides, the salinity in wastewater is also a key parameter for microalgae biomass production, and a higher salinity concentration may hinder the microalgae biomass production [17]. Although Yang et al. evaluated the effect of salinity on Cr(VI) adsorption efficiency by algal-bacterial AGS for the first time, the exposure time was very short (6 h) [15]. To date, the feasibility of Cr(VI) removal (with the co-existence of salt ions) by algal-bacterial AGS system has not yet been examined under a long-term operation condition. Furthermore, other parameters relating to algal-bacterial AGS system performance under the co-existence of Cr(VI) and salinity, such as nutrients removal and granule characteristics, have not been reported. Restated, little information is available regarding the efficiency of Cr(VI) removal by algal-bacterial AGS under a high salinity condition during a long-term operation. Therefore, this study aimed to investigate the effect of salinity on Cr(VI) bioremediation by algal-bacterial AGS in sequencing batch reactors (SBRs) in addition to its nutrients removal performance. Moreover, the effects of the co-existence of salinity and Cr(VI) on the characteristics of algal-bacterial AGS, such as morphological change, granule size, granule settleability, and stability, were also evaluated.

2. Materials and Methods

2.1. Experimental Design and Operation

In this study, two identical cylindrical SBRs with a working volume of 250 mL (D × H = 43 mm × 300 mm) each, namely R1 (control, no additional salinity addition) and R2 (with 1% additional salinity), were used. The reactors were operated at room temperature (25 ± 2 °C) under a 6 h cycle consisting of 7 min of feeding, 60 min of non-aeration, 282 min of aeration, 2 min of settling, 8 min of decanting, and 1 min of idling. The volume exchange rate (VER) was set at 50%, resulting in a hydraulic retention time (HRT) of 12 h. Air was introduced into each reactor from the bottom by an air pump (AK-40, KOSHIN, Japan), and the superficial uplift air velocity was controlled at 0.81 cm/s. The reactors were operated under 24 h illumination, which was provided by two artificial lights (lb055cw, RIKAKEN, Nagoya, Japan) attached to the outside wall of the reactors at an illuminance of around 6000 lux. During the experimental period, the sludge retention time (SRT) was not controlled, which varied between 30–40 days for the two SBRs according to the biomass concentration in the reactors and their effluent biomass concentrations. The whole experiment and test consisted of four stages as listed in Table 1.

Table 1. The 4 operation stages for the two SBRs in this study.

Reactor	Salinity and/or Cr(VI) Exposure	Operation Stage			
		I (Days 1–7)	II (Days 8–14)	III (Days 15–21)	IV (Days 22–28)
R1	NaCl (g/L)	0	0	0	0
	Cr(VI) (mg/L)	0	0	2	0
R2	NaCl (g/L)	0	10	10	10
	Cr(VI) (mg/L)	0	0	2	0

2.2. Synthetic Wastewater and Seed Mature Algal-Bacterial Granules

Synthetic wastewater was prepared and fed to the two SBRs, which consisted of the main components as follows: 300 mg chemical oxygen demand (COD)/L (with NaAc as the sole carbon source), 30 mg NH_4^+-N/L (NH_4Cl), 5 mg PO_4-P/L (KH_2PO_4), 10 mg Ca^{2+}/L ($CaCl_2 \cdot 2H_2O$), 5 mg Fe^{2+}/L ($FeSO_4 \cdot 7H_2O$), 5 mg Mg^{2+}/L ($MgSO_4 \cdot 7H_2O$), and 1 mL/L of trace elements solution. The specific composition of the trace elements solution was described in a previous study [18]. In addition, 50 mg/L $NaHCO_3$ was added to maintain the influent pH. In our previous study [15], the algal-bacterial AGS system was found to tolerate relatively low salinity <5 g/L (or 0.5%) during the 6 h Cr(VI) biosorption, which would be negatively affected when exposed to higher salinity levels. As the major purpose of this study was to examine the effect of the co-existing Cr(VI) and salinity on algal-bacterial AGS during a long-term operation, approximately 1% salinity was initially applied in the experiments. The results from the operation under higher salinity levels or real wastewaters will be included in the follow-up reports. The 1% salinity (with conductivity of about 17,000 µS/cm) in wastewater was generated by adding NaCl at a concentration of 10 g/L. The 2 mg/L Cr(VI) in the influent wastewater was obtained by diluting a 1000 mg/L stock Cr(VI) solution prepared from $K_2Cr_2O_7$. Mature algal-bacterial AGS was used in this study, which was developed from bacterial AGS. More specifically, the bacterial AGS was first sampled from the long-term bacterial AGS system operated by Wang et al. [19]. Then, the bacterial AGS was cultivated in another lab-scale SBR (D × H = 200 mm × 800 mm with a working volume of 15.7 L) under an artificial light illumination system of approximately 15k lux on the water surface and 40 k lux on the reactor wall (light on/off = 12/12 h). This lab-scale SBR was stably operated for 6 months at a mixed liquor suspended solids (MLSS) concentration of 3.4 g/L with sludge volume index in 30 min (SVI_{30}) of about 70 mL/g.

Moreover, prior to the 4-stage operation in this study, the algal-bacterial AGS was again cultivated in R1 and R2 for about 14 days by feeding the same influent wastewater without supplementation of salt and Cr(VI). Both achieved the same stable system performance before starting Stage I.

2.3. Analytical Methods

All water samples were first collected at the end of the third operation cycle (19:00) every day, then filtered through 0.22 µm membrane filters before further analysis. Ammonia nitrogen (NH_4^+-N), nitrite nitrogen (NO_2^--N), nitrate nitrogen (NO_3^--N), phosphate phosphorus (PO_4^{3-}-P), mixed liquor (volatile) suspended solids (ML(V)SS), and sludge volume index (SVI) were measured according to the standard methods [20]. In this study, total inorganic nitrogen (TIN) was calculated as the sum of NH_4^+-N, NO_2^--N, and NO_3^--N. Dissolved organic carbon (DOC), dissolved inorganic carbon (DIC), and dissolved total carbon (DTC) in the effluent samples were analyzed by a total organic carbon (TOC) analyzer (TOC-VCSN, Shimadzu, Kyoto, Japan). The concentration of Cr(VI) was measured by a UV-vis spectrophotometer (UV 1800, Shimadzu, Japan), which has been described in detail elsewhere [14]. Chlorophyll *a* (Chl-*a*) content in the granules was extracted and analyzed by the method described previously [21]. All the other assays, such as the granule size and distribution, granule strength (expressed as integrity coefficient or IC), extraction of extracellular polymeric substances (EPS) (including loosely bound EPS (LB-EPS) and tightly bound EPS (TB-EPS)), and determination of proteins (PN) and polysaccharides (PS), were the same as in a previous study [22].

2.4. Statistical Analysis

All the analyses were performed in triplicate and the results were expressed as mean or mean ± standard deviation (SD). One-way analysis of variance (ANOVA) was conducted to compare the performance of the two SBRs and test stages using Microsoft Excel 2010. Statistically significant difference was assumed if $p < 0.05$.

3. Results and Discussion

3.1. Changes in Characteristics of Algal-Bacterial AGS in the Two Reactors

3.1.1. Morphology and Granule Size

In this study, changes in morphology and size of the granules from the two reactors were observed and recorded during the four stages' operation (Figure 1). Initially, the mature algal-bacterial AGS from both reactors was relatively regular in shape and dark green in color, demonstrating a dense and compact structure. After Stage I, there was not much change in the morphology of the granules because the salinity and Cr(VI) were not supplemented to the influent of the two reactors. Starting from Stage II, when 1% salinity was added to the influent wastewater of R2, some difference was clearly observed in the granules from the two reactors, especially the growth of filamentous bacteria outside the granules. Specifically, under the 1% salinity condition (R2), bacterial growth may have been inhibited, and further inhibition was observed under the prolonged exposure to salinity (to the end of Stage IV). In contrast, the granules in R1 (control) demonstrated the rapid growth of filamentous bacteria on their surface along with the operation. In addition, the presence of Cr(VI) in the influent to the two reactors (in Stage III) led to the appearance of specific colored precipitates (in gray and bright yellow) surrounding the surface of the granules from R1 and R2, respectively.

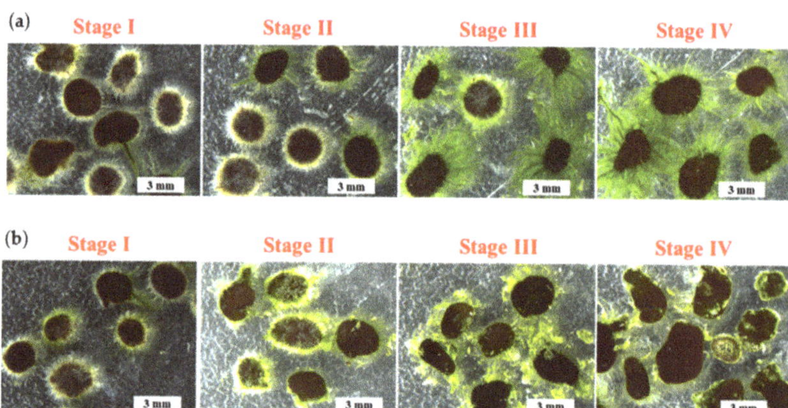

Figure 1. Microscopic images of algal-bacterial AGS from the two reactors during the four stages' operation. (**a**) R1; (**b**) R2.

Figure 2 demonstrates the changes in granular size and its distribution in the two reactors (R1 and R2). Generally, granules ranging between 1.5–2 mm and 2–3 mm were predominant in both R1 and R2 during the test period. In addition, an opposite trend was observed in the occupying ratio of the smallest (≤0.5 mm, decreased) and largest (>3 mm, increased) granules in both reactors. This trend was more obvious in R2, possibly due to the salinity exposure. Regarding the average diameter of granules, there was a gradual increase over time in both reactors, from the initial 1.92 ± 0.87 (R1) and 1.90 ± 0.87 (R2) mm to 2.58 ± 1.39 (R1) and 2.62 ± 1.11 (R2) mm at the end of test, respectively.

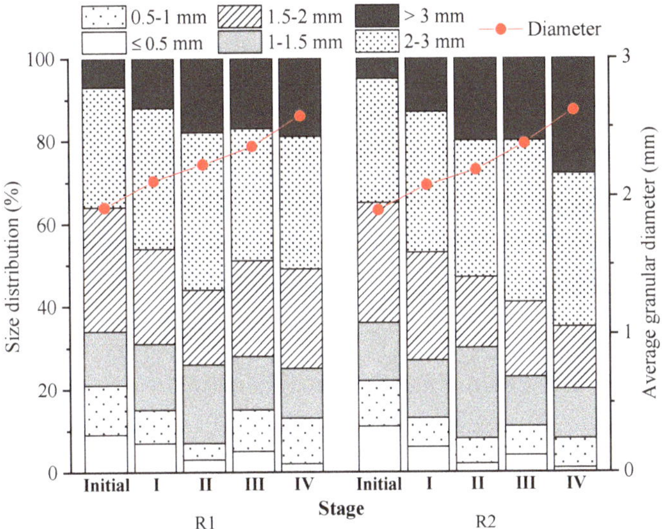

Figure 2. Changes in the diameter and its size distribution of algal-bacterial AGS in R1 and R2.

3.1.2. Granular Biomass Growth and Settleability

Figure 3a shows the changes of the MLSS and MLVSS/MLSS ratio in the two reactors during the four operation stages. Obviously, the MLSS concentration in the two reactors increased gradually over time, from the initial 3.40 to 6.67 g/L in R1 and 9.08 g/L in R2, respectively. A larger increase in MLSS was detected in R2, which is probably attributable to the effect of salinity exposure. The larger increase in biomass under the salinity exposure condition (R2 in this study) is consistent with previous studies [23,24], possibly due to the increased accumulation of inorganic materials into the granules. In addition, the presence of Cr(VI) in wastewater (Stage III) also triggered the increase in biomass growth in R1. In terms of MLVSS/MLSS, this ratio in R1 fluctuated between 80–92% during the experiment, while that in R2 decreased from 86% (Stage I) to 52% after exposure to 1% NaCl (Stage II), and then stabilized around 89% during the rest period of test.

On the other hand, salinity exposure also enhanced the granule settleability (Figure 3b), as the SVI_5 of granules from R2 decreased from 62.0 mL/g (Stage I) to 26.4 mL/g (Stage IV). This finding agreed well with other previous works [23,25]. In addition, Dong et al. [25] attributed the excellent settleability of algal-bacterial AGS under salinity conditions to its inhibitory effect on the growth of filamentous bacteria.

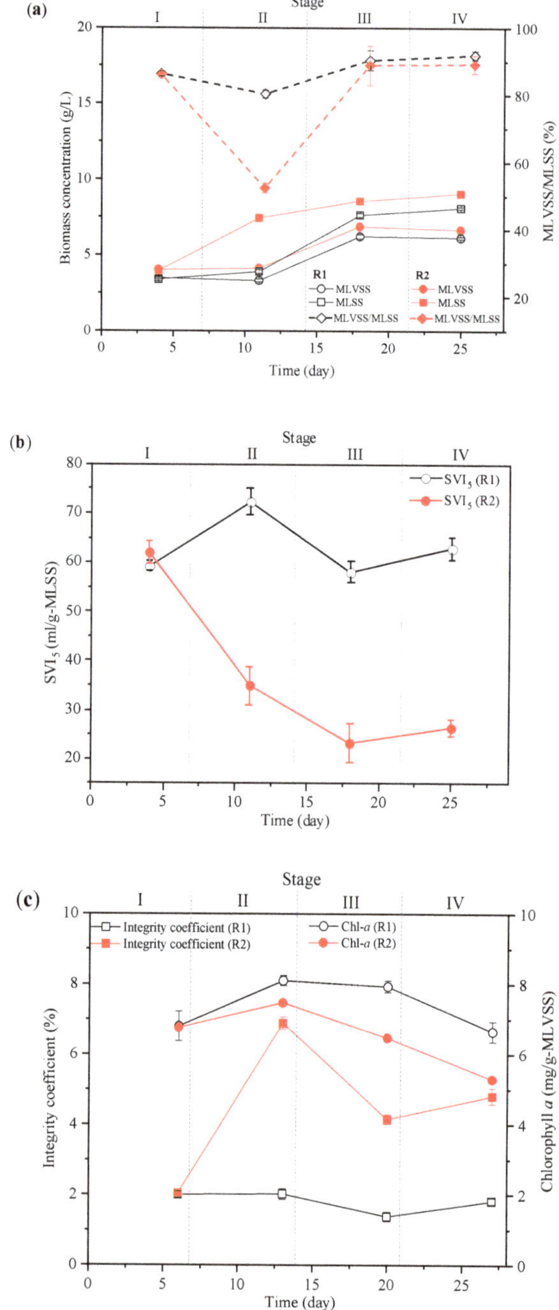

Figure 3. Characteristic changes of algal-bacterial AGS in R1 and R2 during the 28 days' operation. (**a**) ML(V)SS and MLVSS/MLSS ratio; (**b**) SVI$_5$; (**c**) integrity coefficient and chlorophyll *a* content.

3.1.3. Granular Stability and Chl-*a* Content

In this study, the integrity coefficient (IC) and Chl-*a* represent the granular stability (or granular strength) and algal content in granules, respectively. From Figure 3c, during Stage I with no salinity exposure and Cr(VI) addition, the IC values of granules from R1 and R2 were similar at 2.01% and 2.04%, respectively. However, in the presence of 10 g/L of NaCl (during Stage II), the IC of algal-bacterial AGS in R2 was consistently higher than that in R1 (no salinity exposure), indicating the decrease in granular strength in R2. The negative effect of salinity on granular strength in this study is consistent with previous studies [25,26]. Interestingly, when 2 mg/L Cr(VI) (Stage III) was co-existing, the IC values of granules in both reactors slightly decreased in comparison to their previous stage, and then slightly increased to 1.82% (R1) and 4.82% (R2), respectively, when Cr(VI) exposure ceased (Stage IV), implying a positive effect of Cr(VI) on granule stability. As a whole, a very slight change in IC value was observed in the granules from both reactors during the test period, probably due to the very compact structure of algal-bacterial AGS and the low concentrations of NaCl and/or Cr(VI) tested.

As seen from Figure 3c, after the exposure to 1% salinity (Stage II), the Chl-*a* content in the granules from R2 was consistently lower than that from R1, implying the negative influence of salinity on the growth of microalgae. On the other hand, in the presence of Cr(VI), the Chl-*a* contents tended to decrease in the granules from both reactors, which continued to decline even after Cr(VI) exposure ceased (Stage IV), to 6.65 mg/g-VSS and 5.30 mg/g-VSS in the granules from R1 and R2, respectively.

3.2. Performance of the Two Reactors

3.2.1. Organics Removal

The changes in effluent DIC, DOC concentrations, and DOC removal efficiency are shown in Figure 4. As seen, little impact was observed on organics removal under the salinity exposure and/or Cr(VI) presence, achieving, on average, >93% of DOC removal by the two reactors ($p = 0.152 \gg 0.05$). Specifically, the effluent DOC concentrations from the two reactors were consistently lower than 18 mg/L, with an average DOC removal efficiency of $96.7 \pm 1.1\%$ by R1 and $96.5 \pm 0.8\%$ by R2 during Stage I, respectively. After the 1% salinity exposure (Stage II), the effluent DOC concentration from R2 began to increase and exceeded that from R1 (no salinity exposure) by 10.52 mg/L and 6.70 mg/L, respectively. Then, both DOC concentrations tended to increase after the exposure to 2 mg/L Cr(VI) (Stage III). When the exposure to Cr(VI) ceased during Stage IV, a slight decrease in DOC was detected in the effluents from both reactors, suggesting that both reactors could recover their DOC removal capacity. The above results imply the limited negative effect of salinity and Cr(VI) on algal-bacterial AGS with respect to DOC removal performance. The relatively low salinity level (1%) may lead to a result of "no significant difference" in DOC removal between the two reactors.

Compared with the effluent DIC concentration from R1, that from R2 was slightly higher. This phenomenon may be associated with the less algae growth in R2, which can be clearly reflected from the lower Chl-*a* content in the granules from R2 after salinity exposure (from Stage II to the end of test, Figure 3c). Thus, salinity exposure may partially contribute to the increase of effluent DIC concentration. For instance, according to Dong et al. [25], high salinity (1% and 3%) conditions can negatively affect the algal growth in algal-bacterial AGS system, resulting in high DIC concentrations in the effluent of SBR. Interestingly, after exposure to 2 mg/L Cr(VI) (Stage III), both reactors produced a lower DIC concentration effluent, and then became relatively stable when the Cr(VI) exposure ceased.

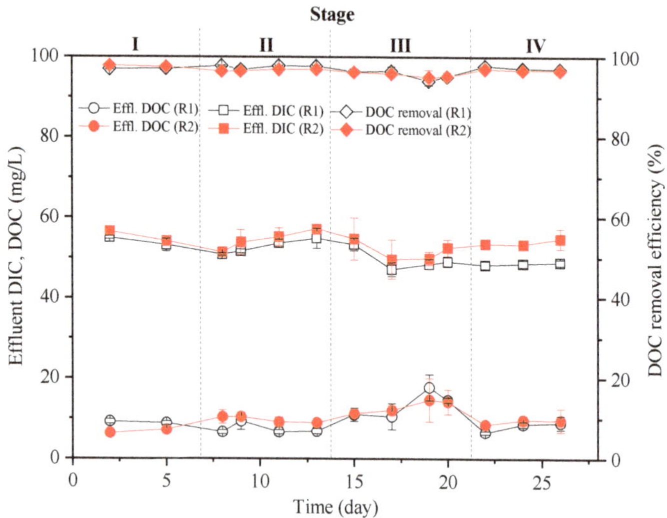

Figure 4. Variations of the effluent DOC and DIC concentrations from R1 and R2, and their DOC removals during the 28 days' operation.

3.2.2. Nitrogen Removal

The changes in effluent N species (including NH_4^+-N, NO_2^--N, and NO_3^--N) and total inorganic nitrogen (TIN) removal are shown in Figure 5. Among the variations of the three N species, the concentrations of effluent NH_4^+-N from the two reactors showed the most apparent fluctuations. More specifically, the effluent NH_4^+-N concentration from R2 gradually increased after the exposure to 1% salinity began (Stage II). This increasing trend continued when the granules were exposed to 2 mg/L Cr(VI) (Stage III), while the NH_4^+-N concentration steadily decreased when the exposure ceased from Stage IV. In contrast, the effluent NH_4^+-N concentration from R1 (no salinity exposure) gradually increased from the beginning of exposure to Cr(VI) (Stage III), but decreased during the later period of Stage III. Being different from the apparent fluctuation in effluent NH_4^+-N concentrations, the effluent NO_2^--N concentrations remained at a relatively constant low level, averagely 0.61 mg/L and 1.22 mg/L from R1 and R2, respectively. As for NO_3^--N, Cr(VI) exposure did not exert a noticeable effect on the granules in the two reactors. In addition, as can be seen from Figure 5, the effluent NO_3^--N concentrations from R2 were generally lower than those from R1 during most of the test period, to some extent reflecting the negative effect of salinity exposure on nitrification.

Based on the changes of TIN removal during the 28 days' operation of R1 and R2, the main findings can be summarized as follows: (i) In the test SBR (R2), nitrification process was inhibited when exposed to 1% salinity, which became worse after the granules were exposed to 2 mg/L Cr(VI) together. The granules could be recovered when the exposure to Cr(VI) ceased. (ii) In the control SBR (R1), Cr(VI) exposure also exhibited an inhibitory effect on nitrification, and the granules could be acclimated and recovered along with the operation. Once the Cr(VI) exposure ceased, the granules could quickly be recovered to the initial level.

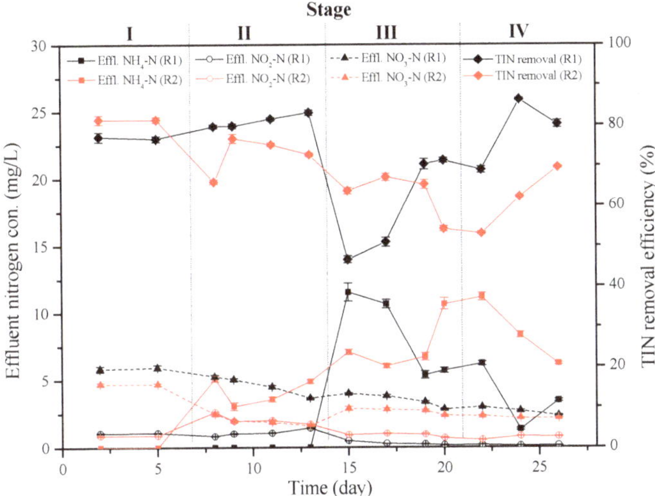

Figure 5. Profiles of the effluent nitrogen species from R1 and R2, and their TIN removals during the 28 days of operation.

The above results indicate that exposure to 1% salinity and/or 2 mg/L Cr(VI) may negatively affect nitrogen removal performance to different extents. This observation is probably associated with the inhibitory effects of the above harsh environment on the growth of algae or bacteria that are responsible for nitrogen uptake and conversion, in agreement with previous works [25,27].

3.2.3. Phosphorus Removal

Figure 6 shows the effluent TP concentration and TP removal efficiency of the two SBRs during the 28 days of operation. At Stage I with no exposure to 1% salinity and 2 mg/L Cr(VI), the TP removal efficiency was averagely kept at about 68.8% by R1 and 80.8% by R2, respectively, and this difference was mainly due to their different SRTs (no SRT control). After that, an obvious increase in the effluent TP concentration from R2 was detected, ranging from 0.97 mg/L to 3.59 mg/L when the granules were exposed to 1% salinity (Stage II), which led to a dramatic reduction in TP removal to the lowest level of 28.1%. However, this P removal performance was quickly recovered over the operation time. When the granules were exposed 2 mg/L Cr(VI) (Stage III), the effluent TP concentrations from R1 and R2 increased from 0.73 mg/L and 1.26 mg/L to 2.34 mg/L and 1.99 mg/L, corresponding to the reductions in TP removal efficiency to 53.3% and 60.3%, respectively. Even so, their TP removal efficiencies could be rapidly recovered, reaching 97.2% and 78.1% by R1 and R2 when the exposure to Cr(VI) ceased (Stage IV). This means that the algal-bacterial granules had a good TP removal capability and recoverability under fluctuated wastewater quality and variable environmental conditions, which could quickly recover from the short-term inhibition of 1% salinity and/or 2 mg/L Cr(VI) exposure. This observation is different from a previous study [25], i.e., the salt-stressed algal-bacterial AGS (under sudden salinity disturbance conditions) could not recover its original TP removal rate. The different response of algal-bacterial AGS in these two works are most probably attributable to the differences in SBR operating conditions (dimension and size, cycle time, aeration time, operating duration, etc.), light condition, salinity exposure procedure, and the different characteristics of the initial granules used.

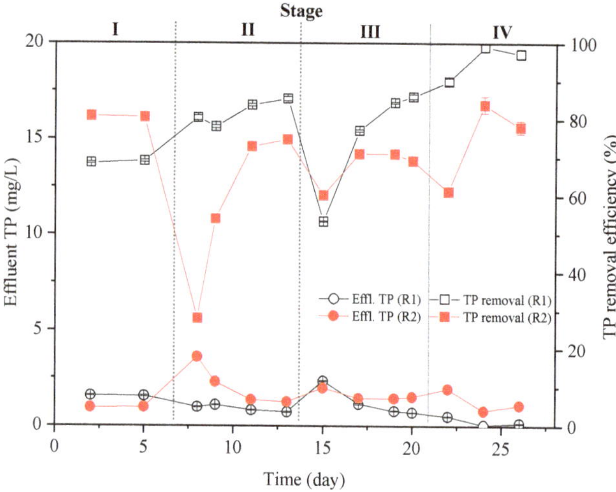

Figure 6. Profiles of the effluent P concentrations from R1 and R2, and their TP removals during the 28 days of operation.

Further studies are still necessary on the effect of salinity and/or Cr(VI) exposure on the anaerobic P release rate and aerobic P uptake rate, thereby shedding light on the mechanism regarding the varied P removal performance of the algal-bacterial AGS system.

3.2.4. Chromium Removal

In order to assess the effect of salinity on Cr(VI) removal efficiency by the algal-bacterial AGS system, the effluent Cr(VI) and Cr(III) concentrations, as well as the Cr(VI) removal efficiency, were monitored during Stage III (Figure 7). It is noteworthy that the Cr(VI) removal rate of R2 was generally lower than that of R1 in Stage III, possibly due to the slightly negative effect of salinity exposure. Specifically, the Cr(VI) removal efficiencies of R1 and R2 fluctuated between 31–51% (37.3% on average) and 28–48% (34.8% on average), respectively ($p = 0.005 < 0.05$). This observation might be attributable to the inhibitory impact of salinity on microbial activity in the AGS system [28], leading to the decline in Cr(VI) removal efficiency. In addition, the increase in ionic strength under the 1% salinity condition (R2) might have some shielding effect on the surface electrostatic field, thus suppressing the adsorption of chromium species on the active sites of algal-bacterial AGS [29].

The effluent Cr(III) concentrations from the two reactors could be obtained by subtracting the effluent Cr(VI) concentration from the effluent total Cr concentration. During Stage III, the effluent Cr(III) concentrations from R1 and R2 were very low, fluctuating between 0.01–0.04 mg/L and 0.01–0.06 mg/L, respectively. Still, it is necessary to further analyze the total Cr (Cr(III) + Cr(VI)) content in the granules, as well as in the EPS samples, to evaluate the effect of salinity on the Cr accumulation capability by the algal-bacterial granules.

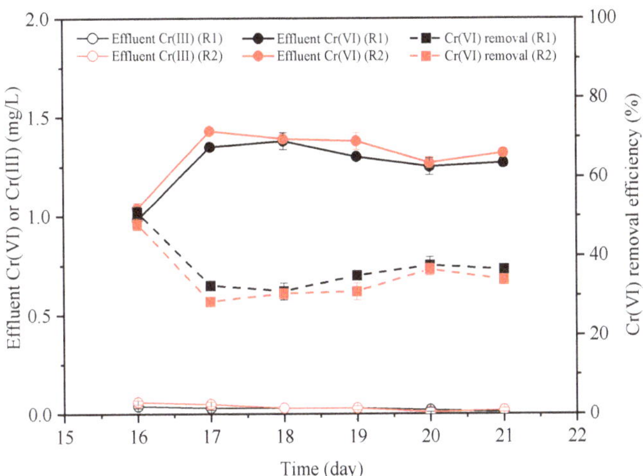

Figure 7. Profiles of the effluent chromium species from R1 and R2, and their Cr(VI) removals during Stage III.

Considering the Cr bioremediation mechanisms, the chemical fractionation of Cr in the algal-bacterial AGS (shown in Supplementary Figure S2), as well as surface functional groups (from FTIR spectra in Supplementary Figure S3), were also examined. For the granules in R1 (control, no salinity exposure), organic bound Cr contributed to the largest fraction, about 51.6%, indicating that metal complexation might be the main mechanism involved in Cr(VI) removal by algal-bacterial AGS in this study. This observation is consistent with previous studies [30,31], in which bacterial AGS and algal-bacterial AGS were used to adsorb Cr(III) and Cr(VI), respectively.

As for the granules in R2 (1% salinity exposure), the Fe/Mn oxides bound Cr fraction accounted for the highest percentage, about 52.0%, implying that Fe/Mn oxides in granules play a more important role in binding and adsorption of Cr than other fractions. Compared with the fractionation results of the granules from R1, under 1% salinity, some influence on Cr removal could be discerned, as the granules in R2 exhibited a higher chromium removal and immobilization by Fe/Mn oxides, and lower organic bound and residual fractions. It is possible that under some harsh condition (such as the 1% salinity condition in this study), the two most stable fractions (residual and organic matter bound parts) could be converted to less stable fractions, resulting in the increase in Cr bioavailability. Further studies are necessary on the leaching risk of Cr accumulated in algal-bacterial AGS with prolonged salinity exposure to better understand the effects of Cr-loaded granules on the environment. Moreover, the results from FTIR analysis (Supplementary Figure S3) and the morphological changes on granule surfaces (Supplementary Figure S1) partly confirmed that the functional groups such as carboxyl and phosphate played an important role in the metal complexation and/or chemical precipitation.

Based on the above results, the algal-bacterial AGS cultivated in the SBR system can potentially treat wastewaters containing both Cr(VI) and high salinity during the long operation period. Still, future works are necessary for a clear and insightful explanation of the mechanisms involved in the effects of co-existing Cr(VI) and salinity, which are also included in our follow-up research. For instance, further work on the changes of microbial composition and algae species during the long-term operation of algal-bacterial AGS system could help to clearly elucidate their roles in Cr(VI) removal.

3.3. EPS Secretion from Algal-Bacterial Granules

Algal-bacterial granule samples used for EPS extraction and EPS content determination were obtained from the two SBRs at the end of each operation stage. EPS are mainly composed of PN and PS that are secreted by microorganisms, which play an essential role in granular structure and stability [32,33]. The changes in PN and PS contents under different test stages are illustrated in Figure 8.

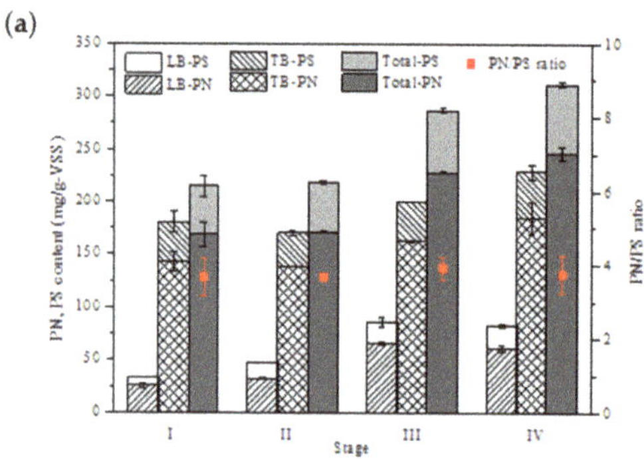
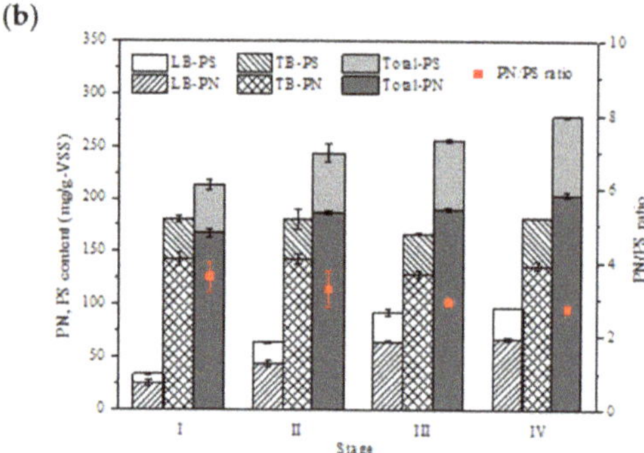

Figure 8. Variation of the EPS content (LB-EPS, TB-EPS, and total EPS), composition (PN and PS), and the PN/PS ratios of algal-bacterial AGS at the end of each test stage during the 28 days of operation. (**a**) R1; (**b**) R2.

As shown in Figure 8a, the total EPS content in the granules from R1 remained stable during Stages I and II. Meanwhile, for the granules in R2, after exposure to 1% salinity (Stage II), a slight increase in the EPS content, from 213.82 90 mg/g-MLVSS to 243.90 mg/g-MLVSS, was detected (Figure 8b). These results indicate that algal-bacterial AGS can produce more EPS to protect themselves from the increasing osmotic pressure due to the increase of influent salinity, in agreement with previous studies using granular sludge to treat saline wastewater [24,25,34,35]. For instance, Corsino and coworkers [35] found that the osmotic pressure generated by the increasing salinity can stimulate the production of exopolymers by microorganisms. EPS might mitigate the osmotic pressure on the cell wall of bacteria because of their gelatinous matrix.

As the influent Cr(VI) concentration increased from 0 mg/L to 2 mg/L (Stage III), the total EPS content in granules from both reactors increased. However, the extent of their increase was different. Specifically, after exposure to 2 mg/L Cr(VI), the granules from R1 excreted more EPS, which was remarkably increased to 286.64 mg/g-MLVSS. This increase trend continued, reaching 311.53 mg/g-MLVSS even after Cr(VI) exposure ceased (Figure 8a). On the other hand, in R2, the granular EPS content rose slightly to 259.39 mg/g-MLVSS and 278.73 mg/g-MLVSS during Stages III and IV, respectively. The significant increase of EPS in granules from R1 during Stage III might help protect microorganisms from harsh external condition (e.g., Cr(VI) presence in the influent) [27,32,36]. The granules from R2 may have better tolerance of Cr(VI) under the co-existence of 1% salinity.

From Figure 8a, the TB-EPS content showed a similar trend to the total EPS content during the four operation stages, while the LB-EPS content displayed a somewhat different variation trend. This finding implies that TB-EPS is an essential part affecting the algal-bacterial granule structure and stability under Cr(VI) exposure condition. On the contrary, the variation of LB-EPS content better matched with the total EPS change than the TB-EPS content in the granules from R2. Therefore, the LB-EPS played a more important role in algal-bacterial granule structure and stability in the presence of salinity, which is in line with a recent study [25].

As shown in Figure 8b, the PN/PS ratio gradually decreased over time under the presence or absence of Cr(VI), suggesting that 1% salinity exposure may trigger more excretion of PS than PN, somewhat helping the granules maintain a more stable structure to resist saline conditions. This observation is consistent with the finding from a previous work [25]. One possible explanation for this phenomenon is that PS contains abundant polar functional groups that are often strongly water-binding and can thicken the hydration layer around the cells, mitigating the adverse effect of osmotic pressure generated by salinity [37]. Meanwhile, for the granules from R1, the PN/PS ratio remained stable at 3.66 and 3.68 during Stages I and II, respectively (Figure 8a). However, after exposure to 2 mg/L Cr(VI) (Stage III), their PN/PS ratio increased to 3.92, which fell to 3.76 when Cr(VI) exposure ceased (Stage IV). More excretion of PN than PS was detected from the granules in R1 during Stage III when compared to Stage II, which is probably associated with some exoenzymes contained in EPS that could reduce heavy metal ions [38] as enzymes are mainly composed of proteins. Interestingly, during the four test stages, the granules from both R1 and R2 contained consistently higher PN content than PS content. To summarize, salinity exposure promoted algal-bacterial AGS to secrete more PS, while Cr(VI) exposure stimulated more PN excretion.

4. Conclusions

The present study examined the effects of salinity on Cr(VI) bioremediation and nutrients removal performance by algal-bacterial AGS. The results from this work showed that algal-bacterial AGS reflected less filamentous and a rougher surface under salinity exposure, with little influence on organics removal being observed. However, the removals of total inorganic nitrogen and total phosphorus were noticeably impacted at the 1% salinity condition, and were further decreased with the co-existence of 2 mg/L Cr(VI). The Cr(VI) removal efficiency, on the other hand, was 31–51% in the control SBR and 28–48% in the test SBR, respectively, indicating that salinity exposure may slightly influence Cr(VI) bioremediation. Furthermore, in this study, it was found that 1% salinity exposure stimulated more PS excretion from algal-bacterial AGS, while PN excretion was enhanced with the introduction of 2 mg/L Cr(VI).

Supplementary Materials: The followings are available online at https://www.mdpi.com/article/10.3390/pr9081400/s1. Figure S1: Morphological changes of algal-bacterial AGS from R1 and R2 at the beginning and the end of the test, respectively. Figure S2: Chemical fractionation of different metals extracted from the Cr-loaded granules. Figure S3: FTIR spectrum of algal-bacterial AGS at the beginning (a) and (b) the end of test from R1 (without salinity addition), and (c) the end of the test from R2 (with 1% salinity addition).

Author Contributions: Conceptualization, B.V.N., X.Y., and Z.L.; Methodology, B.V.N., X.Y., J.W., Z.Z. (Ziwen Zhao), Z.L., and S.H.; Software, B.V.N. and X.Y.; Validation, B.V.N. and X.Y.; Formal analysis, B.V.N. and X.Y.; Investigation, B.V.N., X.Y., J.W., Z.Z. (Ziwen Zhao), and S.H.; Data curation, B.V.N., X.Y., J.W., Z.Z. (Ziwen Zhao), Z.L., S.H., and S.X.L.; Writing–original draft preparation, B.V.N. and X.Y.; Writing–review and editing, B.V.N., X.Y., Z.L., K.S., Z.Z. (Zhenya Zhang), and S.H.; Supervision, Z.L., K.S., and Z.Z. (Zhenya Zhang); Project administration, Z.L.; Funding acquisition, Z.L. All authors have read and agreed to the published version of the manuscript.

Funding: This work was supported by JSPS KAKENHI Grant Number JP18H03403.

Institutional Review Board Statement: Not applicable.

Informed Consent Statement: Not applicable.

Data Availability Statement: The data presented in this study are available on request from the corresponding author.

Acknowledgments: The first author, Bach Van Nguyen, would like to thank the financial and technical support by the Project for Human Resource Development Scholarship by Japanese Grant Aid (JDS), Japan International Cooperation Agency (JICA), and the Institute of Marine Environment and Resources (IMER), Vietnam Academy of Science and Technology (VAST), Vietnam for supporting his study in University of Tsukuba, Japan.

Conflicts of Interest: The authors declare no conflict of interest.

References

1. Sartore, L.; Dey, K. Preparation and heavy metal ions chelating properties of multifunctional polymer-grafted silica hybrid materials. *Adv. Mater. Sci. Eng.* **2019**, *2019*, e7260851. [CrossRef]
2. Gibb, H.J.; Lees, P.S.; Pinsky, P.F.; Rooney, B.C. Lung cancer among workers in chromium chemical production. *Am. J. Ind. Med.* **2000**, *38*, 115–126. [CrossRef]
3. Zhitkovich, A. Chromium in drinking water: Sources, metabolism, and cancer risks. *Chem. Res. Toxicol.* **2011**, *24*, 1617–1629. [CrossRef]
4. Tumolo, M.; Ancona, V.; De Paola, D.; Losacco, D.; Campanale, C.; Massarelli, C.; Uricchio, V.F. Chromium pollution in european water, sources, health risk, and remediation strategies: An overview. *Int. J. Environ. Res. Public. Health* **2020**, *17*, 5438. [CrossRef]
5. Mitra, S.; Sarkar, A.; Sen, S. Removal of chromium from industrial effluents using nanotechnology: A review. *Nanotechnol. Environ. Eng.* **2017**, *2*, 11. [CrossRef]
6. Brasili, E.; Bavasso, I.; Petruccelli, V.; Vilardi, G.; Valletta, A.; Bosco, C.D.; Gentili, A.; Pasqua, G.; Di Palma, L. Remediation of hexavalent chromium contaminated water through zero-valent iron nanoparticles and effects on tomato plant growth performance. *Sci. Rep.* **2020**, *10*, 1920. [CrossRef] [PubMed]
7. Fu, F.; Wang, Q. Removal of heavy metal ions from wastewaters: A review. *J. Environ. Manag.* **2011**, *92*, 407–418. [CrossRef] [PubMed]
8. Pushkar, B.; Sevak, P.; Parab, S.; Nilkanth, N. Chromium pollution and its bioremediation mechanisms in bacteria: A review. *J. Environ. Manag.* **2021**, *287*, 112279. [CrossRef] [PubMed]
9. Joseph, L.; Jun, B.-M.; Flora, J.R.V.; Park, C.M.; Yoon, Y. Removal of heavy metals from water sources in the developing world using low-cost materials: A review. *Chemosphere* **2019**, *229*, 142–159. [CrossRef]
10. Zabochnicka-Świątek, M.; Krzywonos, M. Potentials of biosorption and bioaccumulation processes for heavy metal removal. *Pol. J. Environ. Stud.* **2014**, *23*, 551–561.
11. Ashraf, S.; Naveed, M.; Afzal, M.; Ashraf, S.; Rehman, K.; Hussain, A.; Zahir, Z.A. Bioremediation of tannery effluent by cr- and salt-tolerant bacterial strains. *Environ. Monit. Assess.* **2018**, *190*, 716. [CrossRef]
12. Zeng, Q.; Hu, Y.; Yang, Y.; Hu, L.; Zhong, H.; He, Z. Cell envelop is the key site for Cr(VI) reduction by oceanobacillus oncorhynchi w4, a newly isolated Cr(VI) reducing bacterium. *J. Hazard. Mater.* **2019**, *368*, 149–155. [CrossRef]
13. U.S. Environmental Protection Agency. *Engineered Approaches to in Situ Bioremediation of Chlorinated Solvents: Fundamentals and Field Applications*; U.S. Environmental Protection Agency: Washington, DC, USA, 2000.
14. Yang, X.; Zhao, Z.; Yu, Y.; Shimizu, K.; Zhang, Z.; Lei, Z.; Lee, D.-J. Enhanced biosorption of Cr(VI) from synthetic wastewater using algal-bacterial aerobic granular sludge: Batch experiments, kinetics and mechanisms. *Sep. Purif. Technol.* **2020**, *251*, 117323. [CrossRef]
15. Yang, X.; Zhao, Z.; Zhang, G.; Hirayama, S.; Nguyen, B.V.; Lei, Z.; Shimizu, K.; Zhang, Z. Insight into Cr(VI) biosorption onto algal-bacterial granular sludge: Cr(VI) bioreduction and its intracellular accumulation in addition to the effects of environmental factors. *J. Hazard. Mater.* **2021**, *414*, 125479. [CrossRef]
16. Stasinakis, A.S.; Thomaidis, N.S.; Mamais, D.; Papanikolaou, E.C.; Tsakon, A.; Lekkas, T.D. Effects of chromium (VI) addition on the activated sludge process. *Water Res.* **2003**, *37*, 2140–2148. [CrossRef]
17. Minhas, A.K.; Hodgson, P.; Barrow, C.J.; Adholeya, A. A review on the assessment of stress conditions for simultaneous production of microalgal lipids and carotenoids. *Front. Microbiol.* **2016**, *7*, 546. [CrossRef] [PubMed]
18. Huang, W.; Cai, W.; Huang, H.; Lei, Z.; Zhang, Z.; Tay, J.H.; Lee, D.-J. Identification of inorganic and organic species of phosphorus and its bio-availability in nitrifying aerobic granular sludge. *Water Res.* **2015**, *68*, 423–431. [CrossRef]
19. Wang, J.; Lei, Z.; Tian, C.; Liu, S.; Wang, Q.; Shimizu, K.; Zhang, Z.; Adachi, Y.; Lee, D.-J. Ionic response of algal-bacterial granular sludge system during biological phosphorus removal from wastewater. *Chemosphere* **2021**, *264*, 128534. [CrossRef] [PubMed]
20. APHA. *Standard Methods for the Examination of Water and Wastewater*; American Public Health Association, American Water Work Association and Water Environment Federation: Washington, DC, USA, 2012.
21. Wang, Q.; Shen, Q.; Wang, J.; Zhang, Y.; Zhang, Z.; Lei, Z.; Shimizu, K.; Lee, D.-J. Fast cultivation and harvesting of oil-producing microalgae ankistrodesmus falcatus var. Acicularis fed with anaerobic digestion liquor via biogranulation in addition to nutrients removal. *Sci. Total Environ.* **2020**, *741*, 140183. [CrossRef]
22. Zhao, Z.; Yang, X.; Cai, W.; Lei, Z.; Shimizu, K.; Zhang, Z.; Utsumi, M.; Lee, D.-J. Response of algal-bacterial granular system to low carbon wastewater: Focus on granular stability, nutrients removal and accumulation. *Bioresour. Technol.* **2018**, *268*, 221–229. [CrossRef] [PubMed]
23. Li, X.; Luo, J.; Guo, G.; Mackey, H.R.; Hao, T.; Chen, G. Seawater-based wastewater accelerates development of aerobic granular sludge: A laboratory proof-of-concept. *Water Res.* **2017**, *115*, 210–219. [CrossRef]
24. Meng, F.; Liu, D.; Huang, W.; Lei, Z.; Zhang, Z. Effect of salinity on granulation, performance and lipid accumulation of algal-bacterial granular sludge. *Bioresour. Technol. Rep.* **2019**, *7*, 100228. [CrossRef]
25. Dong, X.; Zhao, Z.; Yang, X.; Lei, Z.; Shimizu, K.; Zhang, Z.; Lee, D.-J. Response and recovery of mature algal-bacterial aerobic granular sludge to sudden salinity disturbance in influent wastewater: Granule characteristics and nutrients removal/accumulation. *Bioresour. Technol.* **2021**, *321*, 124492. [CrossRef]

26. Meng, F.; Huang, W.; Liu, D.; Zhao, Y.; Huang, W.; Lei, Z.; Zhang, Z. Application of aerobic granules-continuous flow reactor for saline wastewater treatment: Granular stability, lipid production and symbiotic relationship between bacteria and algae. *Bioresour. Technol.* **2020**, *295*, 122291. [CrossRef]
27. Wang, X.; Dai, H.; Zhang, J.; Yang, T.; Chen, F. Unraveling the long-term effects of Cr(VI) on the performance and microbial community of nitrifying activated sludge system. *Water* **2017**, *9*, 909. [CrossRef]
28. Bassin, J.P.; Pronk, M.; Muyzer, G.; Kleerebezem, R.; Dezotti, M.; van Loosdrecht, M.C.M. Effect of elevated salt concentrations on the aerobic granular sludge process: Linking microbial activity with microbial community structure. *Appl. Environ. Microbiol.* **2011**, *77*, 7942–7953. [CrossRef] [PubMed]
29. Shukla, A.; Zhang, Y.-H.; Dubey, P.; Margrave, J.L.; Shukla, S.S. The role of sawdust in the removal of unwanted materials from water. *J. Hazard. Mater.* **2002**, *95*, 137–152. [CrossRef]
30. Yao, L.; Ye, Z.; Tong, M.; Lai, P.; Ni, J. Removal of Cr3+ from aqueous solution by biosorption with aerobic granules. *J. Hazard. Mater.* **2009**, *165*, 250–255. [CrossRef] [PubMed]
31. Yang, X.; Zhao, Z.; Nguyen, B.V.; Hirayama, S.; Tian, C.; Lei, Z.; Shimizu, K.; Zhang, Z. Cr(VI) bioremediation by active algal-bacterial aerobic granular sludge: Importance of microbial viability, contribution of microalgae and fractionation of loaded Cr. *J. Hazard. Mater.* **2021**, *418*, 126342. [CrossRef]
32. Fang, J.; Su, B.; Sun, P.; Lou, J.; Han, J. Long-term effect of low concentration Cr(VI) on p removal in granule-based enhanced biological phosphorus removal (EBPR) system. *Chemosphere* **2015**, *121*, 76–83. [CrossRef]
33. Lin, H.; Ma, R.; Hu, Y.; Lin, J.; Sun, S.; Jiang, J.; Li, T.; Liao, Q.; Luo, J. Reviewing bottlenecks in aerobic granular sludge technology: Slow granulation and low granular stability. *Environ. Pollut.* **2020**, *263*, 114638. [CrossRef]
34. Wan, C.; Yang, X.; Lee, D.-J.; Liu, X.; Sun, S.; Chen, C. Partial nitrification of wastewaters with high nacl concentrations by aerobic granules in continuous-flow reactor. *Bioresour. Technol.* **2014**, *152*, 1–6. [CrossRef]
35. Corsino, S.F.; Capodici, M.; Torregrossa, M.; Viviani, G. Physical properties and extracellular polymeric substances pattern of aerobic granular sludge treating hypersaline wastewater. *Bioresour. Technol.* **2017**, *229*, 152–159. [CrossRef] [PubMed]
36. Wang, Z.; Gao, M.; Wang, S.; Xin, Y.; Ma, D.; She, Z.; Wang, Z.; Chang, Q.; Ren, Y. Effect of hexavalent chromium on extracellular polymeric substances of granular sludge from an aerobic granular sequencing batch reactor. *Chem. Eng. J.* **2014**, *251*, 165–174. [CrossRef]
37. Li, J.; Ye, W.; Wei, D.; Ngo, H.H.; Guo, W.; Qiao, Y.; Xu, W.; Du, B.; Wei, Q. System performance and microbial community succession in a partial nitrification biofilm reactor in response to salinity stress. *Bioresour. Technol.* **2018**, *270*, 512–518. [CrossRef] [PubMed]
38. Wang, Z.; Gao, M.; Wei, J.; Ma, K.; Zhang, J.; Yang, Y.; Yu, S. Extracellular polymeric substances, microbial activity and microbial community of biofilm and suspended sludge at different divalent cadmium concentrations. *Bioresour. Technol.* **2016**, *205*, 213–221. [CrossRef] [PubMed]

Article

Long-Term Stability of Nitrifying Granules in a Membrane Bioreactor without Hydraulic Selection Pressure

Zhaohui An [1], Xueyao Zhang [1], Charles B. Bott [2] and Zhi-Wu Wang [1,*]

[1] Department of Civil and Environmental Engineering, Virginia Polytechnic Institute and State University, Blacksburg, VA 24061, USA; za1@vt.edu (Z.A.); xueyao@vt.edu (X.Z.)

[2] Operations Department, Hampton Roads Sanitation District, Virginia Beach, VA 23455, USA; cbott@hrsd.com

* Correspondence: wzw@vt.edu

Abstract: To understand the long-term stability of nitrifying granules in a membrane bioreactor (GMBR), a membrane module was submerged in an airlift reactor to eliminate the hydraulic selection pressure that was believed to be the driving force of aerobic granulation. The long-term monitoring results showed that the structure of nitrifying granules could remain stable for 305 days in the GMBR without hydraulic selection pressure; however, the majority of the granule structure was actually inactive due to mass diffusion limitation. As a consequence, active biomass free of mass diffusion limitation only inhabited the top 60–80 μm layer of the nitrifying granules. There was a dynamic equilibrium between bioflocs and membrane, i.e., 25% of bioflocs attached on the membrane surface within the last nine days of the backwash cycle in synchronization with the emergence of a peak of soluble extracellular polymeric substances (sEPS), with a concentration of around 47 mg L^{-1}. Backwash can eventually detach and return these bioflocs to the bulk solution. However, the rate of membrane fouling did not change with and without the biofloc attachment. In a certain sense, the GMBR investigated in this study functioned in a similar fashion as an integrated fixed-film activated sludge membrane bioreactor and thus defeated the original purpose of GMBR development. The mass diffusion problem and sEPS production should be key areas of focus in future GMBR research.

Keywords: granules; hydraulic selection pressure; continuous flow; MBR

Citation: An, Z.; Zhang, X.; Bott, C.B.; Wang, Z.-W. Long-Term Stability of Nitrifying Granules in a Membrane Bioreactor without Hydraulic Selection Pressure. *Processes* **2021**, *9*, 1024. https://doi.org/10.3390/pr9061024

Academic Editors: Yongqiang Liu and Zhiya Sheng

Received: 13 May 2021

Accepted: 8 June 2021

Published: 10 June 2021

Publisher's Note: MDPI stays neutral with regard to jurisdictional claims in published maps and institutional affiliations.

1. Introduction

Aerobic granulation technology has been developed for two decades for the advantage of granular sludge over activated sludge in terms of settleability, filterability, biomass retention, and resistance to shock loading [1]. Since the debut of aerobic granular sludge as an alternative catalyst to activated sludge for advanced wastewater treatment, its formation mechanism was attributed to the hydraulic selection pressure [1]. Theoretically, granules with mass diffusion limitation are unable to compete with bioflocs in terms of growth without the facilitation of hydraulic selection pressure [2]. For this reason, it was reported that aerobic granules could not remain structurally stable for more than 80 days within MBRs operated without hydraulic selection pressure because the submerged membrane module retained both granules and bioflocs within the same reactor without selection [3–5]. However, our recent study demonstrated that nitrifying granules have remained stable within a completely mixed airlift reactor for 340 days under minimum hydraulic selection pressure in terms of the surface overflow rate (SOR) as low as 0.04 m h^{-1} [6]. Technically, such a minimum hydraulic selection pressure should be insufficient to keep the long-term stability of nitrifying granules because it is even lower than the SOR commonly used in full-scale wastewater treatment clarifiers that only retain activated sludge [7]. Therefore, it is reasonable to hypothesize that the existing granulation mechanism for nitrifying granules should be reconsidered. To test this hypothesis, an insightful experiment was designed to provide a mechanistic explanation of the phenomenon contradictory to the conventional pressure-driven aerobic granulation theory [1]. The experiment in this study eliminated all hydraulic selection pressure by inserting an ultrafiltration membrane into the airlift reactor to turn it into a granule membrane bioreactor (GMBR), which was operated for another 305 days to check whether nitrifying granules can still remain structurally stable. This study presented our observation over this extended experimental duration with insightful data analysis to reveal the mechanic understanding behind the phenomenon. It is our intention to use this study to verify the validity of conventional aerobic granulation theory and also discuss the perspective of GMBR that has been pursued in recent years.

2. Materials and Methods

2.1. Reactor Setup

The airlift reactor (ALR) used in our previous long-term nitrifying granule stability study was used in this study (Figure 1A) [6]. The only change made was by replacing the previous overflow design with a membrane inserted for completely retaining all bioparticles inside the ALR (Figure 1B). Briefly, the ALR comes with a 1.5-L working volume and a 3-cm diameter internal riser for hydraulic mixing. The reactor diameter was 6 cm, and the height was 44.5 cm. The SOR was controlled at around 0.04 m h^{-1}. The ALR was operated at 20–23 °C in a temperature-controlled room. The air was introduced to the bottom of the ALR at an aeration rate of 0.3 L min^{-1} using an air pump. The substrate containing 50–60 mg L^{-1} ammonium-nitrogen (NH$_4^+$-N) along with other nutrients as described in a recipe of the previous study [6] was continuously pumped into the ALR at a flow rate of 0.3 L h^{-1}, giving a hydraulic retention time of 5.2 h. The seed granules were the anammox granules collected from a DEMON® reactor operated by Hampton Roads Sanitation District, Virginia Beach, VA. As reported previously, the ALR was operated for 340 days without any granule stability issue [6]. After that, a membrane module (ZENON Environmental, ZeeWeed, Burlington, ON, Canada) was inserted into the ALR to turn it into a GMBR to totally eliminate the hydraulic selection pressure by retaining all biomass in the GMBR. The membrane module consisted of 30 hollow fibers with a pore size of 0.04 μm and a total effective area of 268.5 cm^2. The effluent was withdrawn from the membrane by a peristaltic pump at a flow rate of 0.3 L h^{-1}, which gives a flux around 11 L m^{-2} h^{-1}. The transmembrane pressure (TMP) was monitored by using a pressure gauge (Winters Instruments, PFQ790, Buffalo, NY, USA). The membrane module was backwashed in the GMBR for 5 min by pumping back the filtrate when the TMP reached 45 kPa. All detached biomass from the membrane during backwash was retained in the reactor by the membrane without any wasting, giving rise to a nearly infinite solids retention time. The period between two backwashes was defined as one backwash cycle. No relaxation time was applied during the membrane operation.

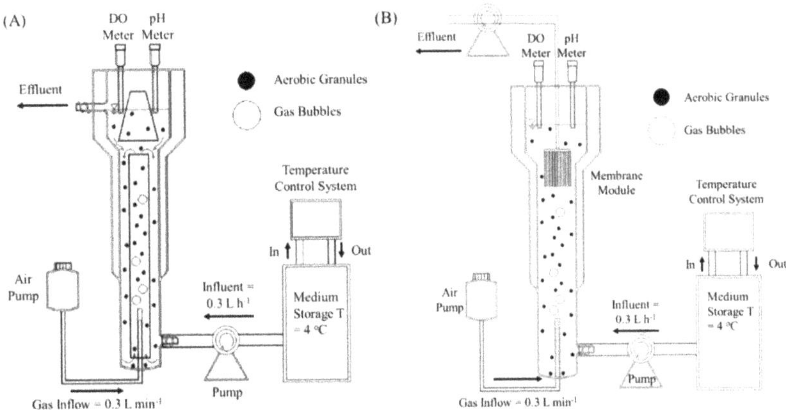

Figure 1. Reactor setup of the ALR (**A**) without membrane and the GMBR (**B**) with membrane.

2.2. Analytical Methods

The dissolved oxygen (DO) concentration of the reactor was monitored online with a 30-min interval by a submerged luminescent DO probe (Hach, Intellical™ LDO101, Loveland, CO, USA). The pH was monitored with a pH electrode (Hach, Intellical™ PHC101, Loveland, CO, USA). The TMP was measured by an inline pressure gauge. The mixed liquor suspended solids (MLSS), mixed liquor volatile suspended solids (MLVSS), settling velocity, ash content, and oxygen utilization rate (OUR) were all analyzed by using standard methods [8]. The nitrogen species, i.e., NH_4^+-N, nitrite-nitrogen (NO_2^--N), and nitrate-nitrogen (NO_3^--N), were measured using the Hach salicylate/cyanurate, NitriVer 3, and Tentpoles 835 kits. The turbidity of influent and effluent was quantified by a turbidity meter (Hach, 2100N, Loveland, CO, USA). The supernatant turbidity in the GMBR was measured after settling all granules. The granule size distribution was analyzed using a laser scattering particle size analyzer (Horiba, LA-950, Kyoto, Japan). Mixed liquor samples were taken from the GMBR for the petri-dish photos. The granule cross-section was stained using LIVE/DEAD BacLight Bacterial Viability Kit L-7012 for microscopy and quantitative assays (Molecular Probes, Eugene, OR, USA). The samples were stained and incubated in the darkroom at room temperature for 15 min. A microscope used for fluorescence visualization (Nikon, Eclipse E200, Melville, NY, USA) was equipped with a camera (Sony, Exmor CMOS, New York, NY, USA) for image capture. Soluble extracellular polymeric substances (sEPS) were extracted from the supernatant of samples after centrifugation (Thermo Scientific, Sorvall Legend X1, Waltham, MA, USA) at $4000\times g$ for 20 min at 4 °C and measured based on the method described previously [9]. Dialysis Kits (Spectrum, Labs Spectra/Por™ 3500 D MWCO Standard RC Pre-Treated, Waltham, MA, USA) were utilized to recover polymers with a molecular weight greater than 3500 Da. A phenol–sulfuric acid method was adopted for carbohydrates (PS) analysis (Nielsen, 2010), and Pierce™ modified Lowry protein assays were used for the protein (PN) analysis (Thermo Scientific, Waltham, MA, USA). The sum of PS and PN concentrations was regarded as sEPS concentration.

2.3. Model Simulation

To understand the substrate utilization profiles in the granules, the mathematical model developed and calibrated in previous studies was adopted in this work [6,10]. Briefly, the model was set up by taking both the Monod equation and Fick's law into consideration. The values of all parameters were kept the same or adjusted to different temperatures according to the Arrhenius equation as calibrated in the previous study [10]. The model was validated in a previous study [6]. The input of the model was the steady-state operation parameters of GMBR listed in Table 1.

Table 1. Reactor performance in a GMBR at steady state on the 305th day.

Parameters	Unit	Influent	In Reactor	Effluent
DO	mg L^{-1}	-	4.63 ± 0.84	-
pH	-	8.22 ± 0.14	7.23 ± 0.23	7.24 ± 0.22
NH_4^+-N	mg L^{-1}	49.70 ± 2.50	3.26 ± 4.17	2.66 ± 5.83
NO_2^--N	mg L^{-1}	0.28 ± 0.21	3.42 ± 4.53	3.85 ± 5.83
NO_3^--N	mg L^{-1}	0.79 ± 0.53	45.56 ± 6.73	45.93 ± 5.97

Each parameter was measured in triplicate and expressed as mean $\pm$ standard deviation

3. Results

3.1. Characteristics of Seed Granules

The seed sludge used for the study was steady-state nitrifying granules after 340 days of cultivation in a previous ALR without membrane insertion [6]. Figure 2A shows that these seed granules possess the typical morphology of nitrifying granules reported in the literature, i.e., smooth surface, round shape, and brown color [11–13]. The seed granule characteristics and performance are summarized in Table 2. It should be pointed out that the seed reactor overflow turbidity was only 0.77 ± 0.31 NTU, indicating excellent structural stability of these seed granules. The settling velocity of seed granules in ALR was as high as 54.2 m h^{-1}, which was comparable with those reported in previous studies, i.e., 18–60 m h^{-1} [14–16]. Table 2 shows that the nitrification efficiency was as high as 97%, indicating the excellent performance of these nitrifying granules. The average particle size of granules was 1.8 mm in Table 2, with 80% of particles above 0.2 mm, as shown in Figure 3. A similar range of particle size distribution was also reported in previous nitrifying granulation studies [6,12,17].

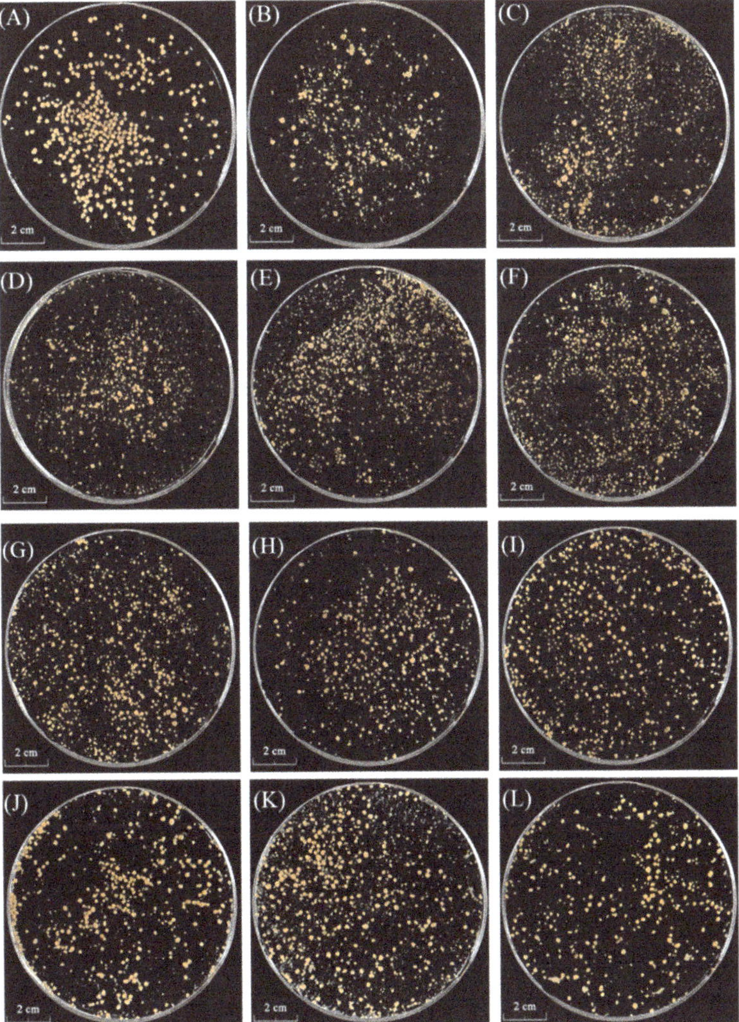

Figure 2. Morphology of nitrifying granules in the GMBR on day 0 (**A**), 37 (**B**), 79 (**C**), 100 (**D**), 130 (**E**), 150 (**F**), 180 (**G**), 210 (**H**), 240 (**I**), 260 (**J**), 280 (**K**) and 305 (**L**).

Table 2. Characteristics of seed granule and seed ALR without membrane.

Parameters	Unit	Granules	Influent	Effluent
Particle size	mm	1.8 ± 0.6	-	-
Settling velocity	$m\ h^{-1}$	54.2 ± 4.4	-	-
DO	$mg\ L^{-1}$	-	4.11 ± 0.72	-
pH	-	-	8.34 ± 0.16	7.26 ± 0.19
Turbidity	NTU	-	2.12 ± 1.12	0.77 ± 0.31
NH_4^+-N	$mg\ L^{-1}$	-	50.74 ± 5.36	1.70 ± 1.62
NO_2^--N	$mg\ L^{-1}$	-	0.80 ± 0.28	0.55 ± 0.32
NO_3^--N	$mg\ L^{-1}$	-	0.65 ± 0.19	49.83 ± 5.07

Each parameter was measured in triplicate and expressed as mean $\pm$ standard deviation.

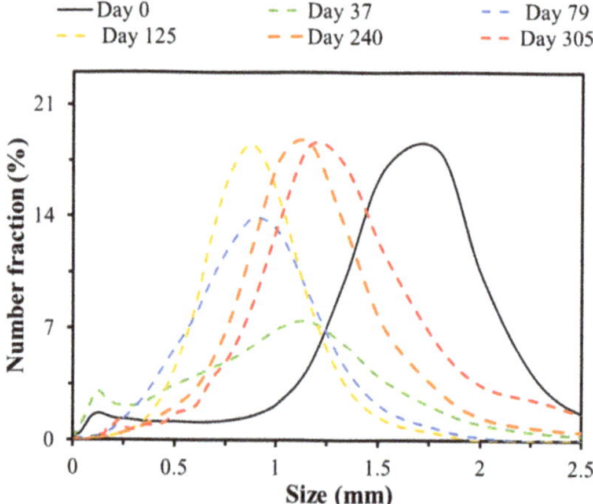

Figure 3. Particle size distribution in the GMBR in the course of 305 days of operation.

3.2. Response of Seed Granules to Membrane Insertion

A membrane module was inserted in the seed ALR to totally eliminate the hydraulic selection pressure in order to test whether these nitrifying granules can still remain long-term stable. It should be realized that the GMBRs such as this were studied previously, but the longest operation time was only 120 days, which is relatively too short to reflect their long-term stability [4,5,18]. Figures 2 and 3 show the granule morphology and particle size evolution over 305 days of the GMBR operation. As can be seen, the round shape seed granules collected have disintegrated into bioflocs with small and irregular structures after 37 days of cultivation in the GMBR, and the average particle sizes dropped from 1.8 mm to 0.8 mm (Figures 2B and 3). Meanwhile, there was also a sharp increase in the number of small particles with a size of around 0.2 mm (Figure 3) as a result of the fine bioflocs formation in Figure 2B. Accordingly, the turbidity of the supernatant of GMBR after 5 min sharply settling increased from 1.03 to 19.4 NTU (Figure 4A). After acclimation to the GMBR operation for 42 days, the peak of fine bioflocs that showed up on day 37 disappeared on day 79, and the particles with a size larger than 0.5 mm increased from 48% to 74% (Figure 3). Meanwhile, the turbidity also dropped to 1.61 NTU (Figure 4A). After around 240 days of operation, the granule morphology appeared to be much more uniform (Figure 2B–I). Although there were still some fine bioflocs shown in Figure 2J,K, probably due to the membrane backwash, the granule structure became quite stable again on day 305 (Figure 2L) in view of the similar particle size distribution on days 240 and 305, as shown in Figure 3. The supernatant turbidity in GMBR also dropped to a level as low as 0.97 NTU. The consistently low turbidity in the membrane effluent in Figure 4A shows that the membrane functioned well over 305 days with the average effluent turbidity under 0.5 NTU. Regardless of the granule morphology evolution, the nitrification efficiency in the GMBR has remained as high as 99% throughout the 305 days of operation without compromise (Figure 4B). Thereby, it can be concluded that nitrifying granules responded to the insertion of a submerged membrane module mainly by disintegration and then reaggregation (Figures 2–4). It is noteworthy that the granules stabilized in the GMBR were not as large as the seed sludge that was stabilized within ALR without membrane (Figure 3). Similar observations were also reported in previous studies [4,18,19]. This could have resulted from the substrate competition by bioflocs that should have been washed out if the hydraulic selection pressure were still in place.

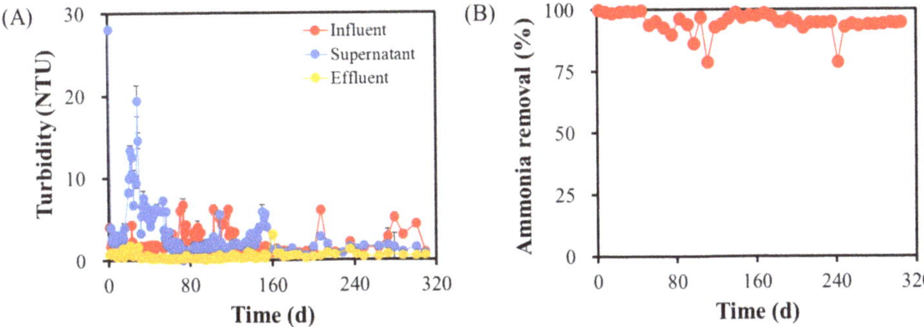

Figure 4. Turbidity in the influent, supernatant, and effluent (**A**) of the GMBR, as well as the ammonia removal (**B**) in the course of 305 days of operation. The error bar refers to the standard deviation from triplicate measurement.

3.3. Characteristics of the Nitrifying Granules Stabilized in the GMBR after 305 Days

After 305 days of operation, the biomass in the GMBR was separated into granules and bioflocs with a 200 μm sieve as defined in this study. The MLSS concentration in Figure 5A shows that 85% of the biomass stabilized in the GMBR was in the form of granules, which was six times that of the bioflocs. However, the MLVSS of granules was only three times that of the bioflocs because the ash content in granules was as high as 64% (Figure 5B), which was two times that in the bioflocs. The live/dead staining in Figure 6 exhibited that only the top 60–80 μm layer of these nitrifying granules was active, with a majority of the granule structure remaining inactive. The mathematical model calibrated in our previous study was applied to simulate the mass diffusion and substrate utilization profiles along the radius of the nitrifying bacteria [6,10]. All the model inputs were listed in Table 1. It can be seen that the 4.63 mg L^{-1} bulk DO was only able to penetrate around 75 μm into the biomass (Figure 7). This means it can penetrate the entire bioflocs with a diameter smaller than 150 μm (Figure 7A). However, the majority of the granule structure was under DO deficiency (Figure 7B). As a consequence, it can be predicted that the inner granule structure beneath the top 75 μm surface layer was decaying under starvation over the 305 days of cultivation. Therefore, it is not surprising to see high ash content resulted in these granules (Figure 5B). In a certain sense, the GMBR might have been converted into an integrated fixed-film activated sludge membrane bioreactor (IFAS-MBR), in which bacteria were either attached on the surface of abiotic carriers in the form of traditional biofilm or suspended in the bulk solution in the form of bioflocs.

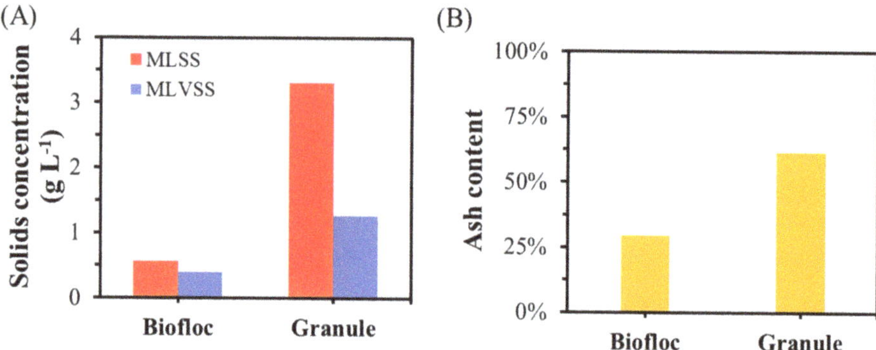

Figure 5. MLSS and MLVSS distribution (**A**) and ash contents (**B**) of bioflocs and granules after 305 days of cultivation in the GMBR.

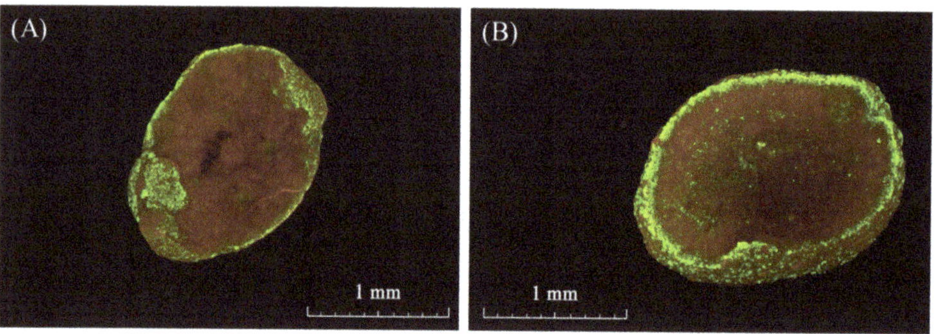

Figure 6. Live/dead stained cross-section of 1.1 mm (**A**) and 1.8 mm (**B**) nitrifying granules after 305 days of cultivation in the GMBR.

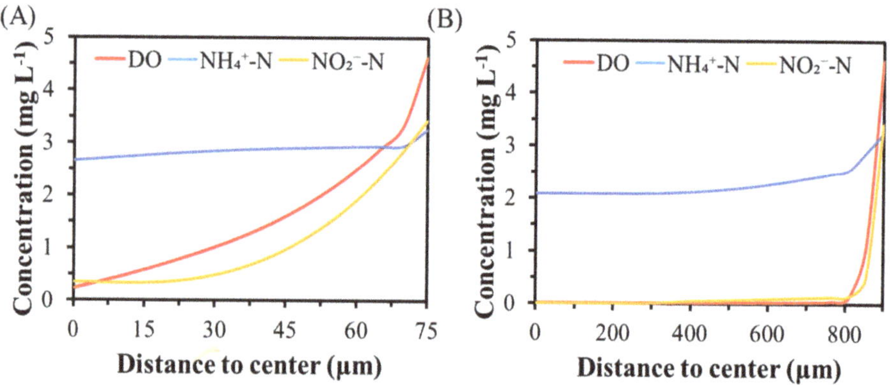

Figure 7. Model simulation of substrate utilization profiles in bioflocs (150 μm diameter, (**A**)) and granules (1800 μm diameter, (**B**)).

3.4. Dynamic Equilibrium within a Backwash Cycle of GMBR at the Steady State

Knowing the nitrifying granules were long-term stabilized in equilibrium with suspended bioflocs, it is our interest to understand how such a dynamic equilibrium was maintained within the GMBR. Therefore, the data in a steady-state backwash cycle (18 d) were analyzed. The OUR measurement in Figure 8A shows that the total activity of nitrifying granules was nearly three times that of the bioflocs, which is consistent with MLVSS distribution measured in Figure 5A, giving almost equal SOURs of 169.7 and 185.8 mg O_2 g MLVSS^{-1} h^{-1} for granules and bioflocs, respectively. This similarity between SOURs of granules and bioflocs implies the lack of mass diffusion limitation into the MLVSS of nitrifying granules, which is consistent with the thin surface layer distribution of the active biomass in nitrifying granules as visualized in Figure 6. The decline of biofloc OUR along with the corresponding membrane OUR increase in Figure 8A show that the biofloc attachment on the membrane did not start until the 9th day following membrane backwash. The emergence of a new peak of particle size distribution between 100 µm and 200 µm right after the membrane backwash evidenced such a dynamic of bioflocs detachment from the membrane in the course of each backwash cycle of the GMBR (Figure 8B). Interestingly, the sEPS profile during the same cycle revealed that the biofloc attachment on the membrane did not start until the sEPS accumulated to a maximum level around 47 mg L^{-1} (Figure 9A). As a matter of fact, it was recognized that sEPS plays a medium role between the bound EPS on the bacteria surface and the fouling surface of the membrane [20]. This may explain why the biofloc attachment on the membrane surface did not start until a relatively high sEPS concentration was accumulated in the GMBR (Figure 9A). It should be realized that the TMP has linearly increased with the membrane operation time (Figure 10A) regardless of the biofloc attachment measured in Figure 8A. Therefore, the membrane fouling before the 9th day of GMBR operation might be due to pore fouling by sEPS even without the biofloc attachment. Similar phenomena were repeatedly reported elsewhere [21,22]. The fact that the sEPS in the membrane effluent was only 6% and 3% those in the GMBR mixed liquor before and after the 9th day of a backwash cycle actually indicated such a pore fouling to cake-layer fouling shift within the GMBR backwash cycle (Figure 9B) and corroborated the TMP increase before the biofloc attachment (Figure 10A).

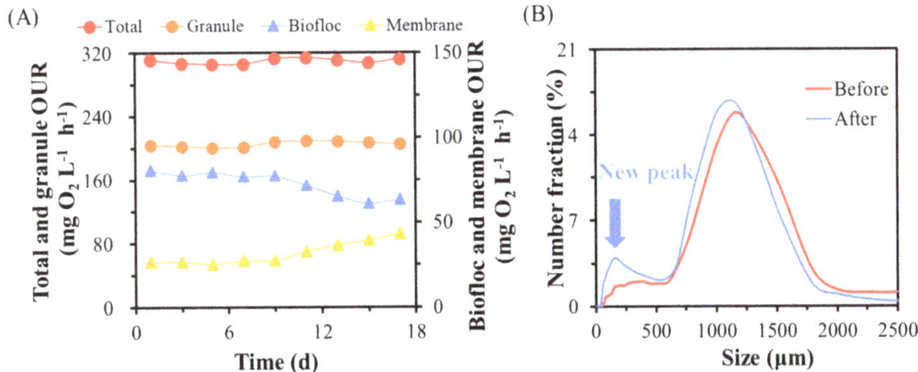

Figure 8. OUR distribution between total, granules, bioflocs, and membrane within one backwash cycle (**A**) and particle size distribution before and after backwash of the membrane (**B**).

3.5. Contributions of Granules and Bioflocs to Membrane Fouling

The rates of TMP increase when the membrane was exposed to only granules or only bioflocs were measured in Figure 10B. It did show that bioflocs caused TMP to increase at a rate four times faster than granules, even though the bioflocs concentration was three times lower than that of granules (Figure 5A). This finding is consistent with previous GMBR studies showing a low trendy of membrane fouling when granules were used in place of activated sludge in MBRs [23]. However, it should be realized that the granules still caused membrane fouling even with their large particle size (Figure 10B). sEPS production, as shown in Figure 9, might be the reason behind this, i.e., both granules and bioflocs can produce sEPS, which was regarded as a primary contributor to membrane fouling [20,22].

Although sEPS was not measured in the batch experiment in Figure 10, the low fouling rates of granules even at a much higher MLVSS indicated that the sEPS production from granules might be lower than that from bioflocs [23,24].

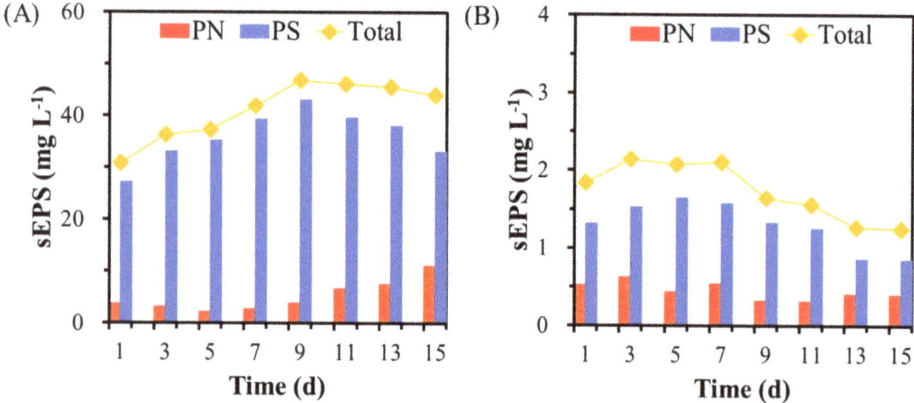

Figure 9. sEPS concentration in the GMBR mixed liquor (**A**) and membrane effluent (**B**) within a backwash cycle.

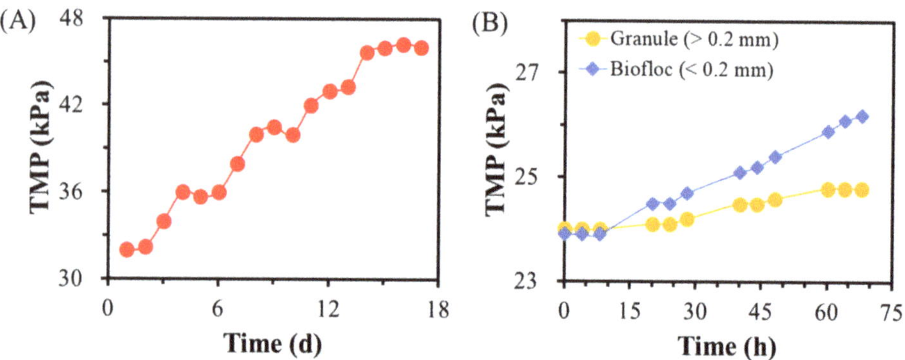

Figure 10. TMP profile of the membrane inserted in the mixed liquor of GMBR within a backwash cycle (**A**), and TMP profile of the membrane when only bioflocs or granules were loaded in the GMBR, respectively (**B**).

4. Discussion

In this study, the nitrifying granules were stabilized in a continuous flow GMBR for 305 days without hydraulic selection pressure (Figures 2–4). However, their size was much smaller than when the membrane was not used (Figure 3). Technically, the growth of granules cannot compete with that of the bioflocs for their severe mass diffusion limitation (Figure 7), which is why a hydraulic selection pressure had to be applied to facilitate the competition of granules over bioflocs. However, experimental results shown in this study seem contradictory to the classic selection pressure-driven aerobic granulation theory, i.e., nitrifying granules were able to remain stable in a GMBR without hydraulic selection pressure for 305 days. The thin layer of active biomass on the surfaces of these high ash content-containing nitrifying granules might have provided an explanation of the unconventional observation (Figures 5 and 6). Specifically, the active biomass actually existed in the form of conventional biofilms attaching to the surfaces of the fossilized inner structure of granules, which functions just like abiotic biofilm carriers (Figure 6). As a matter of fact, the high ash contents distributed in the inner structure of aerobic granules as a result of the mass diffusion limitation were broadly reported [2,25]. It was also recognized that the granule structure could remain stable even under long-term starvation [26], which explains why granules can remain structurally stable even when a majority of their structures were fossilized (Figure 6). However, a GMBR with such high ash and low active biomass contents just defeated the purpose of using GMBR because IFAS-MBR can literally perform the same job with fewer concerns of structural stability [27].

The GMBR was anticipated to become a novel approach to controlling membrane fouling [1,4,23]. This is because the granules' large size and rigid structure are expected to reduce cake-layer formation, pore blocking, and surface deposition on the membrane surface [23,28]. Although the better performance of GMBR over MBR was reported in a number of studies [4,5], the long-term system operation instability of GMBR was identified as a concern [1]. This study further revealed that, although nitrifying granules were capable of remaining structurally stable for the long term, their activity has dropped to extremely low levels in GMBR due to the lack of effective mass diffusion into the inner structure (Figures 6 and 7). Technically, this challenge cannot be resolved in completely stirred tank reactors, in which bulk substrate concentration is expected to be low. Instead, plug flow reactors, in which deeper mass diffusion can be facilitated by a relatively higher initial substrate concentration, should be considered in future GMBR research [16]. Besides this mass diffusion problem, special attention should also be paid to the high sEPS concentration accumulated in the GMBR [22]. The results in Figure 9, together with other recent studies, indicated that it was the sEPS but not the bioflocs that was the primary cause of the membrane fouling [20,22,23]. The sEPS can be produced by both granules and bioflocs, even though the former appeared to show a less fouling tendency than the latter (Figure 10B). Although there was no organic carbon provided for autotrophic bacteria in nitrifying granules, quite high sEPS concentration was apparently produced in the GMBR (Figure 9), which was also reported in other nitrifying bioreactors [29,30]. Therefore, the cause and control of sEPS production is a key area requiring further research in the field of GMBR study.

5. Conclusions

The structure of nitrifying granules can remain long-term stable in a GMBR without hydraulic selection pressure; however, the majority of the granule structure was inactive due to mass diffusion limitation. As a consequence, active biomass free of mass diffusion limitation only inhabited the top 60–80 μm layer of the nitrifying granules. There was a dynamic equilibrium between suspended bioflocs and the membrane. Twenty-five percent of bioflocs shifted to attached living on the membrane surface within the last nine days of the backwash cycle when sEPS concentration peaked at 47 mg L^{-1}. Backwash can eventually detach and return these bioflocs to a bulk solution. In a certain sense, the GMBR investigated in this study functioned in a similar fashion as IFAS-MBR and thus defeated the original purpose of GMBR development. The mass diffusion problem and sEPS production should be key areas of focus in future GMBR research.

Author Contributions: Conceptualization, C.B.B. and Z.-W.W.; Experimental Design, C.B.B. and Z.-W.W.; Data Collection and Analysis, Z.A., C.B.B., and Z.-W.W.; Writing—Review and Editing, Z.A., X.Z., C.B.B., and Z.-W.W. All authors have read and agreed to the published version of the manuscript.

Funding: This research received no external funding.

Institutional Review Board Statement: Not applicable.

Informed Consent Statement: Not applicable.

Data Availability Statement: All data is contained within the article, in Section 3.

Acknowledgments: Authors appreciate VT's OASF support in publishing this article.

Conflicts of Interest: The authors declare that they have no known competing financial interests or personal relationships that could have appeared to influence the work reported in this paper.

References

1. Kent, T.R.; Bott, C.B.; Wang, Z.-W. State of the art of aerobic granulation in continuous flow bioreactors. *Biotechnol. Adv.* **2018**, *36*, 1139–1166. [CrossRef]
2. Liu, Y.-Q.; Lan, G.-H.; Zeng, P. Excessive precipitation of $CaCO_3$ as aragonite in a continuous aerobic granular sludge reactor. *Appl. Microbiol. Biotechnol.* **2015**, *99*, 8225–8234. [CrossRef]
3. Corsino, S.; Campo, R.; Di Bella, G.; Torregrossa, M.; Viviani, G. Study of aerobic granular sludge stability in a continuous-flow membrane bioreactor. *Bioresour. Technol.* **2016**, *200*, 1055–1059. [CrossRef] [PubMed]
4. Sajjad, M.; Kim, I.S.; Kim, K.S. Development of a novel process to mitigate membrane fouling in a continuous sludge system by seeding aerobic granules at pilot plant. *J. Membr. Sci.* **2016**, *497*, 90–98. [CrossRef]
5. Chen, C.; Bin, L.; Tang, B.; Huang, S.; Fu, F.; Chen, Q.; Wu, L.; Wu, C. Cultivating granular sludge directly in a continuous-flow membrane bioreactor with internal circulation. *Chem. Eng. J.* **2017**, *309*, 108–117. [CrossRef]
6. An, Z.; Kent, T.R.; Sun, Y.; Bott, C.B.; Wang, Z.-W. Free ammonia resistance of nitrite-oxidizing bacteria developed in aerobic granular sludge cultivated in continuous upflow airlift reactors performing partial nitritation. *Water Environ. Res.* **2021**, *93*, 421–432. [CrossRef] [PubMed]
7. Martin, K.; Shaw, A.; de Clippeleir, H.; Sturm, B. "Accidental granular sludge?": Understanding process design and operational conditions that lead to low SVI-30 values through a survey of full scale facilities in North America. *Proc. Water Environ. Fed.* **2016**, *2016*, 3385–3394. [CrossRef]
8. Baird, R.; Bridgewater, L. *Standard Methods for Examination of Water and Wastewater*, 23rd ed.; American Public Health Association: Washington, DC, USA, 2017.
9. Liu, H.; Fang, H.H. Extraction of extracellular polymeric substances (EPS) of sludges. *J. Biotechnol.* **2002**, *95*, 249–256. [CrossRef]
10. Kent, T.R.; Sun, Y.; An, Z.; Bott, C.B.; Wang, Z.-W. Mechanistic understanding of the NOB suppression by free ammonia inhibition in continuous flow aerobic granulation bioreactors. *Environ. Int.* **2019**, *131*, 105005. [CrossRef] [PubMed]
11. Zhang, B.; Chen, Z.; Qiu, Z.; Jin, M.; Chen, Z.; Chen, Z.; Li, J.; Wang, X.; Wang, J. Dynamic and distribution of ammonia-oxidizing bacteria communities during sludge granulation in an anaerobic–aerobic sequencing batch reactor. *Water Res.* **2011**, *45*, 6207–6216.
12. Jin, R.-C.; Zheng, P.; Mahmood, Q.; Zhang, L. Performance of a nitrifying airlift reactor using granular sludge. *Sep. Purif. Technol.* **2008**, *63*, 670–675. [CrossRef]
13. Shi, X.-Y.; Yu, H.-Q.; Sun, Y.-J.; Huang, X. Characteristics of aerobic granules rich in autotrophic ammonium-oxidizing bacteria in a sequencing batch reactor. *Chem. Eng. J.* **2009**, *147*, 102–109. [CrossRef]
14. Zheng, Y.-M.; Yu, H.-Q.; Sheng, G.-P. Physical and chemical characteristics of granular activated sludge from a sequencing batch airlift reactor. *Process Biochem.* **2005**, *40*, 645–650. [CrossRef]
15. Liu, Y.; Wang, Z.-W.; Liu, Y.Q.; Qin, L.; Tay, J.H. A generalized model for settling velocity of aerobic granular sludge. *Biotechnol. Progr.* **2005**, *21*, 621–626. [CrossRef]
16. Sun, Y.; Angelotti, B.; Wang, Z.-W. Continuous-flow aerobic granulation in plug-flow bioreactors fed with real domestic wastewater. *Sci. Total Environ.* **2019**, *688*, 762–770. [CrossRef]
17. Tsuneda, S.; Nagano, T.; Hoshino, T.; Ejiri, Y.; Noda, N.; Hirata, A. Characterization of nitrifying granules produced in an aerobic upflow fluidized bed reactor. *Water Res.* **2003**, *37*, 4965–4973. [CrossRef] [PubMed]
18. Wang, J.; Wang, X.; Zhao, Z.; Li, J. Organics and nitrogen removal and sludge stability in aerobic granular sludge membrane bioreactor. *Appl. Microbiol. Biotechnol.* **2008**, *79*, 679–685. [CrossRef]
19. Li, X.; Li, Y.; Liu, H.; Hua, Z.; Du, G.; Chen, J. Characteristics of aerobic biogranules from membrane bioreactor system. *J. Membr. Sci.* **2007**, *287*, 294–299. [CrossRef]
20. Geng, Z.; Hall, E.R. A comparative study of fouling-related properties of sludge from conventional and membrane enhanced biological phosphorus removal processes. *Water Res.* **2007**, *41*, 4329–4338. [CrossRef] [PubMed]
21. Liu, Q.; Yao, Y.; Xu, D. Mechanism of Membrane Fouling Control by HMBR: Effect of Microbial Community on EPS. *Int. J. Environ. Res. Public Health* **2020**, *17*, 1681. [CrossRef]
22. Zuthi, M.; Ngo, H.; Guo, W. Modelling bioprocesses and membrane fouling in membrane bioreactor (MBR): A review towards finding an integrated model framework. *Bioresour. Technol.* **2012**, *122*, 119–129. [CrossRef] [PubMed]
23. Wang, J.; Qiu, Z.; Chen, Z.; Li, J.; Zhang, Y.; Wang, X.; Zhang, B. Comparison and analysis of membrane fouling between flocculent sludge membrane bioreactor and granular sludge membrane bioreactor. *PLoS ONE* **2012**, *7*, e40819.
24. Wang, X.; Zhang, B.; Shen, Z.; Qiu, Z.; Chen, Z.; Jin, M.; Li, J.; Wang, J. The EPS characteristics of sludge in an aerobic granule membrane bioreactor. *Bioresour. Technol.* **2010**, *101*, 8046–8050.
25. Liu, Y.-Q.; Lan, G.-H.; Zeng, P. Size-dependent calcium carbonate precipitation induced microbiologically in aerobic granules. *Chem. Eng. J.* **2016**, *285*, 341–348. [CrossRef]
26. Wang, X.; Zhang, H.; Yang, F.; Wang, Y.; Gao, M. Long-term storage and subsequent reactivation of aerobic granules. *Bioresour. Technol.* **2008**, *99*, 8304–8309. [CrossRef]
27. De la Torre, T.; Rodriguez, C.; Gomez, M.; Alonso, E.; Malfeito, J. The IFAS-MBR process: A compact combination of biofilm and MBR tecnnology as RO pretreatment. *Desalin. Water Treat.* **2013**, *51*, 1063–1069. [CrossRef]
28. Iorhemen, O.T.; Hamza, R.A.; Tay, J.H. Membrane bioreactor (MBR) technology for wastewater treatment and reclamation: Membrane fouling. *Membranes* **2016**, *6*, 33. [CrossRef]

29. Thanh, B.X.; Visvanathan, C.; Aim, R.B. Fouling characterization and nitrogen removal in a batch granulation membrane bioreactor. *Int. Biodeterior. Biodegrad.* **2013**, *85*, 491–498. [CrossRef]
30. Li, X.; Du, R.; Peng, Y.; Zhang, Q.; Wang, J. Characteristics of sludge granulation and EPS production in development of stable partial nitrification. *Bioresour. Technol.* **2020**, *303*, 122937. [CrossRef]

 *processes*

Review

Review on Digestibility of Aerobic Granular Sludge

Mohamed S. Zaghloul, Asmaa M. Halbas, Rania A. Hamza * and Elsayed Elbeshbishy *

Environmental Research Group for Resource Recovery, Department of Civil Engineering, Faculty of Engineering, Architecture and Science, Toronto Metropolitan University, 350 Victoria Street, Toronto, ON M5B 2K3, Canada
* Correspondence: rhamza@torontomu.ca (R.A.H.); elsayed.elbeshbishy@torontomu.ca (E.E.)

Abstract: Full-scale wastewater treatment plants utilizing aerobic granular sludge technology are being built in many countries worldwide. As with all biological wastewater treatment plants, the produced waste biomass must be stabilized to protect the population, wildlife, and the environment. Digestion is usually used to break down the complex organics in the waste sludge; however, the digestibility of aerobic granular sludge still needs to be fully understood compared to the conventional activated sludge. This paper reviews the studies published on the digestibility of waste aerobic granular sludge to date. Studies comparing aerobic granular sludge and activated sludge in terms of composition, properties, and digestibility are highlighted. The impact of biological composition and physical properties on the digestibility of sludge is reviewed in terms of biomethane production and biodegradability. The effect of pre-treatment is also covered. Areas for future research are presented.

Keywords: aerobic granular sludge; wastewater treatment; digestibility; pre-treatment; biomethane production

Citation: Zaghloul, M.S.; Halbas, A.M.; Hamza, R.A.; Elbeshbishy, E. Review on Digestibility of Aerobic Granular Sludge. *Processes* **2023**, *11*, 326. https://doi.org/10.3390/pr11020326

Academic Editors: Yongqiang Liu and Yingnan Yang

Received: 22 December 2022
Revised: 11 January 2023
Accepted: 14 January 2023
Published: 19 January 2023

1. Introduction

Aerobic granular sludge (AGS) is becoming more popular as an alternative to activated sludge (AS) [1,2]. This is mainly due to the advantages of (AGS) over floccular sludge, such as higher density, microbial diversity leading to simultaneous removal of nutrients and organics, resilience, a 75% smaller footprint, and a 50% reduction in energy demand [3]. The AGS technology is operated in a sequencing batch reactor (SBR) without the need for primary sedimentation tanks. The high concentration of solids in the influent, coupled with the shear force due to the aeration and wastewater up-flow in the SBR, triggers the granulation of the biomass. Controlling the settling time of the SBR cycle allows the reactor to retain the faster-settling granules and wash out the slower-settling biomass flocs, leading to the accumulation of the denser granules inside the reactor [3].

Sludge wasting in biological wastewater treatment is performed primarily to control the solids retention time (SRT) and maintain the biomass concentration in the bioreactor [4]. In the AS process, waste sludge can be taken from the aeration tank's mixed liquor directly or from the return activated sludge (RAS) line, typically the preferred wasting location, as it minimizes the volume of sludge discharged and handled. Most AGS-based bioreactors operate as SBRs, where aeration and settling occur in the same tank, and thus RAS is eliminated. Waste AGS is usually withdrawn between the settling and filling phases to obtain the highest concentrated sludge [5]. Another portion of sludge is removed unintentionally during the decanting phase, which does not fall below the decant port during the settling phase. This process is called the selection pressure of AGS, and it is regarded as one of the main factors that help with granule formation by naturally selecting the faster-settling granules [6]. The other main factor that triggers granule formation is the shear force induced by the air bubbles on the biomass [3,7]. This sludge is usually more floccular and is separated from the effluent using an effluent equalization tank [8]. It has been reported that particle size distribution of the wasted sludge showed that more than 90% of the AGS wasted intentionally (i.e., selective discharge) was larger than 500 μm, while the sludge wasted with the effluent wastewater during decanting had an average particle size smaller than 500 μm [8,9].

The quantity of wasted AGS is determined using Equation (1) [10]. The SRT is a control parameter chosen by the plant designers to maintain the required mixed liquor volatile suspended solids (MLVSS) concentrations, allow for new biological growth, and optimize nutrient removal [11]. The mass of volatile suspended solids (VSS) in the waste sludge is controlled by the operators, where a certain amount of sludge is withdrawn after settling and before filling. The mass of VSS in the effluent is more challenging to measure due to the inconsistency of solids concentration in the effluent stream [8].

$$SRT = \frac{Mass\ of\ MLVSS\ in\ the\ reactor}{Mass\ of\ VSS\ in\ effluent + Mass\ of\ VSS\ in\ Waste\ Sludge} \tag{1}$$

The washed-out sludge leaves the reactor as the effluent begins to be decanted, then once the sludge-liquid interface passes the effluent port, the effluent becomes clear. Theoretically, the only way to accurately sample the VSS in the effluent is by placing the effluent wastewater in a completely stirred reactor and sampling from there, which is not feasible in full-scale applications. Therefore, SRT estimation is often not accurate in AGS bioreactors. In full-scale wastewater treatment plants, however, it was reported that both the conventional AS (i.e., flocculent sludge) and the AGS processes have similar waste sludge production rates. The AGS process at the Nereda® plant in Garmerwolde produced 0.23 kgVSS/kgCOD of waste AGS with an SRT of 28 days, while the AS process at Harnaschpolder produced 0.25 kgVSS/kgCOD at an SRT of 24 days [8]. In Lubawa, Poland, a full-scale AGS facility operated at an SRT of 30 days produced 0.6 kgVSS/kgCOD of biomass [12]. Using synthetic wastewater at a C:N ratio of 100:5, the biomass yield was 0.499 kgVSS/kgCOD [13].

This mini-review article compiled the research published to date on the digestibility of AGS. The literature on the digestibility of AGS in the last ten years is still very limited, and thus, a summary is provided to motivate further research in this area. A recent review article, covering the AGS technology as well as its digestibility has also found that the research is limited [14]. In this article, the biological and chemical characteristics of waste AGS in the reviewed AGS digestibility studies were discussed. The pre-treatment and digestion processes of waste AGS in these studies were summarized. The comparisons made in these studies between the digestion of AGS, waste-activated sludge (WAS), and primary sludge (PS) were highlighted. The effect of the type of AGS and its biological and microbial composition on the pre-treatment and digestibility of waste AGS were also reviewed.

2. Aerobic Granular Sludge Physical Properties

The physical properties of waste AGS can vary according to the type of wastewater treated and the control of the SBR reactor. Wastewater with higher readily biodegradable organics, such as municipal wastewater with high volatile fatty acid (VFA) content, leads to faster granule formation with larger particle sizes and lower density due to the rapid formation of new biomass and larger amounts of slow-settling filamentous growth [15]. Higher ratios of slowly biodegradable to readily biodegradable organics, such as wastewater from the food industry, produce less floccular growth. Denser and faster-settling granules are obtained using the selection pressure by controlling the SBR cycle times, where the settling time is reduced to only allow the denser AGS to remain in the reactor while the floccular sludge is washed out [6].

In a steady-state AGS bioreactor, the washed-out floccular sludge is the amount of biomass growth during each cycle that did not granulate and is usually not used in controlling the reactor SRT [8]. The SRT is controlled by withdrawing settled sludge with fully formed granules. However, washed-out floccular sludge is still collected and requires stabilization similar to the withdrawn waste AGS. The structural morphology and microbial composition differences between the two waste sludge streams of AGS reactors result in differences in their digestibility behavior. Sludge with intact granules contains complex biopolymers that maintain the structural integrity of the granules and have hydrophobic characteristics, which require homogenization and pre-treatment to release the organic contents of the granules. Floccular washed-out sludge, on the other hand, is composed of filamentous bacteria as well as cellulose-like fibers, which are highly biodegradable, which leads to faster digestion and less requirement for pre-treatment [8,16].

The average particle size of aerobic granules in AGS reactors varies according to the type of wastewater treated and the process operation. The aerobic granules, by definition, should have an average particle size of 200 μm [17]. However, a reactor with a granulation percentage of 100% is theoretically not possible. Typically, AGS reactors are hybrid systems of co-existing flocculent and granular biomass [18]. It has been reported that SVI_{30}/SVI_5 can represent the granulation percent, where granular settling would be that of discrete settling rather than flocculent or zone settling [19,20]. During start-up, the AGS reactor is seeded with WAS, which then transforms into aerobic granules. The particle size distribution progression for the first 60 days of operation was plotted to show that the biomass average particle size increases from around 100 μm to above 1700 μm [21]. In comparison to AS, the average particle size at an OLR of 0.93–0.95 gCOD/L.d was 800 μm for AGS and 290 μm for AS, where the AGS was cultivated in the lab, and the AS was collected from a full-scale treatment plant [22]. Table 1 shows the average particle size reported in AGS digestion studies. Other studies have reported that the activated sludge average particle size was 114 μm while AGS was above 1600 μm in both lab and full-scale applications [16,21,23].

Table 1. Average particle size in AGS and AS in various studies.

Sludge Type	Particle Size (mm)	Reference
AGS	1.4–1.7	[23]
AGS	1.3–1.6	[24]
AGS	1	[25]
AGS	1.74	[26]
AGS	0.4–0.8	[22]
AS	0.29	
AGS	1.75	[21]
AGS	1.79	[16]
WAS	0.114	
AGS	0.09–0.35 mm	[27]

The settleability of AGS is typically better than AS due to the larger particle size and faster particle settling velocity. A well-operated granular reactor would have biomass with SVI values below 80 mL/g and an average particle size higher than 200 μm [20,28]. Thermal pre-treatment has been shown to reduce the settleability of AGS due to the transformation of the biomass to a gelatinous consistency with a higher viscosity than the original AGS. It was reported that thermal pre-treatment at temperatures above 115 °C–125 °C drastically increased the granular biomass viscosity, while no change was observed at temperatures less than 100 °C [29,30]. It was shown that this behavior was the opposite of AS under thermal pre-treatment, where the SVI decreased with raising temperature [29,31–33]. The reason for the transformation of AGS to a gel consistency and increased viscosity under thermal pre-treatment is still not understood.

3. Aerobic Granular Sludge Biochemical Properties

The type of influent wastewater affects the fractions of readily to slowly biodegradable organics in the aerobic AGS. Little research is available in the literature on the biodegradability of AGS, but some comparisons were made on the organic composition of AGS cultivated using swine and municipal wastewater [30]. Val del Río et al. [30] observed minimal improvement in biodegradability after the thermal pre-treatment (170 °C–210 °C) of AGS cultivated using municipal wastewater, while AGS cultivated using swine wastewater improved by up to 88%. One of the possible justifications of this finding is that the high content of complex organics in swine AGS was broken down under thermal pre-treatment leading to noticeable digestibility improvement, while municipal AGS already contained a higher fraction of readily biodegradable organics, so thermal treatment had little effect. Table 2 shows the concentrations of chemical oxygen demand (COD), total solids (TS), and volatile solids (VS) in AGS and AS obtained from full-scale and lab-scale reactors reported in different studies. More research is needed to gain further insight into how the wastewater type can influence the composition of complex organics in AGS.

Table 2. Characteristics of AGS and AS from different sources.

Source of Sludge	Type of Wastewater	Sludge Type	COD	TS	VS	Reference
	Swine manure	AGS	39.7 g/L	29.6 g/L	27.3 g/L	[30]
Pilot plant (100 L)	Synthetic wastewater	AGS	85.7 g/L	106 g/L	60.1 g/L	
Pilot plant	Brewery wastewater	AGS	8–31.1 g/L	14–21.1 g/L	7.3–15.9 g/L	[23]
WWTP Calo-Milladoiro, Spain	Municipal wastewater	AS	29 g/L	15.7 g/L	10.9 g/L	
Pilot (100 L)	Swine manure	AGS	7.7–27.5 g/L	9.2–21.1 g/L	8.4–19.2 g/L	[24]
Lab scale SBR	-	AGS	Soluble: 403 (mg/gVSS)	-	-	[34]
Lab scale SBR	-	AS	Soluble: 329 (mg/gVSS)	-	-	
Lab scale SBR (4.5 L)	-	AGS	-	2.28 (%)	1.47 (%)	[25]
WWTP Olsztyn, Poland	Municipal wastewater	AS	-	4.51 (%)	3.46 (%)	
WWTP Olsztyn, Poland	Municipal wastewater	PS	-	1.71 (%)	1.33 (%)	
Lab scale SBR	-	AGS	17.21 g/L	15.3 (gTSS/L)	12.8 (gVSS/L)	[26]
Lab scale SBR (6–8 L)	-	AGS	-	-	-	[22]
Municipal WWTP	Municipal wastewater	AS	-	-	-	
Nereda® plant Garmerwolde	Municipal wastewater	AGS wasted	71.3 (g/L)	6.1 (%)	4.9 (%)	[8]
	Municipal wastewater	AGS Washed out	79.1 (g/L)	6.6 (%)	5.1 (%)	
WWTP Harnaschpolder	Municipal wastewater	WAS	72.4 (g/L)	6.2 (%)	5 (%)	
	Municipal wastewater	PS	77.8 (g/L)	6.4 (%)	5 (%)	
Nereda® plant Garmerwolde	Municipal wastewater	AGS	-	5.41 (%)	4.25 (%)	[16]
WWTP in Harnaschpolder, the Netherlands	Municipal wastewater	WAS	-	5.16 (%)	4.14 (%)	
WWTP in Lubawa (Poland)	Municipal wastewater	AGS	-	1.55 (%)	1.24 (%)	[27]

Additionally, AGS that is wasted after settling was reported to have a different organic composition from washed-out AGS during the decant phase. Guo, van Lier, et al., [8], tested the waste AGS, washed-out AGS, WAS, and PS for three types of fibers: cellulose-like, hemicellulose-like, and lignin-like fibers, as well as proteins, carbohydrates, and lipids. They found that the waste AGS was very similar in composition to the WAS, while the washed-out AGS was similar to the PS. These fibers constitute 30–50% of suspended solids (SS) in wastewater and are typically removed in primary settling in the AS process [35]. However, in the AGS processes, no primary settling is done where raw wastewater enters the AGS reactors directly. Guo, van Lier, et al., [8], found that the fibrous content (as VS) in waste AGS and washed-out AGS was 32.5% and 38%, but the cellulose fraction in the washed-out AGS was double that of the waste AGS. Pronk et al. [9] indicated that fibers do not attach to mature granules but rather remain with the floccular sludge that gets washed out during the AGS process. Guo, van Lier, et al., [8], also indicated that the carbohydrate, lipid, and VFA contents of the washed-out sludge were about double that of the waste AGS, while the proteins were higher in the waste AGS. The difference in organic composition and structural morphology between the two types of sludge directly impacts their biodegradability, where the BMP of washed-out AGS was 1.5 times that of the waste AGS due to the abundance of the biodegradable cellulose-like fibers [8].

The AGS structural integrity relies heavily on a matrix of extracellular polymeric substances (EPS) in which they are entrapped. EPS is a complex mixture of polysaccharides, proteins, nucleic acids, lipids, and humic substances, which provides the cross-linkage between microbial cells in the biofilm matrix in which they are self-immobilized. There is a consensus that EPS synthesis and regulation are linked with microbial interactions, which are further affected by wastewater composition and operating conditions [18,36,37]. EPS production is also highly influenced by environmental stress conditions: shear forces, salinity, and the presence of toxic compounds such as contaminants of emerging concern (CECs) and heavy metals, all of which stimulate EPS production as a defense mechanism. The presence of toxins has been found to increase protein content but to have less effect on polysaccharide content. Although these findings might suggest that the extracellular proteins produced would stimulate the granular formation by promoting nucleation, in fact, excessive EPS could adversely affect granular system performance [38].

The biodegradability of EPS is a key factor in the ease of anaerobic digestion of un-homogenized aerobic granules. Wang et al. [39] indicated that the outer shell of granules that contain aerobic heterotrophic microbes was composed of hydrophobic non-readily biodegradable EPS, while the inner core with anaerobic microbes contained readily biodegradable and soluble EPS. The hydrophobic non-readily biodegradable EPS is responsible for the structural integrity of the granules [40,41]. Homogenization then becomes needed to break the cohesion of the granules to expose the biodegradable organic content, and thermal hydrolysis was used to break down the complex biopolymers into more readily biodegradable compounds. Structural EPS, or only those polymers within the EPS able to form hydrogels, have also been identified as the main component that distinguishes and explains the resistance to degradation of waste AGS compared to AS [8,16]. Table 3 shows the organic biopolymer composition of AGS and AS in different studies.

The selection pressure applied during the operation of the AGS reactors to promote the growth of fast-settling granules also plays a role in the biodegradability of AGS and differentiates it from CAS [8]. The waste AGS extracted to maintain the required SRT is intentionally removed from the settled sludge, which means the granules are more compact and affected by the presence of EPS, while slower-settling sludge is washed out of the reactor. This selective separation leads to the presence of a much higher content of slowly settling readily biodegradable cellulose-like fibers in the washed-out AGS sludge and their absence from the intact granules, further lowering the biodegradability of the waste AGS [9].

Sludge generated from wastewater treatment also contains nutrients and inorganic components. Nutrients are ammonia, nitrogen, and phosphorus, which are used to manufacture plant fertilizer and can cause eutrophication if released untreated into the environment [42]. Inorganic components include several ions, as shown in Table 4. The most commonly present elements, other than carbon, oxygen, nitrogen, and hydrogen, are potassium, calcium, magnesium, and iron.

Table 3. Organic biopolymers composition of AGS and AS from different sources.

Source of Sludge	Type of Wastewater	Sludge Type	Carbohydrates	Proteins (mg/gVS)	Polysaccharides	Lipids (mg/gVS)	VFAs (mg/gVS)	PHA	References
Pilot plant (100 L)	Swine manure	AGS	3.6 g/L	16.6 g/L	-	0.05 g/L	1.4 g/L	0.8 g/L	[30]
	Synthetic wastewater	AGS	6.9 g/L	26.9 g/L	-	0.013 g/L	7.5 g/L	5.5 g/L	
Pilot plant	Brewery wastewater	AGS	-	-	-	-	ND	-	[23]
WWTP Calo-Milladoiro, Spain	Municipal wastewater	AS	-	-	-	-	0.27 (g/L)	-	
Pilot (100 L)	Swine manure	AGS	-	-	-	-	ND–1 g/L	-	[24]
Lab scale SBR	-	AGS	Soluble: 42 (mg/gVSS)	Soluble: 82.7 (mg/gVSS)	-	-	355	-	[34]
Lab scale SBR	-	AS	Soluble: 39.1 (mg/gVSS)	Soluble: 63 (mg/gVSS)	-	-	352	-	
Lab scale SBR (4.5 L)	-	AGS	0.002 (%)	0.926 (%)	-	0.008 (%)	-	-	[25]
WWTP Olsztyn, Poland	Municipal wastewater	AS	0.095 (%)	2.121 (%)	-	0.04 (%)	-	-	
WWTP Olsztyn, Poland	Municipal wastewater	PS	0.162 (%)	0.292 (%)	-	0.01 (%)	-	-	
Lab scale SBR	-	AGS	113.4 (mg/gVSS)	701.6 (mg/gVSS)	-	-	-	-	[26]
Lab scale airlift SBR	-	AGS	-	365 mg/gVSS	135 mg/gVSS	-	-	-	[21]
WWTP in Hong Kong	Municipal wastewater	AS	-	265 mg/gVSS	245 mg/gVSS	-	-	-	
Nereda® plant Garmerwolde	Municipal wastewater	AGS wasted	217 (mg glucose/g sludge)	498	-	37	4.6	-	[8]
	Municipal wastewater	AGS Washed out	429 (mg glucose/g sludge)	301	-	60	9.7	-	
	Municipal wastewater	WAS	190 (mg glucose/g sludge)	389	-	35	5.6	-	
WWTP Harnaschpolder	Municipal wastewater	PS	464 (mg glucose/g sludge)	248	-	73	8.6	-	

Table 4. Inorganic composition of AGS and AS from different sources.

Reference	[34]		[25]			[26]	[21]		[29]	[27]
Sludge Type	AGS	AS	AGS	AS	PS	AGS	AGS	AS	AGS	AGS
NH_4^+	0.7 (mg/gVSS)	1.8 (mg/gVSS)	-	-	-	-	-	-	-	-
TN	39.4 (mg/gVSS)	38.3 (mg/gVSS)	-	-	-	-	-	-	-	-
PO_4^{3-}	9.7 (mg/gVSS)	10.2 (mg/gVSS)	-	-	-	-	-	-	-	-
TP	39.3 (mg/gVSS)	26.7 (mg/gVSS)	-	-	-	31.5 (mg/gVSS)	-	-	9.21–56.51 (g/kgTS)	-
K	18.2 (mg/gVSS)	17 (mg/gVSS)	-	-	-	-	0.34%	0.27%	2.62–3.3 (g/kgTS)	-
Ca	-	-	-	-	-	60.3 (mg/gVSS)	5.34%	0.31%	19.77–188.1 (g/kgTS)	-
Na	-	-	-	-	-	-	5.11%	2.29%	5.42–5.87 (g/kgTS)	-
Mg	10.4 (mg/gVSS)	7.6 (mg/gVSS)	-	-	-	9.1 (mg/gVSS)	0.19%	0.40%	1.9–3.05 (g/kgTS)	-
Mn	-	-	-	-	-	-	-	-	0.025–0.018 (g/kgTS)	-
Al	-	-	-	-	-	2.4 (mg/gVSS)	ND	0.08%	-	-
Zn	-	-	-	-	-	-	0.05%	0.01%	0.009–0.004 (g/kgTS)	295.6 (mg/kgTSS)
Fe	-	-	-	-	-	6.3 (mg/gVSS)	0.02%	0.16%	0.34–0.21 (g/kgTS)	-
Cu	-	-	-	-	-	-	ND	0.01%	0.086–0.058 (g/kgTS)	268 (mg/kgTSS)
Co	-	-	-	-	-	-	-	-	0.009–0.021 (g/kgTS)	-
Ni	-	-	-	-	-	-	-	-	0.009 (g/kgTS)	24 (mg/kgTSS)
Hg	-	-	-	-	-	-	-	-	-	0.0098 (mg/kgTSS)
Cr	-	-	-	-	-	-	-	-	-	39.7 (mg/kgTSS)
Cd	-	-	-	-	-	-	-	-	-	1.6 (mg/kgTSS)
Pb	-	-	-	-	-	-	-	-	-	15.1 (mg/kgTSS)
C	-	-	49.80%	57%	60%	-	38.62%	40.80%	-	-
O	-	-	1.76%	7.56%	9.50%	-	-	-	-	-
N	-	-	7.90%	7%	2.80%	-	9.38%	8.54%	-	-
H	-	-	5%	5.20%	5.60%	-	5.81%	6.41%	-	-
S	-	-	-	-	-	-	0.76%	2.12%	-	-

4. Sludge Preparation Prior to Digestion

Pre-treatment of organic feedstock prior to anaerobic digestion has been proven to improve its biodegradability by breaking down the complex organics in the biological material, which constitute the majority of the WWTP waste sludge components [43–45]. In AGS, the abundance of EPS and other biopolymers, such as alginates, results in more compact granules with hydrophobic properties, which complicates the digestion process [46]. Therefore, pre-treatment may appear necessary to achieve two outcomes: breaking down the compact granular structure and breaking down the complex biopolymers and hydrocarbons. Mechanical or ultrasound homogenization was used to break the structural cohesion of the granules, while thermal pre-treatment was used to break down the complex organics [27,30]. Very scarce research has been done to explore the pre-treatment of AGS for digestion. However, the current literature reported that pre-treatment only improved the methane production rates and had little effect on the final total yield [8]. Zou and Li [34] also reported that thermal pre-treatment improves the recovery of carbon, phosphorus, and hydrogen during anaerobic fermentation.

Mechanical and ultrasound homogenization was used by Cydzik-Kwiatkowska et al. [27] on AGS collected from a full-scale municipal wastewater treatment plant. It was found that the ultrasound was much more effective than mechanical homogenization and resulted in higher methane content in the produced biogas. It was also reported that the digestion of the ultrasound-homogenized AGS was 25% faster than raw AGS. Mechanical crushing was used to homogenize sludge samples in a comparison between AGS and WAS [8]. A household blender was used at 1000 RPM for 5 min. The raw AGS sludge particle size distribution was 70% above 2000 μm, while the raw WAS was 20% between 50 and 100 μm, and 65% between 100 and 500 μm. The homogenized samples were similar in particle size, with around 30% between 0 and 20 μm, 35% between 20 and 50 μm, 20% between 50 and 100 μm, and 15% between 100 and 500 μm. The biogas production analysis also showed a minimal effect of homogenization on the overall yield in AGS, but the rate was faster by about five days to reach the plateau in the accumulated methane production curve. Homogenizing WAS did not have any effect on biogas production. The rate of degradation of readily biodegradable fractions of AGS increased by 24% due to crushing.

The type of wastewater treated was shown to have a significant impact on the effectiveness of thermal pre-treatment. A study by Val del Río et al. [30] compared the digestion of AGS cultivated using municipal and swine wastewater. It was found that the raw AGS cultivated using municipal wastewater was more biodegradable than that cultivated using swine wastewater, with biodegradability of 49% and 33%, respectively. Thermal pre-treatment at 60 °C and 170 °C was found to improve the biodegradability of the swine-based AGS by 20 and 88%, respectively. The more biodegradable municipal AGS was pre-treated at 190 °C and 210 °C with a lower improvement in biodegradability of 14% and 18%, respectively [30].

The compact granular structure was found to have little impact on the final biodegradability and total biogas yield of the AGS, where the overall biodegradability of the AGS was found to be similar to CAS under similar conditions between 30% to 50% [47]. A study used different pre-treatment methods on the same type of AGS and showed that the raw sludge was 44% biodegradable and the anaerobic digestion of the untreated AGS had a 32% solids reduction [24]. Thermal pre-treatment at 133 °C resulted in a 47% improvement in biodegradability. Mixing the untreated AGS with CAS improved the solids reduction during the anaerobic digestion. Val Del Río et al. [24] also compared the biodegradability of CAS with that of AGS and found no major differences in overall biogas yield.

Steam explosion has been found to be an effective pre-treatment method for the anaerobic digestion of AGS cultivated in mineral-rich wastewater [29]. The mineral content inhibits the digestibility of the sludge, leading to lower biogas yields. Steam explosion at 170 °C for 30 min led to a 20% improvement in BMP over CAS under similar conditions [29].

5. Digestion of Aerobic Granular Sludge

The digestibility of aerobic AGS has been studied to investigate its biodegradability and its methane production. It was generally found that the biodegradability of AGS was lower than that of AS due to the granules' morphology and organic biopolymer composition [8,16,34]. Other studies have shown different behavior where both AGS and AS had similar methane production when digested under similar conditions [22,23]. Thermal pre-treatment has been found to improve biomethane production in AGS, where a study compared AGS and AS digestion with and without thermal pre-treatment at 135 °C [24]. Without pre-treatment, the methane production in AS was 20% higher than that of AGS. After pre-treatment, the methane production of AGS was 8% higher than that of AS. It was also found that mixing PS with AGS improved methane production by 16%, but it reduced the methane production of AS by 12% [25].

To the knowledge of the authors, only ten manuscripts were published between 2011 and 2022 that reported the digestibility performance of AGS. Five of these studies used AGS obtained from municipal wastewater treatment, two used AGS obtained from the liquid portion of swine farms wastewater treatment, two used AGS obtained from synthetic wastewater treatment, and one used AGS obtained from brackish wastewater treatment. Table 5 shows the summary of the reported data on the digestion of AGS and the comparison with activated and PS in these studies. It was also noted that all published digestion studies so far were performed as lab-scale experiments with digester volumes between 0.4 and 5 L. Some of these digesters were semi-continuous, while some were just batch experiments. Additionally, only four out of ten studies experimented with the pre-treatment of AGS, and three of them involved thermal pre-treatment. One study only tested ultrasonic homogenization [27]. In the current state of research, the digestion of AGS is still not fully understood. The reported increased viscosity of AGS with heat treatment needs further analysis to identify its reasons and ways to overcome this phenomenon in order to benefit from pre-treatment without increasing the mixing energy demand. Continuous pilot and full-scale digestibility studies should also be performed to validate the experimental results, especially with AGS treatment technology becoming more popular in full-scale applications for wastewater treatment. Further, AGS from different sources, such as food industry wastewater, should be studied to compare their digestibility to municipal AGS.

Table 5. Anaerobic digestion of different AGS and AS substrates.

Type of Sludge	Type of Source Wastewater	Reactor Type	Rector Volume	SRT	OLR	Pre-Treatment	Methane Production (mL/g VS)	Biodegrdability	References
AGS	Swine manure	Batch	0.4 L	-	-	Thermal: 170 °C	337 mL/gVS	62%	[30]
AGS	Synthetic wastewater	Batch	0.4 L	-	-	Thermal 210 °C	404 mL/gVS	58%	
AGS	Brackish wastewater	CSTR	5 L	-	0.4–1.6 (gCOD/L/d)	-	78–136 mL/gCOD	23–42%	[23]
AS		CSTR	5 L	-	1.5 (gCOD/L/d)	-	94 mL/gCOD	27%	
AGS	Swine manure	CSTR–Semi-continuous	5 L	10 days	0.4–1.4 g/L	-	208 mL/gVS	44	[24]
WAS						-	254 mL/gVS	50	
AGS						Thermal: 135 °C	309 mL/gVS	58	
WAS						Thermal: 135 °C	285 mL/gVS	54	
AGS	Synthetic wastewater	Batch	0.6 L	7 days	-	-	0–0.1 mL/gVSS	-	[34]
AS		Batch	0.6 L	7 days	-	-	0–6 mL/gVSS	-	
AGS		Batch	-	21 days	2–6 (kgVS/m^{3}d)	-	492.5 m^3/kgVS	-	
AS	Municipal	-	-	-	2–6 (kgVS/m^{3}d)	-	1178.5 m^3/kgVS	-	[25]
PS:AGS (2:1)		-	-	-	2–6 (kgVS/m^{3}d)	-	574.5 m^3/kgVS	-	
PS: AS (2:1)		-	-	-	2–6 (kgVS/m^{3}d)	-	1035.4 m^3/kgVS	-	
AGS	Municipal	Semi-continuous	0.525 L	20	0.68–0.98 gCOD/Ld	-	285 mL/gVSS	-	[22]
AS		Semi-continuous	0.525 L	22	0.93 gCOD/Ld	-	245 mL/gVSS	-	
AGS	Synthetic	-	-	-	-	-	235 mL/gVS	-	[29]
						Steam Explosion: 170 °C	370–400 mL/gVS	-	
AGS wasted		-	-	-	-	-	296.5 mL/gVS	-	
AGS washed-out	Municipal	-	-	-	-	-	192.9 mL/gVS	-	[8]
WAS		-	-	-	-	-	231.8 mL/gVS	-	
PS		-	-	-	-	-	313.5 mL/gVS	-	
AGS	Municipal	Batch CSTR	2 L	44 days	-	-	197 mL/gVS	-	[16]
WAS		Batch CSTR	2 L	44 days	-	-	242 mL/gVS	-	
AGS	Municipal	Batch	1 L	21 days	-	-	215 (mL/g VS)	-	[27]
						Ultrasound homogenization	300 (mL/g VS)	-	

6. Impact of Microbial Community Structure on AGS Digestibility

Research has shown that the microbial community of AGS was different from that of the flocculent sludge owing to the sludge structure despite being cultivated under the conditions and using the same feed sludge [34], where the microbial community of AGS evolved during the granulation process, and major shifts in the dominated species were observed [48]. The presence of anaerobic and anoxic zones in AGS was reported to promote the survival of anaerobes, which were found to be over 35% more compared to flocculent sludge. These anaerobes facilitated the hydrolysis and acidification during fermentation and thus resulted in more hydrogen production during the early stages of fermentation [34]. The fermentation of nitrifying AGS showed that more soluble chemical oxygen demand (SCOD) and total volatile fatty acids (TVFAs) were released compared to flocculent sludge fermentation, which was attributed to the higher amounts of EPS in AGS and the presence of hydrolytic-acidogenic bacteria [26]. Alkaline fermentation of AGS was reported to favor carbon recovery, whereas acidic fermentation promoted the release of phosphorus in the form of apatite [26].

7. Utilization of Modelling Tools for Prediction of AGS Digestibility

Modeling is a valuable tool that can be used to explain the dynamics of the digestion process of AGS and the factors that influence its rate and methane yields. A partial least square model was used to estimate the anaerobic biodegradability of AGS using the relationship between the initial sludge composition and its BMP [30]. Macroscopic parameters, such as the soluble organic carbon and the COD: TOC ratio, and biochemical parameters, such as the carbohydrate, protein, and lipid concentrations, were used as inputs to the model. The model was used to predict the biodegradability of two AGS samples from different origins (type of influent wastewater) after thermal pre-treatment at various temperatures. A total of 12 samples were modeled, and the predicted biodegradability was within an error of 2–18%. The model was used to quantify the contribution of each of the AGS components, such as the proteins, carbohydrates, lipids, COD/TOC ratio, and soluble organics, to the final biodegradability and add more insight into the behavior of different types of sludge, where the focus was on the particulate and soluble components.

An analysis of the biogas production kinetics using first-order kinetics and non-linear regression analysis was used to compare and understand the dynamics of biodegradation of AGS, WAS, and PS [25]. The models were used to simulate biogas production with an R2 between 88% and 98%. The model was used to estimate the rate, rate constant, and theoretical yield of biogas for each sludge type, as well as mixtures of sludge samples, to simulate co-digestion. Using a set of physicochemical tests, they were able to link the biogas production rates with the sludge characteristics. The impact of lignin content and the TS:VS ratio was found to have a significant impact on the rate and yield of biomethane production. They were also able to identify the benefits of co-digesting AGS with PS rather than WAS.

8. Conclusions and Potential Areas of Research

Studies in the literature have found that the digestibility of AGS is similar to that of AS, making it a viable feedstock for digestion. The AGS morphology and structural polymers slowed down the biomethane production rates, but the overall yield was on par with AS. Thermal pre-treatment was found to improve the methane digestibility of AGS, but it was also found that it can raise the biomass viscosity at temperatures above 125 °C.

The AGS technology is starting to become established as a full-scale wastewater treatment technology, but it is still relatively new compared to the conventional AS technology. Therefore, the available literature on the digestibility of waste AGS is still scarce compared to WAS. There are several gaps in the current literature that allow for future research to be in this area. The first gap is that all the published literature on the digestion of AGS is based on lab-scale digestibility experiments, to the knowledge of the authors, even though full-scale plants already exist in several countries across the world.

In the available literature, only anaerobic digestion was used for AGS. In AS WWTPs, both anaerobic and aerobic digestion provide the same level of stabilization of sludge. Anaerobic digestion has the advantage of the recovery of methane, while aerobic digestion is faster and requires less capital investment than anaerobic digestion. There were no reported studies that investigated the use of aerobic digestion of AGS. It is recommended to compare the aerobic and anaerobic digestion of AGS and study the breakdown of the structural biopolymers.

Additionally, thermal pre-treatment has been used in three out of the four studies that reported using pre-treatment of the AGS. The fourth was ultrasonic homogenization. The impact of pre-treatment on AGS is still not fully understood and requires further analysis. The lack of information in the literature makes it difficult to provide definite conclusions on the efficiency of pre-treatment methods under different conditions.

Author Contributions: Conceptualization, E.E. and R.A.H.; Methodology, M.S.Z., E.E., and R.A.H.; Formal Analysis, M.S.Z.; Investigation, M.S.Z. and R.A.H.; Resources, E.E. and R.A.H.; Data Curation, M.S.Z. and A.M.H.; Writing—Original Draft Preparation, M.S.Z. and R.A.H.; Writing—Review & Editing, M.S.Z., R.A.H., and E.E.; Supervision, E.E. and R.A.H.; Project Administration, E.E. and R.A.H.; Funding Acquisition, E.E., R.A.H., and M.S.Z. All authors have read and agreed to the published version of the manuscript.

Funding: This research received funding from the Natural Science and Engineering Research of Canada: NSERC Discovery Grant (No. RGPIN-2022-03825)-NSERC Postdoctoral Fellowship for Dr. Mohamed S. Zaghloul.

Data Availability Statement: Not applicable.

Acknowledgments: The authors would like to acknowledge the support of the Natural Science and Engineering Research of Canada for the funding provided and the Faculty of Engineering and Architectural Sciences at Toronto Metropolitan University.

Conflicts of Interest: The authors declare no conflict of interest.

References

1. Nancharaiah, Y.V.; Sarvajith, M. Aerobic granular sludge process: A fast growing biological treatment for sustainable wastewater treatment. *Curr. Opin. Environ. Sci. Health* **2019**, *12*, 57–65. [CrossRef]
2. Pronk, M.; Giesen, A.; Thompson, A.; Robertson, S.; Van Loosdrecht, M. Aerobic granular biomass technology: Advancements in design, applications and further developments. *Water Pract. Technol.* **2017**, *12*, 987–996. [CrossRef]
3. Hamza, R.; Rabii, A.; Ezzahraoui, F.Z.; Morgan, G.; Iorhemen, O.T. A review of the state of development of aerobic granular sludge technology over the last 20 years: Full-scale applications and resource recovery. *Case Stud. Chem. Environ. Eng.* **2022**, *5*, 100173. [CrossRef]
4. Hreiz, R.; Latifi, M.A.; Roche, N. Optimal design and operation of activated sludge processes: State-of-the-art. *Chem. Eng. J.* **2015**, *281*, 900–920. [CrossRef]
5. Bassin, J.P.; Winkler, M.K.H.; Kleerebezem, R.; Dezotti, M.; van Loosdrecht, M.C.M. Improved phosphate removal by selective sludge discharge in aerobic granular sludge reactors. *Biotechnol. Bioeng.* **2012**, *109*, 1919–1928. [CrossRef] [PubMed]
6. Qin, L.; Tay, J.H.; Liu, Y. Selection pressure is a driving force of aerobic granulation in sequencing batch reactors. *Process Biochem.* **2004**, *39*, 579–584. [CrossRef]
7. Tay, J.H.; Liu, Q.S.; Liu, Y. The effect of upflow air velocity on the structure of aerobic granules cultivated in a sequencing batch reactor. *Water Sci. Technol.* **2004**, *49*, 35–40. [CrossRef] [PubMed]
8. Guo, H.; van Lier, J.B.; de Kreuk, M. Digestibility of waste aerobic granular sludge from a full-scale municipal wastewater treatment system. *Water Res.* **2020**, *173*, 115617. [CrossRef]
9. Pronk, M.; de Kreuk, M.K.; de Bruin, B.; Kamminga, P.; Kleerebezem, R.; van Loosdrecht, M.C.M. Full scale performance of the aerobic granular sludge process for sewage treatment. *Water Res.* **2015**, *84*, 207–217. [CrossRef]
10. Metcalf & Eddy Inc. *Wastewater Engineering Treatment and Resource Recovery*; Tchobanoglous, G., Stensel, D., Tsuchihashi, R., Burton, F., Abu-Orf, M., Bowden, G., Pfrang, W., Eds.; McGraw Hill: New York, NY, USA, 2014.
11. Castellanos, R.M.; Dias, J.M.R.; Bassin, I.D.; Dezotti, M.; Bassin, J.P. Effect of sludge age on aerobic granular sludge: Addressing nutrient removal performance and biomass stability. *Process Saf. Environ. Prot.* **2021**, *149*, 212–222. [CrossRef]
12. Świątczak, P.; Cydzik-Kwiatkowska, A. Performance and microbial characteristics of biomass in a full-scale aerobic granular sludge wastewater treatment plant. *Environ. Sci. Pollut. Res.* **2018**, *25*, 1655–1669. [CrossRef]
13. Hamza, R.A.; Zaghloul, M.S.; Iorhemen, O.T.; Sheng, Z.; Tay, J.H. Optimization of organics to nutrients (COD:N:P) ratio for aerobic granular sludge treating high-strength organic wastewater. *Sci. Total Environ.* **2019**, *650*, 3168–3179. [CrossRef]

14. Kazimierowicz, J.; Dębowski, M. Aerobic Granular Sludge as a Substrate in Anaerobic Digestion—Current Status and Perspectives. *Sustainability* **2022**, *14*, 904. [CrossRef]

15. Haaksman, V.A.; Mirghorayshi, M.; van Loosdrecht, M.C.M.; Pronk, M. Impact of aerobic availability of readily biodegradable COD on morphological stability of aerobic granular sludge. *Water Res.* **2020**, *187*, 116402. [CrossRef]

16. Guo, H.; Felz, S.; Lin, Y.; van Lier, J.B.; de Kreuk, M. Structural extracellular polymeric substances determine the difference in digestibility between waste activated sludge and aerobic granules. *Water Res.* **2020**, *181*, 115924. [CrossRef]

17. Hamza, R.A.; Iorhemen, O.T.; Zaghloul, M.S.; Tay, J.H. Rapid formation and characterization of aerobic granules in pilot-scale sequential batch reactor for high-strength organic wastewater treatment. *J. Water Process Eng.* **2018**, *22*, 27–33. [CrossRef]

18. Aqeel, H.; Weissbrodt, D.G.; Cerruti, M.; Wolfaardt, G.M.; Wilén, B.-M.; Liss, S.N. Drivers of bioaggregation from flocs to biofilms and granular sludge. *Environ. Sci. Water Res. Technol.* **2019**, *5*, 2072–2089. [CrossRef]

19. de Sousa Rollemberg, S.L.; De Oliveira, L.Q.; De Barros, A.N.; Firmino, P.I.M.; Dos Santos, A.B. Pilot-scale aerobic granular sludge in the treatment of municipal wastewater: Optimizations in the start-up, methodology of sludge discharge, and evaluation of resource recovery. *Bioresour. Technol.* **2020**, *311*, 123467. [CrossRef]

20. Liu, Y.-Q.; Moy, B.; Kong, Y.-H.; Tay, J.-H. Formation, physical characteristics and microbial community structure of aerobic granules in a pilot-scale sequencing batch reactor for real wastewater treatment. *Enzyme Microb. Technol.* **2010**, *46*, 520–525. [CrossRef]

21. Li, X.; Lin, S.; Hao, T.; Khanal, S.K.; Chen, G. Elucidating pyrolysis behaviour of activated sludge in granular and flocculent form: Reaction kinetics and mechanism. *Water Res.* **2019**, *162*, 409–419. [CrossRef]

22. Jahn, L.; Saracevic, E.; Svardal, K.; Krampe, J. Anaerobic biodegradation and dewaterability of aerobic granular sludge. *J. Chem. Technol. Biotechnol.* **2019**, *94*, 2908–2916. [CrossRef]

23. Palmeiro-Sánchez, T.; Val del Río, A.; Mosquera-Corral, A.; Campos, J.L.; Méndez, R. Comparison of the anaerobic digestion of activated and aerobic granular sludges under brackish conditions. *Chem. Eng. J.* **2013**, *231*, 449–454. [CrossRef]

24. Val Del Río, Á.; Palmeiro-Sanchez, T.; Figueroa, M.; Mosquera-Corral, A.; Campos, J.L.; Méndez, R. Anaerobic digestion of aerobic granular biomass: Effects of thermal pre-treatment and addition of primary sludge. *J. Chem. Technol. Biotechnol.* **2014**, *89*, 690–697. [CrossRef]

25. Bernat, K.; Cydzik-Kwiatkowska, A.; Wojnowska-Baryła, I.; Karczewska, M. Physicochemical properties and biogas productivity of aerobic granular sludge and activated sludge. *Biochem. Eng. J.* **2017**, *117*, 43–51. [CrossRef]

26. Zou, J.; Pan, J.; He, H.; Wu, S.; Xiao, N.; Ni, Y.; Li, J. Nitrifying aerobic granular sludge fermentation for releases of carbon source and phosphorus: The role of fermentation pH. *Bioresour. Technol.* **2018**, *260*, 30–37. [CrossRef] [PubMed]

27. Cydzik-Kwiatkowska, A.; Bernat, K.; Zielińska, M.; Gusiatin, M.Z.; Wojnowska-Baryła, I.; Kulikowska, D. Valorization of full-scale waste aerobic granular sludge for biogas production and the characteristics of the digestate. *Chemosphere* **2022**, *303*, 135167. [CrossRef]

28. Dangcong, P.; Bernet, N.; Delgenes, J.-P.; Moletta, R. Aerobic granular sludge—A case report. *Water Res.* **1999**, *33*, 890–893. [CrossRef]

29. Liu, Y.; Nilsen, P.J.; Maulidiany, N.D. Thermal pretreatment to enhance biogas production of waste aerobic granular sludge with and without calcium phosphate precipitates. *Chemosphere* **2019**, *234*, 725–732. [CrossRef]

30. Val del Río, A.; Morales, N.; Isanta, E.; Mosquera-Corral, A.; Campos, J.L.; Steyer, J.P.; Carrère, H. Thermal pre-treatment of aerobic granular sludge: Impact on anaerobic biodegradability. *Water Res.* **2011**, *45*, 6011–6020. [CrossRef]

31. Farno, E.; Baudez, J.C.; Parthasarathy, R.; Eshtiaghi, N. The viscoelastic characterisation of thermally-treated waste activated sludge. *Chem. Eng. J.* **2016**, *304*, 362–368. [CrossRef]

32. Feng, G.; Ma, H.; Bai, T.; Guo, Y. Rheology characteristics of activated sludge and thermal treated sludge at different process temperature. *Water Sci. Technol.* **2018**, *2017*, 229–237. [CrossRef]

33. Bougrier, C.; Delgenès, J.P.; Carrère, H. Effects of thermal treatments on five different waste activated sludge samples solubilisation, physical properties and anaerobic digestion. *Chem. Eng. J.* **2008**, *139*, 236–244. [CrossRef]

34. Zou, J.; Li, Y. Anaerobic fermentation combined with low-temperature thermal pretreatment for phosphorus-accumulating granular sludge: Release of carbon source and phosphorus as well as hydrogen production potential. *Bioresour. Technol.* **2016**, *218*, 18–26. [CrossRef]

35. Champagne, P.; Li, C. Enzymatic hydrolysis of cellulosic municipal wastewater treatment process residuals as feedstocks for the recovery of simple sugars. *Bioresour. Technol.* **2009**, *100*, 5700–5706. [CrossRef]

36. Lin, H.; Ma, R.; Hu, Y.; Lin, J.; Sun, S.; Jiang, J.; Li, T.; Liao, Q.; Luo, J. Reviewing bottlenecks in aerobic granular sludge technology: Slow granulation and low granular stability. *Environ. Pollut.* **2020**, *263*, 114638. [CrossRef]

37. Sepúlveda-Mardones, M.; Campos, J.L.; Magrí, A.; Vidal, G. Moving forward in the use of aerobic granular sludge for municipal wastewater treatment: An overview. *Rev. Environ. Sci. Bio/Technol.* **2019**, *18*, 741–769. [CrossRef]

38. Feng, C.; Lotti, T.; Canziani, R.; Lin, Y.; Tagliabue, C.; Malpei, F. Extracellular biopolymers recovered as raw biomaterials from waste granular sludge and potential applications: A critical review. *Sci. Total Environ.* **2021**, *753*, 142051. [CrossRef]

39. Wang, Z.-W.; Liu, Y.; Tay, J.-H. Distribution of EPS and cell surface hydrophobicity in aerobic granules. *Appl. Microbiol. Biotechnol.* **2005**, *69*, 469–473. [CrossRef]

40. Liu, X.; Pei, Q.; Han, H.; Yin, H.; Chen, M.; Guo, C.; Li, J.; Qiu, H. Functional analysis of extracellular polymeric substances (EPS) during the granulation of aerobic sludge: Relationship among EPS, granulation and nutrients removal. *Environ. Res.* **2022**, *208*, 112692. [CrossRef]

41. Zhang, B.; Lens, P.N.L.; Shi, W.; Zhang, R.; Zhang, Z.; Guo, Y.; Bao, X.; Cui, F. Enhancement of aerobic granulation and nutrient removal by an algal–bacterial consortium in a lab-scale photobioreactor. *Chem. Eng. J.* **2018**, *334*, 2373–2382. [CrossRef]

42. Gherghel, A.; Teodosiu, C.; De Gisi, S. A review on wastewater sludge valorisation and its challenges in the context of circular economy. *J. Clean. Prod.* **2019**, *228*, 244–263. [CrossRef]

43. Jeong, S.Y.; Chang, S.W.; Ngo, H.H.; Guo, W.; Nghiem, L.D.; Banu, J.R.; Jeon, B.H.; Nguyen, D.D. Influence of thermal hydrolysis pretreatment on physicochemical properties and anaerobic biodegradability of waste activated sludge with different solids content. *Waste Manag.* **2019**, *85*, 214–221. [CrossRef] [PubMed]

44. Farhat, A.; Asses, N.; Ennouri, H.; Hamdi, M.; Bouallagui, H. Combined effects of thermal pretreatment and increasing organic loading by co-substrate addition for enhancing municipal sewage sludge anaerobic digestion and energy production. *Process Saf. Environ. Prot.* **2018**, *119*, 14–22. [CrossRef]

45. Zieliński, M.; Dębowski, M.; Kisielewska, M.; Nowicka, A.; Rokicka, M.; Szwarc, K. Comparison of Ultrasonic and Hydrothermal Cavitation Pretreatments of Cattle Manure Mixed with Straw Wheat on Fermentative Biogas Production. *Waste Biomass Valorization* **2019**, *10*, 747–754. [CrossRef]

46. Ding, Z.; Bourven, I.; Guibaud, G.; van Hullebusch, E.D.; Panico, A.; Pirozzi, F.; Esposito, G. Role of extracellular polymeric substances (EPS) production in bioaggregation: Application to wastewater treatment. *Appl. Microbiol. Biotechnol.* **2015**, *99*, 9883–9905. [CrossRef] [PubMed]

47. Sun, D.; Qiao, M.; Xu, Y.; Ma, C.; Zhang, X. Pretreatment of waste activated sludge by peracetic acid oxidation for enhanced anaerobic digestion. *Environ. Prog. Sustain. Energy* **2018**, *37*, 2058–2062. [CrossRef]

48. Li, A.; Yang, S.; Li, X.; Gu, J. Microbial population dynamics during aerobic sludge granulation at different organic loading rates. *Water Res.* **2008**, *42*, 3552–3560. [CrossRef]

MDPI
St. Alban-Anlage 66
4052 Basel
Switzerland
www.mdpi.com

Processes Editorial Office
E-mail: processes@mdpi.com
www.mdpi.com/journal/processes